FUEL CELL FUNDAMENTALS

FUEL CELL FUNDAMENTALS

RYAN P. O'HAYRE
Joint NSF Fellow
Department of Mechanical Engineering
Stanford University
Delft Institute for Sustainable Energy
Delft University of Technology

SUK-WON CHA
School of Mechanical and Aerospace Engineering
Seoul National University

WHITNEY COLELLA
Department of Civil and Environmental Engineering
Stanford University

FRITZ B. PRINZ
R. H. Adams Professor of Engineering
Departments of Mechanical Engineering and Materials Science and Engineering
Stanford University

JOHN WILEY & SONS, INC.

Library of Congress Cataloging-in-Publication Data

Fuel cell fundamentals / Ryan O'Hayre . . . [et al.].
  p. cm.
Includes bibliographical references and index.
ISBN-13 978-0-471-74148-0
ISBN-10 0-471-74148-5 (cloth: alk. paper)
1. Fuel cells—Textbooks. I. O'Hayre, Ryan.
TK2931.F834 2005
621.31′2429—dc22   2005047495

Printed in the United States of America

10 9 8 7 6 5 4 3 2 1

To the parents who nurtured us.
To the teachers who inspired us.

CONTENTS

PREFACE

Imagine driving home in a fuel cell car with nothing but pure water dripping from the tailpipe. Imagine a laptop computer that runs for 30 hours on a single charge. Imagine a world where you plug your *house* into your *car* and power lines are a distant memory. These dreams motivate today's fuel cell research. While some dreams (like powering your home with your fuel cell car) may be distant, others (like a 30-hour fuel cell laptop) may be closer than you think.

By taking fuel cells from the dream world to the real world, this book teaches you the *science* behind the technology. This book focuses on the questions *"how"* and *"why."* Inside you will find straightforward descriptions of *how* fuel cells work, *why* they offer the potential for high efficiency, and *how* their unique advantages can best be used. Emphasis is placed on the fundamental scientific principles that govern fuel cell operation. These principles remain constant and universally applicable, regardless of fuel cell type or technology.

Following this philosophy, the first part, Fuel Cell Principles, is devoted to basic fuel cell physics. Illustrated diagrams, examples, text boxes, and homework questions are all designed to impart a unified, *intuitive* understanding of fuel cells. Of course, no treatment of fuel cells is complete without at least a brief discussion of the practical aspects of fuel cell technology. This is the aim of the second part of the book, Fuel Cell Technology. Informative diagrams, tables, and examples provide an engaging review of the major fuel cell technologies. In this half of the book, you will learn how to select the right fuel cell for a given application and how to design a complete system. Finally, you will learn how to assess the potential environmental impact of fuel cell technology.

Comments or questions? Suggestions for improving the book? Found a typo, think our explanations could be improved, want to make a suggestion about other important concepts to discuss, or have we got it all wrong? Please send us your feedback by emailing us at fcf3@yahoogroups.com. We will take your suggestions into consideration for the next edition. Our website http://groups.yahoo.com/group/fcf3/ posts these discussions and additional educational materials. Thank you.

ACKNOWLEDGMENTS

The authors would like to thank their friends and colleagues at Stanford University and the Rapid Prototyping Laboratory (RPL) for their support, critiques, comments, and enthusiasm. Without you, this text would not have been written!

In particular, we would like to thank the Deans of the School of Engineering, Jim Plummer and Channing Robertson, and John Bravman, Vice Provost Undergraduate Education, for the support that made this book possible. We would also like to acknowledge Honda R&D, its representatives J. Araki, T. Kawanabe, Y. Fujisawa, Y. Kawaguchi, Y. Higuchi, T. Kubota, N. Kuriyama, Y. Saito, J. Sasahara, and H. Tsuru, and the GCEP community under the leadership of Lynn Orr and Chris Edwards for creating an atmosphere conducive to studying and researching new forms of power generation. All members of RPL are recognized for stimulating discussions. Special thanks to Tim Holme for his innumerable contributions, including his careful review of the text, integration work, nomenclature and equation summaries, and the appendixes. Thanks also to Rojana Pornprasertsuk, who developed the wonderful quantum simulation images for Chapter 3 and Appendix D. Juliet Risner deserves gratitude for her beautiful editing job, and Hong Huang deserves thanks for content contribution. Jeremy Cheng, Kevin Crabb, Turgut Gur, Shannon Miller, Massafumi Nakamura, and A. J. Simon also provided significant editorial advice. We would like extend our gratitude to Steven Schneider, Mark Jacobson, Mark Delucchi (University of California at Davis), Michael Mastrandrea, and Gerard Ketafani for their insightful comments on the environmental impact of fuel cells. Additional thanks go to Jay Garaw, Fred Kornbrust, Paul Farris, and Bob Greene of Energy Technology. We also would like to thank Richard Stone (Oxford University), Colin Snowdon (Oxford University), Ali Mani, and Lee Shunn for their perceptive remarks on the design and integration of fuel cell systems.

Fritz B. Prinz wants to thank his wife, Gertrud, and his childern, Marie-Helene and Benedikt, for their love, support, and patience.

Whitney Colella would like to thank her family, Caterina Qualtieri, Eulalia Pandolfi, Tom Judge, and Emily Zedler.

Suk-Won Cha wishes to acknowledge his friends and family for their encouragement and enthusiasm. In particular, he would like to thank Unjung for her support and for letting him skip babysitting William from time to time.

Ryan O'Hayre sends his thanks and gratitude to Lisa for her friendship, encouragement, confidence, support, and love. Thanks also to Kendra, Arthur, and Morgan. Ryan has always wanted to write a book . . . probably something about dragons and adventure. Well, things have a funny way of working out, and although he ended up writing about fuel cells, he had to put the dragons in somewhere . . . .

NOMENCLATURE

Symbol	Meaning	Common Units
A	Area	cm^2
A_c	Catalyst area coefficient	Dimensionless
a	Activity	Dimensionless
ASR	Area specific resistance	$\Omega \cdot cm^2$
C	Capacitance	F
C_{dl}	Double-layer capacitance	F
c^*	Concentration at reaction surface	mol/cm^2
c	Concentration	mol/m^3
c	Constant describing how mass transport affects concentration losses	V
c_p	Heat capacity	$J/mol \cdot K$
D	Diffusivity	cm^2/s
E	Electric field	V/cm
E	Thermodynamic ideal voltage	V
E_{thermo}	Thermodynamic ideal voltage	V
E_T	Temperature-dependent thermodynamic voltage at reference concentration	V
F	Helmholtz free energy	J, J/mol
F	Faraday constant	96,485 C/mol
F_k	Generalized force	N
f	Reaction rate constant	Hz, s^{-1}
f	Friction factor	Dimensionless
G, g	Gibbs free energy	J, J/mol

Symbol	Meaning	Common Units
g	Acceleration due to gravity	m/s^2
$\Delta G^{\ddagger}$	Activation energy barrier	J/mol, J
ΔG_{act}	Activation energy barrier	J/mol, J
H	Heat	J
H, h	Enthalpy	J, J/mol
H_C	Gas channel thickness	cm
H_E	Diffusion layer thickness	cm
h	Planck's constant	6.63×10^{-34} J/s
$\hbar$	Reduced Planck constant, $h/2\pi$	1.05×10^{-34} J/s
h_m	Mass transfer convection coefficient	m/s
i	Current	A
J	Molar flux	$mol/cm^2 \cdot s$
$\hat{J}$	Mass flux	$g/cm^2 \cdot s$, $kg/m^2 \cdot s$
J_C	Convective mass flux	$kg/m^2 \cdot s$
j	Current density	A/cm^2
j_0	Exchange current density	A/cm^2
j_0^0	Exchange current density at reference concentration	A/cm^2
j_L	Limiting current density	A/cm^2
j_{leak}	Fuel leakage current	A/cm^2
k	Boltzmann's constant	1.38×10^{-23} J/K
L	Length	cm
M	Molar mass	g/mol, kg/mol
M	Mass flow rate	kg/s
M_{ik}	Generalized coupling coefficient between force and flux	Varies
m	Mass	kg
mc_p	Heat capacity flow rate	$kW/kg \cdot \,^{\circ}C$
N	Number of moles	Dimensionless
N_A	Avogadro's number	6.02×10^{23} mol^{-1}
n	Number of electrons transferred in the reaction	Dimensionless
n_g	Number of moles of gas	Dimensionless
P	Power or power density	W or W/cm^2
p	Pressure	bar, atm, Pa
Q	Heat	J, J/mol
Q	Charge	C
Q_h	Adsorption charge	C/cm^2
Q_m	Adsorption charge for smooth catalyst surface	C/cm^2
q	Fundamental charge	1.60×10^{-19} C
R	Ideal gas constant	8.314 J/mol $\cdot$ K
R	Resistance	Ω
R_f	Faradaic resistance	Ω
Re	Reynolds number	Dimensionless
S, s	Entropy	J/K, $J/mol \cdot K$
S/C	Steam-to-carbon ratio	Dimensionless

Symbol	Meaning	Common Units
Sh	Sherwood number	Dimensionless
T	Temperature	K, °C
t	Thickness	cm
U	Internal energy	J, J/mol
u	Mobility	$cm^2/V \cdot s$
$\bar{u}$	Mean flow velocity	cm/s, m/s
V	Voltage	V
V	Volume	L, cm^3
v	Reaction rate per unit area	$mol/cm^2 \cdot s$
v	Velocity	cm/s
v	Hopping rate	s^{-1}, Hz
W	Work	J, J/mol
X	Parasitic power load	W
x	Mole fraction	Dimensionless
x_V	Vacancy fraction	mol V/mol
y_x	Yield of element x	Dimensionless
Z	Impedance	Ω
z	Height	m

Greek Symbols

Symbol	Meaning	Common Units
α	Charge transfer coefficient	Dimensionless
α	Coefficient for CO_2 equivalent	Dimensionless
α^*	Channel aspect ratio	Dimensionless
β	Coefficient for CO_2 equivalent	Dimensionless
γ	Activity coefficient	Dimensionless
Δ	Denotes change in quantity	Dimensionless
δ	Diffusion layer thickness	m, cm
ε	Efficiency	Dimensionless
ε_{FP}	Efficiency of fuel processor	Dimensionless
ε_{FR}	Efficiency of fuel reformer	Dimensionless
ε_H	Efficiency of heat recovery	Dimensionless
ε_O	Efficiency overall	Dimensionless
ε_R	Efficiency, electrical	Dimensionless
ε	Porosity	Dimensionless
$\dot{\varepsilon}$	Strain rate	s^{-1}
η	Overvoltage	V
η_{act}	Activation overvoltage	V
η_{conc}	Concentration overvoltage	V
η_{ohmic}	Ohmic overvoltage	V
λ	Stoichiometric coefficient	Dimensionless
λ	Water content	Dimensionless

Symbol	Meaning	Common Units
μ	Viscosity	$kg \cdot m/s$
μ	Chemical potential	J, J/mol
$\tilde{\mu}$	Electrochemical potential	J, J/mol
ρ	Resistivity	Ω cm
ρ	Density	kg/cm^3, kg/m^3
σ	Conductivity	S/cm, $(\Omega \cdot cm)^{-1}$
σ	Warburg coefficient	$\Omega/s^{0.5}$
τ	Mean free time	s
τ	Shear stress	Pa
ϕ	Electrical potential	V
ϕ	Phase factor	Dimensionless
ω	Angular frequency ($\omega = 2\pi f$)	rad/s

Superscripts

Symbol	Meaning
0	Denotes standard or reference state
eff	Effective property

Subscripts

Symbol	Meaning
diff	Diffusion
E, e, elec	Electrical (e.g., P_e, W_{elec})
f	Quantity of formation (e.g., ΔH_f)
(HHV)	Higher heating value
i	Species i
P	Product
P	Parasitic
R	Reactant
rxn	Change in a reaction (e.g., ΔH_{rxn})
SK	Stack
SYS	System

PART I

FUEL CELL PRINCIPLES

CHAPTER 1

INTRODUCTION

You are about to embark on a journey into the world of fuel cells and electrochemistry. This chapter will act as a roadmap for your travels, setting the stage for the rest of the book. In broad terms, this chapter will acquaint you with fuel cells: what they are, how they work, and what significant advantages and disadvantages they present. From this starting point, the subsequent chapters will lead you onward in your journey as you acquire a fundamental understanding of fuel cell principles.

1.1 WHAT IS A FUEL CELL?

You can think of a fuel cell as a "factory" that takes fuel as input and produces electricity as output. (See Figure 1.1.) Like a factory, a fuel cell will continue to churn out product (electricity) as long as raw material (fuel) is supplied. This is the key difference between a fuel cell and a battery. While both rely on electrochemistry to work their magic, a fuel cell is not consumed when it produces electricity. It is really a factory, a *shell*, which transforms the chemical energy stored in a fuel into electrical energy.

Figure 1.1. General concept of a hydrogen–oxygen (H_2–O_2) fuel cell.

Viewed this way, combustion engines are also "chemical factories." Combustion engines also take the chemical energy stored in a fuel and transform it into useful mechanical or electrical energy. So what is the difference between a combustion engine and a fuel cell?

In a conventional combustion engine, fuel is burned, releasing heat. Consider the simplest example, the combustion of hydrogen:

$$H_2 + \tfrac{1}{2}O_2 \rightleftharpoons H_2O \tag{1.1}$$

On the molecular scale, collisions between hydrogen molecules and oxygen molecules result in a reaction. The hydrogen molecules are oxidized, producing water and releasing heat. Specifically, at the atomic scale, in a matter of picoseconds, hydrogen–hydrogen bonds and oxygen–oxygen bonds are broken, while hydrogen-oxygen bonds are formed. These bonds are broken and formed by the transfer of electrons between the molecules. The energy of the product water bonding configuration is lower than the bonding configurations of the initial hydrogen and oxygen gases. This energy difference is released as heat. Although the energy difference between the initial and final states occurs by a re-configuration of electrons as they move from one bonding state to another, this energy is recoverable only as heat because the bonding reconfiguration occurs in picoseconds at an intimate, subatomic scale. (See Figure 1.2.) To produce electricity, this heat energy must be converted into mechanical energy, and then the mechanical energy must be converted into electrical energy. Going through all these steps is potentially complex and inefficient.

Consider an alternative solution: to produce electricity directly from the chemical reaction by somehow harnessing the electrons as they move from high-energy reactant bonds to low-energy product bonds. In fact, this is exactly what a fuel cell does. But the question is how do we harness electrons that reconfigure in picoseconds at subatomic length scales? The answer is to spatially separate the hydrogen and oxygen reactants so that the electron

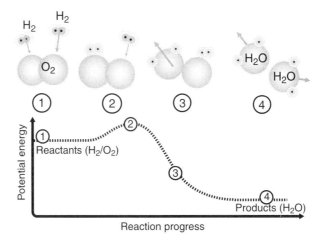

Figure 1.2. Schematic of H_2–O_2 combustion reaction. (Arrows indicate the relative motion of the molecules participating in the reaction.) Starting with the reactant H_2-O_2 gases (1), hydrogen–hydrogen and oxygen–oxygen bonds must first be broken, requiring energy input (2) before hydrogen–oxygen bonds are formed, leading to energy output (3, 4).

BONDS AND ENERGY

Atoms are social creatures. They almost always prefer to be together instead of alone. When atoms come together, they form bonds, lowering their total energy. Figure 1.3 shows a typical energy–distance curve for a hydrogen–hydrogen bond. When the hydrogen atoms are far apart from one another (1), no bond exists and the system has high energy. As the hydrogen atoms approach one another, the system energy is lowered until the most stable bonding configuration (2) is reached. Further overlap between the atoms is energetically unfavorable because the repulsive forces between the nuclei begin to dominate (3). Remember:

- Energy is released when a bond is formed.
- Energy is absorbed when a bond is broken.

For a reaction to result in a net release of energy, the energy released by the formation of the product bonds must be more than the energy absorbed to break the reactant bonds.

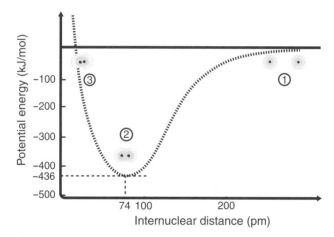

Figure 1.3. Bonding energy versus internuclear separation for hydrogen–hydrogen bond. (1) No bond exists. (2) Most stable bonding configuration. (3) Further overlap unfavorable due to internuclear repulsion.

transfer necessary to complete the bonding reconfiguration occurs over a greatly extended length scale. Then, as the electrons move from the fuel species to the oxidant species, they can be harnessed as an electrical current.

1.2 A SIMPLE FUEL CELL

In a fuel cell, the hydrogen *combustion* reaction is split into two *electrochemical* half reactions:

$$H_2 \rightleftharpoons 2H^+ + 2e^- \qquad (1.2)$$

$$\tfrac{1}{2}O_2 + 2H^+ + 2e^- \rightleftharpoons H_2O \qquad (1.3)$$

By spatially separating these reactions, the electrons transferred from the fuel are forced to flow through an external circuit (thus constituting an electric current) and do useful work before they can complete the reaction.

Spatial separation is accomplished by employing an electrolyte. An electrolyte is a material that allows ions (charged atoms) to flow but not electrons. At a minimum, a fuel cell must posses two electrodes, where the two electrochemical half reactions occur, separated by an electrolyte.

Figure 1.4 shows an example of an extremely simple H_2–O_2 fuel cell. This fuel cell consists of two platinum electrodes dipped into sulfuric acid (an aqueous acid electrolyte). Hydrogen gas, bubbled across the left electrode, is split into protons (H^+) and electrons following Equation 1.2. The protons can flow through the electrolyte (the sulfuric acid is like a "sea" of H^+), but the electrons cannot. Instead, the electrons flow from left to right through a piece of wire that connects the two platinum electrodes. Note that the resulting current, as it is traditionally defined, is in the opposite direction. When the electrons reach the right electrode, they recombine with protons and bubbling oxygen gas to produce water following Equation 1.3. If a load (e.g., a light bulb) is introduced along the path of the electrons, the flowing electrons will provide power to the load, causing the light bulb to glow. Our fuel cell is producing electricity! The first fuel cell, invented by William Grove in 1839, probably looked a lot like the one discussed here.

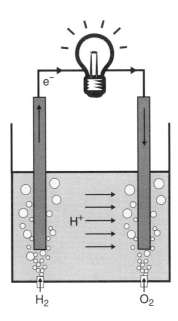

Figure 1.4. A simple fuel cell.

ENERGY, POWER, ENERGY DENSITY, AND POWER DENSITY

To understand how a fuel cell compares to a combustion engine or a battery, several quantitative metrics, or *figures of merit*, are required. The most common figures of merit used to compare energy conversion systems are *power density* and *energy density*.

To understand energy density and power density, you first need to understand the difference between energy and power:

- *Energy* is defined as the ability to do work. Energy is usually measured in joules (J) or calories (cal).
- *Power* is defined as the rate at which energy is expended or produced. In other words, power represents the *intensity* of energy use or production. Power is a rate. The typical unit of power, the watt (W), represents the amount of energy used or produced per second (1 W = 1J/sec).

From the above discussion, it is obvious that energy is the product of power and time:

$$\text{Energy} = \text{power} \times \text{time} \tag{1.4}$$

Although the International System of Units (SI) uses the joule as the unit of energy, you will often see energy expressed in terms of watt-hours (Wh) or kilowatt-hours (kWh). These units arise when the units of power (e.g., watts) are multiplied by a length of time (e.g., hours) as in Equation 1.4. Obviously, watt-hours can be converted to joules or vice-versa using simple arithmetic:

$$1\ \text{Wh} \times 3600\ \text{s/h} \times 1\ (\text{J/s})/\text{W} = 3600\ \text{J} \tag{1.5}$$

Refer to Appendix A for a list of some of the more common unit conversions for energy and power. For portable fuel cells and other mobile energy conversion devices, power density and energy density are more important than power and energy because they provide information about *how big* a system needs to be to deliver a certain amount of energy or power. Power density refers to the amount of power that can be produced by a device per unit mass or volume. Energy density refers to the total energy capacity available to the system per unit mass or volume.

- *Volumetric power density* is the amount of power that can be supplied by a device per unit volume. Typical units are W/cm^3 or kW/m^3.
- *Gravimetric power density (or specific power)* is the amount of power that can be supplied by a device per unit mass. Typical units are W/g or kW/kg.
- *Volumetric energy density* is the amount of energy that is available to a device per unit volume. Typical units are Wh/cm^3 or kWh/m^3.
- *Gravimetric energy density (or specific energy)* is the amount of energy that is available to a device per unit mass. Typical units are Wh/g or kWh/kg.

1.3 FUEL CELL ADVANTAGES

Because fuel cells are "factories" that produce electricity as long as they are supplied with fuel, they share some characteristics in common with combustion engines. Because fuel cells are electrochemical energy conversion devices that rely on electrochemistry to work their magic, they share some characteristics in common with primary batteries. In fact, fuel cells combine many of the advantages of both engines and batteries.

Since fuel cells produce electricity directly from chemical energy, they are often far more efficient than combustion engines. Fuel cells can be all solid state and mechanically ideal, meaning no moving parts. This yields the potential for highly reliable and long-lasting systems. A lack of moving parts also means that fuel cells are silent. Also, undesirable products such as NO_x, SO_x, and particulate emissions are virtually zero.

Unlike batteries, fuel cells allow easy independent scaling between power (determined by the fuel cell size) and capacity (determined by the fuel reservoir size). In batteries, power and capacity are often convoluted. Batteries scale poorly at large sizes while fuel cells scale well from the 1-W range (cell phone) to the megawatt range (power plant). Fuel cells offer potentially higher energy densities compared to batteries and can be quickly recharged by refueling, while batteries must be thrown away or plugged in for a time-consuming recharge.

1.4 FUEL CELL DISADVANTAGES

While fuel cells present intriguing advantages, they also possess some serious disadvantages. Cost represents a major barrier to fuel cell implementation. Because of prohibitive costs, fuel cell technology is currently only economically competitive in a few highly specialized applications (e.g., onboard the Space Shuttle orbiter). Power density is another significant limitation. Power density expresses how much power a fuel cell can produce per unit volume (volumetric power density) or per unit mass (gravimetric power density). Although fuel cell power densities have improved dramatically over the past decades, further improvements are required if fuel cells are to compete in portable and automotive applications. Combustion engines and batteries generally outperform fuel cells on a volumetric power density basis; on a gravimetric power density basis, the race is much closer. (See Figure 1.5.)

Fuel availability and storage pose further problems. Fuel cells work best on hydrogen gas, a fuel which is not widely available, has a low volumetric energy density, and is difficult to store. (See Figure 1.6.) Alternative fuels (e.g., gasoline, methanol, formic acid) are difficult to use directly and usually require reforming. These problems can reduce fuel cell performance and increase the requirements for ancillary equipment. Thus, although gasoline looks like an attractive fuel from an energy density standpoint, it is not well suited to fuel cell use.

Additional fuel cell limitations include operational temperature compatibility concerns, susceptibility to environmental poisons, and durability under start–stop cycling. These significant disadvantages will not be easy to overcome. Fuel cell adoption will be severely limited unless technological solutions can be developed to hurdle these barriers.

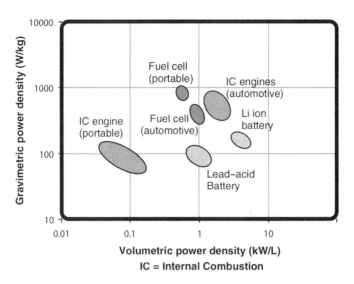

Figure 1.5. Power density comparison of selected technologies (approximate ranges).

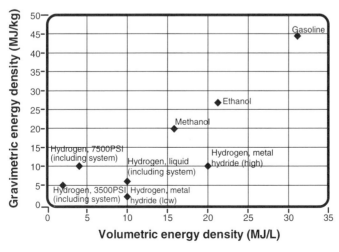

Figure 1.6. Energy density comparison of selected fuels (lower heating value).

1.5   FUEL CELL TYPES

There are five major types of fuel cells, differentiated from one another by their electrolyte:

1. Phosphoric acid fuel cell (PAFC)
2. Polymer electrolyte membrane fuel cell (PEMFC)
3. Alkaline fuel cell (AFC)
4. Molten carbonate fuel cell (MCFC)
5. Solid-oxide fuel cell (SOFC)

While all five fuel cell types are based upon the same underlying electrochemical principles, they all operate at different temperature regimens, incorporate different materials, and often differ in their fuel tolerance and performance characteristics, as shown in Table 1.1. Most of the examples in this book focus on PEMFCs or SOFCs. We will briefly contrast these two fuel cell types.

- PEMFCs employ a thin *polymer* membrane as an electrolyte (the membrane looks and feels a lot like plastic wrap). Protons are the ionic charge carrier in a PEMFC membrane. As we have already seen, the electrochemical half reactions in a H_2–O_2 PEMFC are

$$H_2 \rightarrow 2H^+ + 2e^-$$
$$\tfrac{1}{2}O_2 + 2H^+ + 2e^- \rightarrow H_2O \qquad (1.6)$$

PEMFCs are attractive for many applications because they operate at low temperature and have high power density.

- SOFCs employ a thin *ceramic* membrane as an electrolyte. Oxygen ions (O^{2-}) are the ionic charge carrier in an SOFC membrane. In a H_2–O_2 SOFC, the electrochemical half reactions are

$$H_2 + O^{2-} \rightarrow H_2O + 2e^-$$
$$\tfrac{1}{2}O_2 + 2e^- \rightarrow O^{2-} \qquad (1.7)$$

To function properly, SOFCs must operate at high temperatures ($> 600°C$). They are attractive for stationary applications because they are highly efficient and fuel flexible.

Note how changing the mobile charge carrier dramatically changes the fuel cell reaction chemistry. In a PEMFC, the half reactions are mediated by the movement of protons (H^+) and water is produced at the cathode. In a SOFC, the half reactions are mediated by the motion of oxygen ions (O^{2-}) and water is produced at the anode. Note in Table 1.1 how other fuel cell types use OH^- or CO_3^{2-} as ionic charge carriers. These fuel cell types will also exhibit different reaction chemistries, leading to unique advantages and disadvantages.

Part I of this book introduces the basic underlying principles that govern all fuel cell devices. What you learn here will be equally applicable to a PEMFC, a SOFC, or any other

TABLE 1.1. Description of Major Fuel Cell Types

	PEMFC	PAFC	AFC	MCFC	SOFC
Electrolyte	Polymer membrane	Liquid H_3PO_4 (immobilized)	Liquid KOH (immobilized)	Molten carbonate	Ceramic
Charge carrier	H^+	H^+	OH^-	CO_3^{2-}	O^{2-}
Operating temperature	80°C	200°C	60–220°C	650°C	600–1000°C
Catalyst	Platinum	Platinum	Platinum	Nickel	Perovskites (ceramic)
Cell components	Carbon based	Carbon based	Carbon based	Stainless based	Ceramic based
Fuel compatibility	H_2, methanol	H_2	H_2	H_2, CH_4	H_2, CH_4, CO

fuel cell for that matter. Part II discusses the technology-specific aspects of the five major fuel cell types, while also delving into fuel cell system issues like stacking, fuel processing, control, and environmental impact.

1.6   BASIC FUEL CELL OPERATION

The current (electricity) produced by a fuel cell scales with the size of the reaction area where the reactants, the electrode, and the electrolyte meet. In other words, doubling a fuel cell's area approximately doubles the amount of current produced.

Although this trend seems intuitive, the explanation comes from a deeper understanding of the fundamental principles involved in the electrochemical generation of electricity. As we have discussed, fuel cells produce electricity by converting a primary energy source (a fuel) into a flow of electrons. This conversion necessarily involves an energy transfer step, where the energy from the fuel source is passed along to the electrons constituting the electric current. This transfer has a finite rate and must occur at an interface or reaction surface. Thus, the amount of electricity produced scales with the amount of reaction surface area or interfacial area available for the energy transfer. Larger surface areas translate into larger currents.

To provide large reaction surfaces that maximize surface-to-volume ratios, fuel cells are usually made into thin, planar structures, as shown in Figure 1.7. The electrodes are highly porous to further increase the reaction surface area and ensure good gas access. One side of the planar structure is provisioned with fuel (the anode electrode), while the other side is provisioned with oxidant (the cathode electrode). A thin electrolyte layer spatially separates the fuel and oxidant electrodes and ensures that the two individual half reactions occur

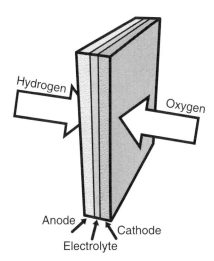

Figure 1.7. Simplified planar anode-electrolyte-cathode structure of a fuel cell.

ANODE = OXIDATION; CATHODE = REDUCTION

To understand any discussion of electrochemistry, it is essential to have a clear concept of the terms *oxidation*, *reduction*, *anode*, and *cathode*.

Oxidation and Reduction

- Oxidation refers to a process where electrons are *removed* from a species. Electrons are *liberated* by the reaction.
- Reduction refers to a process where electrons are *added* to a species. Electrons are *consumed* by the reaction.

For example, consider the electrochemical half reactions that occur in a H_2–O_2 fuel cell:

$$H_2 \rightarrow 2H^+ + 2e^- \tag{1.8}$$

$$\frac{1}{2}O_2 + 2H^+ + 2e^- \rightarrow H_2O \tag{1.9}$$

The hydrogen reaction is an oxidation reaction because electrons are being liberated by the reaction. The oxygen reaction is a reduction reaction because electrons are being consumed by the reaction. The above electrochemical half reactions are therefore known as the *hydrogen oxidation reaction* (*HOR*) and the *oxygen reduction reaction* (*ORR*).

Anode and Cathode

- Anode refers to an electrode where oxidation is taking place. More generally, the anode of any two-port device, such as a diode or resistor, is the electrode where electrons *flow out*.
- Cathode refers to an electrode where reduction is taking place. More generally, the cathode is the electrode where electrons *flow in*.

For a hydrogen–oxygen fuel cell:

- The anode is the electrode where the HOR takes place.
- The cathode is the electrode where the ORR takes place.

Note that the above definitions have nothing to do with which electrode is the positive electrode or which electrode is the negative electrode. *Be careful!* Anodes and cathodes can be either positive or negative. For a *galvanic cell* (a cell that *produces* electricity, like a fuel cell), the anode is the negative electrode and the cathode is the positive electrode. For an *electrolytic cell* (a cell that *consumes* electricity), the anode is the positive electrode and the cathode is the negative electrode.

Just remember anode = oxidation, cathode = reduction, and you will always be right!

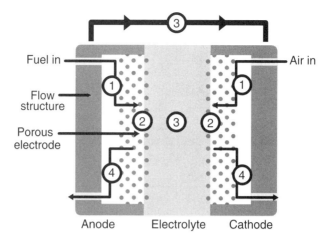

Fuel in

Flow
structure

Porous
electrode

Air in

Anode   Electrolyte   Cathode

Figure 1.8. Cross section of fuel cell illustrating major steps in electrochemical generation of electricity: (1) reactant transport; (2) electrochemical reaction; (3) ionic and electronic conduction; (4) product removal.

in isolation from one another. Compare this planar fuel cell structure with the simple fuel cell discussed earlier in Figure 1.4. While the two devices look quite different, noticeable similarities exist between them.

Figure 1.8 shows a detailed, *cross-sectional view* of a planar fuel cell. Using this figure as a map, we will now embark on a brief journey through the major steps involved in producing electricity in a fuel cell. Sequentially, as numbered on the drawing, these steps are as follows:

1. Reactant delivery (transport) into the fuel cell
2. Electrochemical reaction
3. Ionic conduction through the electrolyte and electron conduction through the external circuit
4. Product removal from the fuel cell

By the end of this book, you will understand the physics behind each of these steps in detail. For now, however, we'll just take a quick tour.

Step 1: Reactant Transport. For a fuel cell to produce electricity, it must be continually supplied with fuel and oxidant. This seemingly simple task can be quite complicated. When a fuel cell is operated at high current, its demand for reactants is voracious. If the reactants are not supplied to the fuel cell quickly enough, the device will "starve." Efficient delivery of reactants is most effectively accomplished by using *flow field plates* in combination with porous electrode structures. Flow field plates contain many fine channels or grooves to carry the gas flow and distribute it over the surface of the fuel cell. The shape, size, and pattern of flow channels can significantly affect the performance of the fuel cell. Understanding how flow structures and porous electrode geometries influence fuel cell performance is an exercise in mass transport, diffusion, and fluid mechanics. The materials aspects of flow structures and electrodes are equally important. Components are held to stringent materials property constraints that include very specific electrical, thermal, mechanical, and corrosion requirements. The details of reactant transport and flow field design are covered in Chapter 5.

Step 2: Electrochemical Reaction. Once the reactants are delivered to the electrodes, they must undergo electrochemical reaction. The current generated by the fuel cell is directly related to how fast the electrochemical reactions proceed. Fast electrochemical reactions result in a high current output from the fuel cell. Sluggish reactions result in low current output. Obviously, high current output is desirable. Therefore, catalysts are generally used to increase the speed and efficiency of the electrochemical reactions. Fuel cell performance critically depends on choosing the right catalyst and carefully designing the reaction zones. Often, the kinetics of the electrochemical reactions represent the single greatest limitation to fuel cell performance. The details of electrochemical reaction kinetics are covered in Chapter 3.

Step 3: Ionic (and Electronic) Conduction. The electrochemical reactions occurring in step 2 either produce or consume ions and electrons. Ions produced at one electrode must be consumed at the other electrode. The same holds for electrons. To maintain charge balance, these ions and electrons must therefore be transported from the locations where they are generated to the locations where they are consumed. For electrons this transport process is rather easy. As long as an electrically conductive path exists, the electrons will be able to flow from one electrode to the other. In the simple fuel cell in Figure 1.4, for example, a wire provides a path for electrons between the two electrodes. For ions, however, transport tends to be more difficult. Fundamentally, this is because ions are much larger and more massive than electrons. An electrolyte must be used to provide a pathway for the ions to flow. In many electrolytes, ions move via "hopping" mechanisms. Compared to electron transport, this process is far less efficient. Therefore, ionic transport can represent a significant resistance loss, reducing fuel cell performance. To combat this effect, the electrolytes in technological fuel cells are made as thin as possible to minimize the distance over which ionic conduction must occur. The details of ionic conduction are covered in Chapter 4.

Step 4: Product Removal. In addition to electricity, all fuel cell reactions will generate at least one product species. The H_2–O_2 fuel cell generates water. Hydrocarbon fuel cells will typically generate water and carbon dioxide (CO_2). If these products are not removed from the fuel cell, they will build up over time and eventually "strangle" the fuel cell, preventing new fuel and oxidant from being able to react. Fortunately, the act of delivering reactants *into* the fuel cell often assists the removal of product species *out of* the fuel cell. The same mass transport, diffusion, and fluid mechanics issues that are important in optimizing reactant delivery (step 1, above) can be applied to product removal. Often, product removal is not a significant problem and is frequently overlooked. However, for certain fuel cells (e.g., PEMFC) "flooding" by product water can be a major issue. Because product removal depends upon the same physical principles and processes that govern reactant transport, it is also treated in Chapter 5.

1.7 FUEL CELL PERFORMANCE

The performance of a fuel cell device can be summarized with a graph of its current–voltage characteristics. This graph, called a current–voltage (i–V) curve, shows the voltage output of the fuel cell for a given current output. An example of a typical i–V curve for a PEMFC is shown in Figure 1.9. Note that the current has been normalized by the area of the fuel cell, giving a current density (in amperes per square centimeter). Because a larger fuel cell can produce more electricity than a smaller fuel cell, i–V curves are normalized by fuel cell area to make results comparable.

An ideal fuel cell would supply any amount of current (as long as it is supplied with sufficient fuel) while maintaining a constant voltage determined by thermodynamics. In practice, however, the actual voltage output of a real fuel cell is less than the ideal thermodynamically predicted voltage. Furthermore, the more current that is drawn from a real fuel cell, the lower the voltage output of the cell, limiting the total power that can be delivered. The power (P) delivered by a fuel cell is given by the product of current and voltage:

$$P = iV \tag{1.10}$$

A fuel cell *power density curve*, which gives the power density delivered by a fuel cell as a function of the current density, can be constructed from the information in a fuel cell i–V curve. The power density curve is produced by multiplying the voltage at each point on the i–V curve by the corresponding current density. An example of combined fuel cell i–V and power density curves is provided in Figure 1.10. Fuel cell voltage is given on the left-hand y-axis while power density is given on the right-hand y-axis.

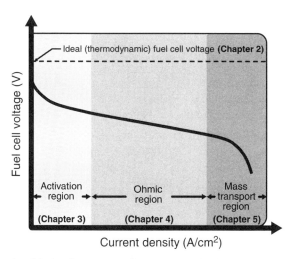

Figure 1.9. Schematic of fuel cell i–V curve. In contrast to the ideal, thermodynamically predicted voltage of a fuel cell (dashed line), the real voltage of a fuel cell is lower (solid line) due to unavoidable losses. Three major losses influence the shape of this i–V curve; they will be described in Chapters 3–5.

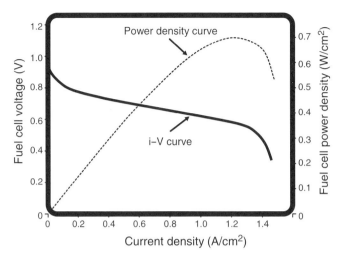

Figure 1.10. Combined fuel cell $i-V$ and power density curves. The power density curve is constructed from the $i-V$ curve by multiplying the voltage at each point on the $i-V$ curve by the corresponding current density. Fuel cell power density increases with increasing current density, reaches a maximum, and then falls at still higher current densities. Fuel cells are designed to operate at or below the power density maximum. At current densities below the power density maximum, voltage efficiency improves but power density falls. At current densities above the power density maximum, both voltage efficiency and power density fall.

The current supplied by a fuel cell is directly proportional to the amount of fuel consumed (each mole of fuel provides n moles of electrons). Therefore, as fuel cell voltage decreases, the electric power produced *per unit of fuel* also decreases. In this way, fuel cell voltage can be seen as a measure of fuel cell efficiency. In other words, you can think of the fuel cell voltage axis as an "efficiency axis." Maintaining high fuel cell voltage, even under high current loads, is therefore critical to the successful implementation of the technology.

Unfortunately, it is hard to maintain a high fuel cell voltage under current load. The voltage output of a real fuel cell is less than thermodynamically predicted voltage output due to irreversible losses. The more current that is drawn from the cell, the greater these losses. There are three major types of fuel cell losses, which give a fuel cell $i-V$ curve its characteristic shape. Each of these losses are associated with one of the basic fuel cell steps discussed in the previous section:

1. Activation losses (losses due to electrochemical reaction)
2. Ohmic losses (losses due to ionic and electronic conduction)
3. Concentration losses (losses due to mass transport)

The real voltage output for a fuel cell can thus be written by starting with thermodynamically predicted voltage output of the fuel cell and then subtracting the voltage drops due to the various losses:

$$V = E_{\text{thermo}} - \eta_{\text{act}} - \eta_{\text{ohmic}} - \eta_{\text{conc}} \qquad (1.11)$$

where

V = real output voltage of fuel cell

E_{thermo} = thermodynamically predicted fuel cell voltage output; this will be the subject of Chapter 2

η_{act} = activation losses due to reaction kinetics; this will be the subject of Chapter 3

η_{ohmic} = ohmic losses from ionic and electronic conduction; this will be the subject of Chapter 4

η_{conc} = concentration losses due to mass transport; this will be the subject of Chapter 5

The three major losses each contribute to the characteristic shape of the fuel cell $i–V$ curve. As shown in Figure 1.9, the activation losses mostly affect the initial part of the curve; the ohmic losses are most apparent in the middle section of the curve, and the concentration losses are most significant in the tail of the $i–V$ curve.

Equation 1.11 sets the stage for the next six chapters of this book. As you progress through these chapters, you will be armed with the tools needed to understand the major losses in fuel cell devices. Using Equation 1.11 as a starting point, you will eventually be able to characterize and model the performance of real fuel cell devices.

1.8 CHARACTERIZATION AND MODELING

Characterization and modeling are pivotal to the development and advancement of fuel cell technology. By assimilating theory and experiment, careful characterization and modeling studies allow us to better understand how fuel cells work, often paving the way toward further improvements.

Because these subjects provide great insight, each has been given a chapter in this book. Fuel cell modeling is covered in Chapter 6. Fuel cell characterization techniques are covered in Chapter 7. These chapters are expected to yield a practical understanding of how fuel cells are tested, how to diagnose their performance, and how to develop simple mathematical models to predict fuel cell behavior.

1.9 FUEL CELL TECHNOLOGY

The majority of this book is devoted to understanding the fundamental principles underlying fuel cells. However, no treatment of fuel cells is complete without a discussion of the practical aspects of fuel cell technology. This is the aim of the second part of this book. A series of chapters will introduce the major considerations for fuel cell stacking and system design, as well as specific technological aspects related to each of the five major fuel cell types. You will gain insight into the state of the art in fuel cell design as well as a historical perspective on the development of practical fuel cell technology.

1.10   FUEL CELLS AND THE ENVIRONMENT

If employed correctly, fuel cells are environmentally friendly. In fact, this may be their single greatest advantage over other energy conversion technologies. However, the environmental impact of fuel cells depends strongly on the context of their use. If they are not deployed wisely, fuel cells may be no better than our current fossil energy conversion system! In the final chapter of this book, you will learn to evaluate possible fuel cell deployment scenarios. Using a technique known as process chain analysis, you will be able to identify promising fuel cell futures.

One such future, referred to as the "hydrogen economy," is illustrated in Figure 1.11. In this figure, H_2 fuel cells are coupled with electrolyzers and renewable energy technologies (such as wind and solar power) to provide a completely closed-loop, pollution-free energy economy. In such a system, fuel cells would play a prominent role, with a primary benefit being their dispatchability. When the sun is shining or the wind is blowing, the electricity produced from solar and wind energy would be used to power cities directly while producing extra hydrogen on the side via electrolysis. Anytime the wind stops or night falls, however, the fuel cells could be dispatched to provide on-demand power by converting the stored hydrogen into electricity. In such a system, fossil fuels are completely eliminated.

Currently, it is unclear when, if ever, the hydrogen economy will become a reality. Various studies have examined the technical and economic hurdles that stand in the way of the hydrogen economy. While many of these studies differ on the details, it is clear that the transition to a hydrogen economy would be difficult, costly, and lengthy. Do not count on it happening anytime soon. In the meantime, we have a fossil fuel world. Even in a fossil fuel world, however, it is important to realize that fuel cells could provide increased efficiency, greater scaling flexibility, reduced emissions, and other advantages compared to conventional power technologies. Fuel cells have found, and will continue to find, niche applications. These applications should continue to drive forward progress for decades to come, with or without the hydrogen economy dream.

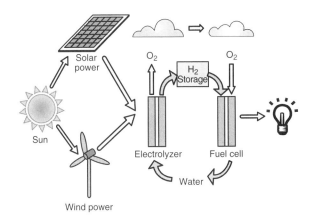

Figure 1.11. Schematic of hydrogen economy dream.

CHAPTER SUMMARY

The purpose of this chapter was to set the stage for learning about fuel cells and to give a broad overview of fuel cell technology.

- A fuel cell is a direct electrochemical energy conversion device. It directly converts energy from one form (chemical energy) into another form (electrical energy) through electrochemistry.
- Unlike a battery, a fuel cell cannot be depleted. It is a "factory" that will continue to generate electricity as long as fuel is supplied.
- At a minimum, a fuel cell must contain two electrodes (an anode and a cathode) separated by an electrolyte.
- Fuel cell power is determined by fuel cell size. Fuel cell capacity (energy capacity) is determined by the fuel reservoir size.
- There are five major fuel cell types, differentiated by their electrolyte.
- Electrochemical systems must contain two coupled half reactions: an oxidation reaction and a reduction reaction. An oxidation reaction liberates electrons. A reduction reaction consumes electrons.
- Oxidation occurs at the anode electrode. Reduction occurs at the cathode electrode.
- The four major steps in the generation of electricity in a fuel cell are (1) reactant transport, (2) electrochemical reaction, (3) ionic (and electronic) conduction, and (4) product removal.
- Fuel cell performance can be assessed by current–voltage curves. Current–voltage curves show the voltage output of a fuel cell for a given current load.
- Ideal fuel cell performance is dictated by thermodynamics.
- Real fuel cell performance is always less than ideal fuel cell performance due to losses. The major types of loss are (1) activation loss, (2) ohmic loss, and (3) concentration loss.

CHAPTER EXERCISES

Review Questions

1.1 List three major advantages and three major disadvantages of fuel cells compared to other power conversion devices. Discuss at least two potential applications where the unique attributes of fuel cells make them attractive.

1.2 In general, do you think a portable fuel cell would be better for an application requiring low power but high capacity (long run time) or high power but small capacity (short run time)? Explain.

1.3 Label the following reactions as oxidation or reduction reactions:
(a) $Cu \rightarrow Cu^{2+} + 2e^-$
(b) $2H^+ + 2e^- \rightarrow H_2$
(c) $O^{2-} \rightarrow \frac{1}{2}O_2 + 2e^-$

(d) $CH_4 + 4O^{2-} \rightarrow CO_2 + 2H_2O + 8e^-$
(e) $O^{2-} + CO \rightarrow CO_2 + 2e^-$
(f) $\frac{1}{2}O_2 + H_2O + 2e^- \rightarrow 2(OH)^-$
(g) $H_2 + 2(OH)^- \rightarrow 2H_2O + 2e^-$

1.4 From the reactions listed in problem 1.3 (or their reverse), write three complete and balanced pairs of electrochemical half reactions. For each pair of reactions, identify which reaction is the cathode reaction and which reaction is the anode reaction.

1.5 Consider the relative volumetric and gravimetric energy densities of 7500 psi compressed H_2 versus liquid H_2. Which would probably be the better candidate for a fuel cell bus? *Hint*: Bus efficiency strongly depends on gross vehicle weight.

1.6 Describe the four major steps in the generation of electricity within a fuel cell. Describe the potential reasons for loss in fuel cell performance for each step.

Calculations

1.7 Energy is released when hydrogen and oxygen react to produce water. This energy comes from the fact that the final hydrogen–oxygen bonds represent a lower total energy state compared to the original hydrogen–hydrogen and oxygen–oxygen bonds. Calculate how much energy (in kilojoules per mole of product) is released by the reaction

$$H_2 + \frac{1}{2}O_2 \rightleftharpoons H_2O \qquad (1.12)$$

at constant pressure and given the following standard bond enthalpies. Standard bond enthalpies denote the enthalpy *absorbed* when bonds are *broken* at standard temperature and pressure (298 K and 1 atm).

Standard Bond Enthalpies

H–H = 432 kJ/mol
O–O = 494 kJ/mol
H–O = 460 kJ/mol

1.8 Consider a fuel cell vehicle. The vehicle draws 30 kW of power at 60 mph and is 40% efficient at rated power. (It converts 40% of the energy stored in the hydrogen fuel to electric power.) You are asked to size the fuel cell system so that a driver can go at least 300 miles at 60 mph before refueling. Specify the minimum volume and weight requirements for the fuel cell system (fuel cell + fuel tank) given the following information:

- Fuel cell power density: 1 kW/L, 500 W/kg
- Fuel tank energy density (compressed hydrogen): 4 MJ/L, 8 MJ/kg

1.9 For the fuel cell $i-V$ curve shown in Figure 1.9, sketch the approximate corresponding current density–power curve.

CHAPTER 2

FUEL CELL THERMODYNAMICS

Thermodynamics is the study of energetics; the study of the transformation of energy from one form to another. Since fuel cells are energy conversion devices, fuel cell thermodynamics is key to understanding the conversion of chemical energy into electrical energy.

For fuel cells, thermodynamics can predict whether a candidate fuel cell reaction is energetically spontaneous. Furthermore, thermodynamics places upper bound limits on the maximum electrical potential that can be generated in a reaction. Thus, thermodynamics yields the theoretical boundaries of what is possible with a fuel cell; it gives the "ideal case."

Any real fuel cell will perform at or below its thermodynamic limit. Understanding real fuel cell performance requires a knowledge of kinetics in addition to thermodynamics. This chapter covers the thermodynamics of fuel cells. Subsequent chapters will cover the major kinetic limitations to fuel cell performance, defining practical performance.

2.1 THERMODYNAMICS REVIEW

This section presents a brief review of the main tenets of thermodynamics. These basic theories are typically taught in an introductory thermodynamics course. Next, these concepts are extended to include parameters which are needed to understand fuel cell behavior. Readers are advised to consult a thermodynamics book if additional review is required.

2.1.1   What Is Thermodynamics?

It is no secret that no one really understands the meaning of popular thermodynamic quantities. For example, Nobel Prize winning physicist Richard Feynman wrote in his *Lectures on Physics*: "It is important to realize that in modern physics today, we have no knowledge of what energy is."[1] We have even less intuition about terms such as *enthalpy* and *free energy*. The fundamental assumptions of thermodynamics are based on human experience. Assumptions are the best we can do. We *assume* that energy can never be created or destroyed (first law of thermodynamics) only because it fits with everything experienced in human existence. Nevertheless, no one knows why it should be so.

If we accept a few of these fundamental assumptions, however, we can develop a self-consistent mathematical description that tells us how important quantities such as energy, temperature, pressure, and volume are related. This is really all that thermodynamics is; it is an elaborate bookkeeping scheme that allows us to track the properties of systems in a self-consistent manner, starting from a few basic assumptions or "laws."

2.1.2   Internal Energy

A fuel cell converts energy stored within a fuel into other, more useful forms of energy. The total intrinsic energy of a fuel (or of any substance) is quantified by a property known as internal energy (U). Internal energy is the energy associated with microscopic movement and interaction between particles on the atomic and molecular scale. It is separated in scale from the macroscopic ordered energy associated with moving objects. For example, a tank of H_2 gas sitting on a table has no apparent energy. However, the H_2 gas actually has significant internal energy (see Figure 2.1); on the microscopic scale it is a whirlwind of molecules traveling hundreds of meters per second. Internal energy is also associated with the *chemical bonds* between the hydrogen atoms. A fuel cell can convert *only a portion* of the internal energy associated with a tank of H_2 gas into electrical energy. The limits on how much of the internal energy of the H_2 gas can be transformed into electrical energy are established by the first and second laws of thermodynamics.

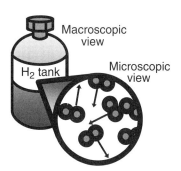

Figure 2.1. Although this tank of H_2 gas has no apparent macroscopic energy, it has significant internal energy. Internal energy is associated with microscopic movement (kinetic energy) and interactions between particles (chemical/potential energy) on the atomic scale.

2.1.3 First Law

The first law of thermodynamics is also known as the law of conservation of energy—energy can never be created or destroyed—as expressed by the equation:

$$d(\text{Energy})_{\text{univ}} = d(\text{Energy})_{\text{system}} + d(\text{Energy})_{\text{surroundings}} = 0 \qquad (2.1)$$

Viewed another way, this equation states that any change in the energy of a system must be fully accounted for by energy transfer to the surroundings:

$$d(\text{Energy})_{\text{system}} = -d(\text{Energy})_{\text{surroundings}} \qquad (2.2)$$

There are two ways that energy can be transferred between a closed system and its surroundings: via *heat* (Q) or *work* (W). This allows us to write the first law in its more familiar form

$$dU = dQ - dW \qquad (2.3)$$

This expression states that the change in the internal energy of a closed system (dU) must be equal to the heat transferred to the system (dQ) minus the work done by the system (dW). To develop this expression from Equation 2.2, we have substituted dU for $d(\text{Energy})_{\text{system}}$; if we choose the proper reference frame, then all energy changes in a system are manifested as internal energy changes. Note that we define positive work as work done *by* the system on the surroundings.

For now, we will assume that only mechanical work is done by a system. Mechanical work is accomplished by the expansion of a system against a pressure. It is given by

$$(dW)_{\text{mech}} = p\, dV \qquad (2.4)$$

where p is the pressure and dV is the volume change. Later, when we talk about fuel cell thermodynamics, we will consider the electrical work done by a system. For now, however, we ignore electrical work. Considering only mechanical work, we can rewrite the expression for the internal energy change of a system as

$$dU = dQ - p\, dV \qquad (2.5)$$

2.1.4 Second Law

The second law of thermodynamics introduces the concept of *entropy*. Entropy is determined by the number of possible microstates accessible to a system, or in other words, the number of possible ways of configuring a system. For this reason, entropy can be thought of as a measure of "disorder," since an increasing entropy indicates an increasing number of ways of configuring a system. For an isolated system (the simplest case)

$$S = k \log \Omega \qquad (2.6)$$

WORK AND HEAT

In contrast to internal energy, work and heat are not properties of matter, or of any particular system (e.g., substance or body). They represent *energy in transit*, in other words, energy that is transferred between substances or bodies.

In the case of work, this transfer of energy is accomplished by the application of a force over a distance. Heat, on the other hand, is transferred between substances whenever they have different thermal energies, as manifested by differences in their temperature.

Due to repercussions of the second law (which we will discuss momentarily), work is often called the most "noble" form of energy; it is the universal donor. Energy, in the form of work, can be converted into any other form of energy at 100% theoretical efficiency. In contrast, heat is the most "ignoble" form of energy; it is the universal acceptor. Any form of energy can eventually be 100% dissipated to the environment as heat, but heat can never be 100% converted back to work.

The nobility of work versus heat illustrates one of the central differences between fuel cells and combustion engines. A combustion engine burns fuel to produce heat and then converts some of this heat into work. Because it first converts energy into heat, the combustion engine destroys some of the work potential of the fuel. This unfortunate destruction of work potential is called the "thermal bottleneck." Because a fuel cell bypasses the heat step, it avoids the thermal bottleneck.

where S is the total entropy of the system, k is Boltzmann's constant, and Ω denotes the number of possible microstates accessible to the system.

Microstates can best be understood with an example. Consider the "perfect" system of 100 identical atoms shown in Figure 2.2*a*. There is only one possible microstate, or *configuration*, for this system. This is because the 100 atoms are exactly identical and indistinguishable from one another. If we were to "switch" the first and the second atoms, the system would look exactly the same. The entropy of this perfect 100-atom crystal is

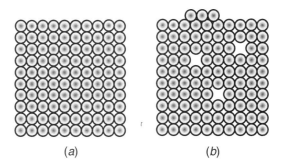

(a)   (b)

Figure 2.2. (a) The entropy of this 100 atom perfect crystal is zero because there is only one possible way to arrange the atoms to produce this configuration. (b) When three atoms are removed from the crystal and placed on the surface, the entropy increases. This is because there are many possible ways to configure a system of 100 atoms where 3 have been removed.

therefore zero ($S = k \log 1 = 0$). Now consider Figure 2.2b, where three atoms have been removed from their original locations and placed on the surface of the crystal. Any three atoms could have been removed from the crystal, and depending on which atoms were removed, the final configuration of the system will be different. In this case, there are many microstates available to the system. (Figure 2.2b represents just one of them.) We can calculate the number of microstates available to the system by evaluating the number of possible ways there are to take N atoms from a total of Z atoms:

$$\Omega = \frac{Z(Z-1)(Z-2)\cdots(Z-N+1)}{N!} = \frac{Z!}{(Z-N)!(N!)} \tag{2.7}$$

In Figure 2.2b, there are 100 atoms. The number of ways to take 3 atoms from 100 is

$$\Omega = \frac{9.3 \times 10^{157}}{(9.6 \times 10^{151})(6)} = 1.7 \times 10^5 \tag{2.8}$$

This yields $S = 7.22 \times 10^{-23}$ J/K.

Except for extremely simple systems like the one in this example, it is impossible to calculate entropy exactly. Instead, a system's entropy is usually *inferred* based on how heat transfer causes the entropy of the system to change. For a reversible transfer of heat at constant pressure, the entropy of a system will change as:

$$dS = \frac{dQ_{\text{rev}}}{T} \tag{2.9}$$

where dS is the entropy change in the system associated with a reversible transfer of heat (dQ_{rev}) at a constant temperature (T). In other words, "dumping" energy, including heat, into a system causes its entropy to increase. Essentially, by providing additional energy to the system, we enable it to access additional microstates, causing its entropy to increase. For an irreversible transfer of heat, the entropy increase will be even larger than that dictated by Equation 2.9. This is a key statement of the second law of thermodynamics.

The most widely known form of the second law acknowledges that the entropy of a system and its surroundings must increase or at least remain zero for any process:

$$dS_{\text{univ}} \geq 0 \tag{2.10}$$

This inequality, when combined with the first law of thermodynamics, allows us to separate thermodynamically "spontaneous" processes from "nonspontaneous" processes.

2.1.5 Thermodynamic Potentials

Based on the first and second laws of thermodynamics, we can write down "rules" to specify how energy can be transferred from one form to another. These rules are called *thermodynamic potentials*. You are already familiar with one thermodynamic potential: the internal energy of a system. We can combine results from the first and the second laws of thermodynamics (Equations 2.3 and 2.9) to arrive at an equation for internal energy which is based on the variation of two independent variables, entropy S and volume V:

$$dU = T\,dS - p\,dV \tag{2.11}$$

Remember, $T\,dS$ represents the reversible heat transfer and $p\,dV$ is the mechanical work. As mentioned above, from this equation we can conclude that U, the internal energy of a system, is a function of entropy and volume:

$$U = U(S, V) \tag{2.12}$$

We can also derive the following useful relations which show how the *dependent* variables (T and p) are related to variations in the *independent* variables (S and V):

$$\left(\frac{dU}{dS}\right)_V = T \tag{2.13}$$

$$\left(\frac{dU}{dV}\right)_S = -p \tag{2.14}$$

Unfortunately, S and V are not easily measurable in most experiments. (There is no such thing as an "entropy meter.") Therefore, a new thermodynamic potential is needed equivalent to U but depending on quantities that are more readily measured than S and V. Temperature T and pressure p fall into this category. Happily, there is a simple mathematical way to accomplish this conversion using a *Legendre transform*. A step-by-step transformation of U begins with defining the new thermodynamic potential $G(T, p)$ as follows:

$$G = U - \left(\frac{dU}{dS}\right)_V S - \left(\frac{dU}{dV}\right)_S V \tag{2.15}$$

Since we know that $(dU/dS)_V = T$ and $(dU/dV)_S = -p$, we obtain

$$G = U - TS + pV \tag{2.16}$$

This function is called the Gibbs free energy. Let us show that G is indeed a function of the temperature and the pressure. The variation of G (mathematically dG) results in

$$dG = dU - T\,dS - S\,dT + p\,dV + V\,dp \tag{2.17}$$

Since we know that $dU = T\,dS - p\,dV$, we can see that

$$dG = -S\,dT + V\,dp \tag{2.18}$$

So, the Gibbs free energy is nothing more than a thermodynamic description of a system that depends on T and p instead of S and V.

What if we want a potential that depends on S and p? No problem! Remember that U is a function of S and V. To get a thermodynamic potential that is a function of S and p, we need only to transform U with respect to V this time. Analogous to Equation 2.15 we define this new thermodynamic potential H as

$$H = U - \left(\frac{dU}{dV}\right)_S V \qquad (2.19)$$

Again, since $(dU/dV)_S = -p$, we obtain

$$H = U + pV \qquad (2.20)$$

where H is called enthalpy. Through differentiation, we can show that H is a function of S and p:

$$dH = dU + p\, dV + V\, dp \qquad (2.21)$$

Again, $dU = T\, dS - p\, dV$, so

$$dH = T\, dS + V\, dp \qquad (2.22)$$

Thus far, we have defined three thermodynamic potentials: $U(S, V)$, $H(S, p)$, and $G(T, p)$. Defining a fourth and final thermodynamic potential which depends on temperature and volume, $F(T, V)$, completes the symmetry:

$$F = U - TS \qquad (2.23)$$

where F is the Helmholtz free energy. We leave it to the reader to show that

$$dF = -S\, dT - p\, dV \qquad (2.24)$$

A summary of these four thermodynamic potentials is provided in Figure 2.3. This mnemonic diagram, originally suggested by Schroeder [2], can help you keep track of the relationships between the thermodynamic potentials. Loosely, the four potentials are defined as follows:

- **Internal Energy** (U). The energy needed to create a system in the absence of changes in temperature or volume.
- **Enthalpy** (H). The energy needed to create a system plus the work needed to make room for it (from zero volume).
- **Helmholtz Free Energy** (F). The energy needed to create a system minus the energy that you can get from the system's environment due to spontaneous heat transfer (at constant temperature).
- **Gibbs Free Energy** (G). The energy needed to create a system and make room for it minus the energy that you can get from the environment due to heat transfer. In other words, G represents the net energy cost for a system created at a constant environmental temperature T from a negligible initial volume after subtracting what the environment automatically supplied.

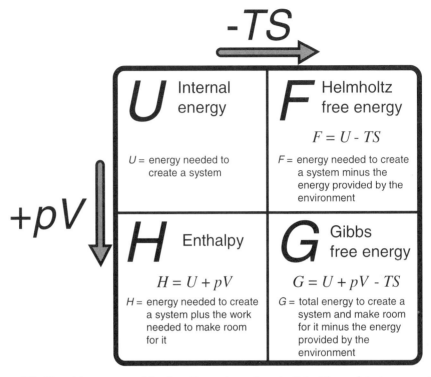

Figure 2.3. Pictorial summary of the four thermodynamic potentials. They relate to one another by offsets of the "energy from the environment" term TS and the "expansion work" term pV. Use this diagram to help remember the relationships. (Fig. 5.2, p. 151, from *An Introduction to Thermal Physics* by Daniel V. Schroeder [2]. Copyright ©2000 by Addison Wesley Longman. Reprinted by permission of Pearson Education, Inc.)

2.1.6 Molar Quantities

Typical notation distinguishes between intrinsic and extrinsic variables. Intrinsic quantities such as temperature and pressure do not scale with the system size; extrinsic quantities such as internal energy and entropy do scale with system size. For example, if the size of a box of gas molecules is doubled and the number of molecules in the box doubles, then the internal energy and entropy double while the temperature and pressure are constant. It is conventional to denote intrinsic quantities with a lowercase letter (p) and extrinsic quantities with an uppercase letter (U).

Molar quantities such as $\hat{u}$, the internal energy per mole of gas (units of kilojoules per-mole), are intrinsic. It is often useful to calculate energy changes due to a reaction on a per-mole basis

$$\Delta \hat{g}_{rxn}, \ \Delta \hat{s}_{rxn}, \ \Delta \hat{v}_{rxn}$$

The Δ symbol denotes a change during a thermodynamic process (such as a reaction), calculated as final state–initial state. Therefore, a negative energy change means energy is released during a process; a negative volume change means the volume decreases during a process. For example, the overall reaction in a H_2–O_2 fuel cell,

$$H_2 + \tfrac{1}{2}O_2 \rightleftharpoons H_2O \tag{2.25}$$

has $\Delta \hat{g}_{rxn} = -237$ kJ/mol H_2 at room temperature and pressure. For every mole of H_2 gas consumed (or every $\tfrac{1}{2}$ mol of O_2 gas consumed or mole of H_2O produced), the Gibbs free energy change is -237 kJ. If 5 mol of O_2 gas is reacted, the *extrinsic* Gibbs free energy change (ΔG_{rxn}) would be:

$$5 \text{ mol } O_2 \times \left(\frac{1 \text{ mol } H_2}{1/(2 \text{ mol } O_2)} \right) \times \left(\frac{-237 \text{ kJ}}{\text{mol } H_2} \right) = -2370 \text{ kJ} \tag{2.26}$$

Of course the *intrinsic* (per-mole) Gibbs free energy of this reaction is still $\Delta \hat{g}_{rxn} = -237$ kJ/mol H_2.

2.1.7 Standard State

Because most thermodynamic quantities depend on temperature and pressure, it is convenient to reference everything to a standard set of conditions. This set of conditions is called the standard-state. Standard-state conditions are taken as room temperature (298.15 K) and atmospheric pressure. (Standard-state pressure is actually defined as 1 bar = 100 kPa. Atmospheric pressure is taken as 1 atm = 101.325 kPa. These slight differences are usually ignored.) The standard-state conditions are also known as standard temperature and pressure, or STP. Standard-state conditions further specify that all reactant and product species are present at unit activity. (Activity is discussed in Section 2.4.3.) Standard-state conditions are designated by a superscript zero. For example, $\Delta \hat{h}^0$ represents an enthalpy change at STP.

2.1.8 Reversibility

We frequently use the term "reversible" when talking about the thermodynamics of fuel cells. Reversible implies equilibrium. A reversible fuel cell voltage is the voltage produced by a fuel cell at thermodynamic equilibrium. A process is thermodynamically reversible when an infinitesimal reversal in the driving force causes it to reverse direction; such a system is always at equilibrium.

Equations relating to reversible fuel cell voltages only apply to equilibrium conditions. As soon as current is drawn from a fuel cell, equilibrium is lost and reversible fuel cell voltage equations no longer apply. To distinguish between reversible and nonreversible fuel cell voltages in this book, we will use the symbols E and V, where E represents a reversible (thermodynamically predicted) fuel cell voltage and V represents an operational (nonreversible) fuel cell voltage.

2.2   HEAT POTENTIAL OF A FUEL: ENTHALPY OF REACTION

Now that we have reviewed general thermodynamics, the exciting work begins. We will now apply what we know about thermodynamics to fuel cells. Remember, the goal of a fuel cell is to extract the internal energy from a fuel and convert it into more useful forms of energy. What is the maximum amount of energy that we can extract from a fuel? The maximum depends on whether we extract energy from the fuel in the form of heat or work. As is shown in this section, the maximum heat energy that can be extracted from a fuel is given by the fuel's enthalpy of reaction (for a constant-pressure process).

Recall the differential expression for enthalpy (Equation 2.22):

$$dH = T\,dS + V\,dp \qquad (2.27)$$

For a constant-pressure process ($dp = 0$), Equation 2.27 reduces to

$$dH = T\,dS \qquad (2.28)$$

Here, dH is the same as the heat transferred (dQ) in a reversible process. For this reason, we can think of enthalpy as a measure of the heat potential of a system under constant-pressure conditions. In other words, for a constant-pressure reaction, the enthalpy change expresses the amount of heat that could be evolved by the reaction. From where does this heat originate? Expressing dH in terms of dU at constant pressure provides the answer:

$$dH = T\,dS = dU + dW \qquad (2.29)$$

From this expression, we see that the heat evolved by a reaction is due to changes in the internal energy of the system, after accounting for any energy that goes toward work. The internal energy change in the system is largely due to the reconfiguration of chemical bonds. For example, as discussed in the previous chapter, burning hydrogen releases heat due to molecular bonding reconfigurations. The product water rests at a lower internal energy state than the initial hydrogen and oxygen reactants. After accounting for the energy that goes toward work, the rest of the internal energy difference is transformed into heat during the reaction. The situation is analogous to a ball rolling down a hill; the potential energy of the ball is converted into kinetic energy as it rolls from the high-potential-energy initial state to the low-potential-energy final state.

The enthalpy change associated with a combustion reaction is called the *heat of combustion*. The name *heat* of combustion indicates the close tie between enthalpy and heat potential for constant-pressure chemical reactions. More generally, the enthalpy change associated with any chemical reaction is called the *enthalpy of reaction* or *heat of reaction*. We use the more general term *enthalpy of reaction* (ΔH_{rxn} or $\Delta \hat{h}_{rxn}$) in this text.

2.2.1   Calculating Reaction Enthalpies

Since reaction enthalpies are associated mainly with the reconfiguration of chemical bonds during reaction, they can be calculated by considering the bond enthalpy differences between the reactants and products. For example, in problem 1.7, we *approximated* how much heat is released in the H_2 combustion reaction by comparing the enthalpies of the reactant O–O and H–H bonds to the product H–O bonds.

Bond enthalpy calculations are somewhat awkward and give only rudimentary approximations. Therefore, enthalpy-of-reaction values are normally calculated by computing the *formation enthalpy* differences between reactants and products. A standard-state formation enthalpy $\Delta \hat{h}_f^0(i)$ tells how much enthalpy is required to form 1 mol of chemical species i at STP from the reference species. For a general reaction

$$aA + bB \rightarrow mM + nN \tag{2.30}$$

where A and B are reactants, M and N are products, and a, b, m, n represent the number of moles of A, B, M, and N, respectively; $\Delta \hat{h}_{rxn}^0$ may be calculated as

$$\Delta \hat{h}_{rxn}^0 = \left[m\,\Delta \hat{h}_f^0(M) + n\Delta \hat{h}_f^0(N) \right] - \left[a\Delta \hat{h}_f^0(A) + b\Delta \hat{h}_f^0(B) \right] \tag{2.31}$$

Thus, the enthalpy of reaction is computed from the difference between the *molar weighted* reactant and product formation enthalpies. Note that enthalpy changes (like all energy changes) are computed in the form of *final state–initial state*, or in other words, *products–reactants*.

An expression analogous to Equation 2.31 may be written for the standard-state entropy of a reaction, $\Delta \hat{s}_{rxn}^0$, using *standard entropy* values $\hat{s}^0$ for the species taking part in the reaction. See Example 2.1 for details.

> **Example 2.1.** A direct methanol fuel cell uses methanol as fuel instead of hydrogen. Calculate the $\Delta \hat{h}_{rxn}^0$ and $\Delta \hat{s}_{rxn}^0$ for the methanol combustion reaction:
>
> $$CH_3OH + \tfrac{3}{2}O_2 \rightarrow CO_2 + 2H_2O_{(liq)} \tag{2.32}$$
>
> *Solution:* From Appendix B, the $\Delta \hat{h}_f^0$ and $\hat{s}^0$ values for CH_3OH, O_2, CO_2, and H_2O are

Chemical Species	$\Delta \hat{h}_f^0$ (kJ/mol)	$\hat{s}^0$[J/(mol · K)]
CH_3OH	−200.95	239.83
O_2	0	205.14
CO_2	−393.51	213.80
$H_2O_{(liq)}$	−285.83	69.95

Following Equation 2.31, the $\Delta \hat{h}_{rxn}^0$ for methanol combustion is calculated as

$$\Delta \hat{h}_{rxn}^0 = \left[2\Delta \hat{h}_f^0(H_2O) + \Delta \hat{h}_f^0(CO_2) \right] - \left[\tfrac{3}{2}\Delta \hat{h}_f^0(O_2) + \Delta \hat{h}_f^0(CH_3OH) \right]$$

$$= [2(-285.83) + (-393.51)] - \left[\tfrac{3}{2}(0) + (-245.98) \right]$$

$$= -719.19 \text{ kJ/mol} \tag{2.33}$$

Similarly, $\Delta \hat{s}_{\text{rxn}}^0$ is calculated as

$$\Delta \hat{s}_{\text{rxn}}^0 = \left[2\hat{s}^0(H_2O) + \hat{s}^0(CO_2)\right] - \left[\tfrac{3}{2}\hat{s}^0(O_2) + \hat{s}^0(CH_3OH)\right]$$

$$= [2(69.95) + (213.8)] - \left[\tfrac{3}{2}(205.14) + (239.83)\right]$$

$$= -193.84 \, \text{J/(mol} \cdot \text{K)} \tag{2.34}$$

2.2.2 Temperature Dependence of Enthalpy

The amount of heat energy that a substance can absorb changes with temperature. It follows that a substance's formation enthalpy also changes with temperature. The variation of enthalpy with temperature is described by a substance's *heat capacity*:

$$\Delta \hat{h}_f = \Delta \hat{h}_f^0 + \int_{T_0}^{T} c_p(T) \, dT \tag{2.35}$$

where $\Delta \hat{h}_f$ is the formation enthalpy of the substance at an arbitrary temperature T, $\Delta \hat{h}_f^0$ is the reference formation enthalpy of the substance at $T_0 = 298.15$ K, and $c_p(T)$ is the constant-pressure heat capacity of the substance (which itself may be a function of temperature). If phase changes occur along the path between T_0 and T, extra caution must be taken to make sure that the enthalpy changes associated with these phase changes are also included.

In a similar manner, the entropy of a substance also varies with temperature. Again, this variation is described by the substance's heat capacity:

$$\hat{s} = \hat{s}^0 + \int_{T_0}^{T} \frac{c_p(T)}{T} \, dT \tag{2.36}$$

From Equations 2.31, 2.35, and 2.36, $\Delta \hat{h}_{\text{rxn}}$ and $\Delta \hat{s}_{\text{rxn}}$ for any reaction at any temperature can be calculated, so long as the basic thermodynamic data ($\Delta \hat{h}_f^0, \hat{s}^0, c_p$) are provided. Appendix B provides a collection of basic thermodynamic data for a variety of chemical species relevant to fuel cells.

Since heat capacity effects are generally minor, $\Delta \hat{h}_f^0$ and $\hat{s}^0$ values are usually assumed to be independent of temperature, simplifying thermodynamic calculations. See Example 2.2 for an illustration.

In a perfect world, we could harness all of the enthalpy released by a chemical reaction to do useful work. Unfortunately, thermodynamics tells us that this is not possible. Only a portion of the energy evolved by a chemical reaction can be converted into useful work. For electrochemical systems (i.e., fuel cells) the Gibbs free energy gives the maximum amount of energy that is available to do electrical work.

2.3   WORK POTENTIAL OF A FUEL: GIBBS FREE ENERGY

Recall from Section 2.1.5 that the Gibbs free energy can be considered to be the net energy required to create a system and make room for it minus the energy received from the environment due to spontaneous heat transfer. Thus, G represents the energy that *you* had to transfer to create the system. (The environment also transferred some energy via heat, but G subtracts this contribution out.) If G represents the net energy you had to transfer to create the system, then G should also represent the maximum energy that you could ever get back out of the system. In other words, the Gibbs free energy represents the exploitable energy potential, or *work potential*, of the system.

2.3.1   Calculating Gibbs Free Energies

Since the Gibbs free energy is the key to the work potential of a reaction, it is necessary to calculate $\Delta \hat{g}_{rxn}$ values as we calculated $\Delta \hat{h}_{rxn}$ and $\Delta \hat{s}_{rxn}$ values. In fact, we can calculate $\Delta \hat{g}_{rxn}$ values *directly* from $\Delta \hat{h}_{rxn}$ and $\Delta \hat{s}_{rxn}$ values. Recalling how G is defined, it is apparent that G already contains H, since $G = U + PV - TS$ and $H = U + PV$. We can therefore write the Gibbs free energy as

$$G = H - TS \qquad (2.37)$$

Differentiating this expression gives

$$dG = dH - T\,dS - S\,dT \qquad (2.38)$$

Holding temperature constant (isothermal process) and writing this relationship in terms of molar quantities gives

$$\Delta \hat{g} = \Delta \hat{h} - T\,\Delta \hat{s} \qquad (2.39)$$

Thus, for an isothermal reaction, we can compute $\Delta \hat{g}$ in terms of $\Delta \hat{h}$ and $\Delta \hat{s}$. The isothermal reaction assumption means that temperature is constant *during* the reaction. However, it is important to realize that we can still use Equation 2.39 to calculate $\Delta \hat{g}$ values at different reaction temperatures.

Example 2.2. Determine the approximate temperature at which the following reaction is no longer spontaneous:

$$CO + H_2O \rightarrow CO_2 + H_2 \qquad (2.40)$$

Solution: To answer this question, we need to calculate the Gibbs free energy for this reaction as a function of temperature, and then solve for the temperature at which the Gibbs free energy for this reaction goes to zero:

$$\Delta \hat{g}_{rxn}(T) = \Delta \hat{h}_{rxn}(T) - T\,\Delta \hat{s}_{rxn}(T) = 0 \qquad (2.41)$$

To get an approximate answer, we can assume that $\Delta\hat{h}_{rxn}$ and $\Delta\hat{s}_{rxn}$ are independent of temperature (heat capacity effects are ignored). In this case, the temperature dependence of $\Delta\hat{g}_{rxn}$ is approximated as:

$$\Delta\hat{g}_{rxn}(T) = \Delta\hat{h}^0_{rxn} - T\,\Delta\hat{s}^0_{rxn} \qquad (2.42)$$

The $\Delta\hat{h}^0_f$ and $\hat{s}^0$ values for CO, CO_2, H_2, and H_2O are given in Appendix B:

Chemical Species	$\Delta\hat{h}^0_f$ (kJ/mol)	$\hat{s}^0$[J/(mol · K)]
CO	−110.54	197.65
CO_2	−393.51	213.80
H_2	0	130.67
$H_2O_{(g)}$	−241.84	188.82

Following Equation 2.31, $\Delta\hat{h}^0_{rxn}$ is calculated as

$$\Delta\hat{h}^0_{rxn} = \left[\Delta\hat{h}^0_f(CO_2) + \Delta\hat{h}^0_f(H_2)\right] - \left[\Delta\hat{h}^0_f(CO) + \Delta\hat{h}^0_f(H_2O)\right]$$
$$= [(-393.51) + (0)] - [(-110.54) + (-241.84)]$$
$$= -41.13 \text{ kJ/mol} \qquad (2.43)$$

Similarly, $\Delta\hat{s}^0_{rxn}$ is calculated as

$$\Delta\hat{s}^0_{rxn} = \left[\hat{s}^0(CO_2) + \hat{s}^0(H_2)\right] - \left[\hat{s}^0(CO) + \hat{s}^0(H_2O)\right]$$
$$= [(213.80) + (130.67)] - [(197.65) + (188.82)]$$
$$= -42.00 \text{ J/(mol · K)} \qquad (2.44)$$

This gives

$$\Delta\hat{g}_{rxn}(T) = -41.13 \text{ kJ/mol} - T\,[-0.042\text{kJ/(mol · K)}] \qquad (2.45)$$

Examining this expression, it is apparent that at low temperatures the enthalpy term will dominate over the entropy term, and the free energy will be negative. However, as the temperature increases, entropy eventually wins and the reaction ceases to be spontaneous. Setting this equation equal to zero and solving for T gives us the temperature where the reaction ceases to be spontaneous:

$$-41.13 \text{ kJ/mol} + T\,[0.042 \text{ kJ/(mol · K)}] = 0 \qquad T \approx 979 \text{ K} \approx 706°C \quad (2.46)$$

This reaction is known as the water–gas shift reaction. It is important for high-temperature internal reforming of direct hydrocarbon fuel cells. These fuel cells run on simple hydrocarbon fuels (such as methane) in addition to hydrogen gas. Since these fuels contain carbon, carbon monoxide is often produced. The water–gas shift reaction allows additional H_2 fuel to be created from the CO stream. However, if the fuel cell is run above 700°C, the water–gas shift reaction is thermodynamically unfavorable. Therefore operating a high-temperature direct hydrocarbon fuel cell requires a delicate balance between the thermodynamics of the reactions (which are more favorable at lower temperatures) and the kinetics of the reactions (which improve at higher temperatures). This balance is discussed in greater detail in Chapter 10.

2.3.2 Relationship between Gibbs Free Energy and Electrical Work

Now that we know how to calculate Δg, we can determine the work potential of a fuel cell. For fuel cells, recall that we are specifically interested in electrical work. Let us find the maximum amount of electrical work that we can extract from a fuel cell reaction.

From Equation 2.17, remember that we define a change in Gibbs free energy as:

$$dG = dU - T\,dS - S\,dT + p\,dV + V\,dp \qquad (2.47)$$

As we have done previously, we can insert the expression for dU based on the first law of thermodynamics (Equation 2.3) into this equation. However, this time we expand the work term in dU to include both mechanical work and electrical work:

$$dU = T\,dS - dW$$
$$= T\,dS - (p\,dV + dW_{\text{elec}}) \qquad (2.48)$$

which yields dG as

$$dG = -S\,dT + V\,dp - dW_{\text{elec}} \qquad (2.49)$$

For a constant-temperature, constant-pressure process ($dT, dp = 0$) this reduces to

$$dG = -dW_{\text{elec}} \qquad (2.50)$$

Thus, the maximum electrical work that a system can perform in a constant-temperature, constant-pressure process is given by the negative of the Gibbs free energy difference for the process. For a reaction using molar quantities, this equation can be written as:

$$W_{\text{elec}} = -\Delta g_{\text{rxn}} \qquad (2.51)$$

Again, remember that the constant temperature, constant pressure assumption used here is not really as restrictive as it seems. The only limitation is that the temperature and pressure do not vary *during* the reaction process. Since fuel cells usually operate at constant temperature and pressure, this assumption is reasonable. *It is important to realize that the expression derived above is valid for different values of temperature and pressure as long as these values are not changing during the reaction.* We could apply this equation for $T = 200$ K and $p = 1$ atm, or just as validly apply it for $T = 400$ K and $p = 5$ atm. Later, we will examine how such steps in temperature and pressure (think of them as changes in the operating conditions from one fixed state to a new fixed state) affect the maximum electrical work available from the fuel cell.

2.3.3   Relationship between Gibbs Free Energy and Reaction Spontaneity

In addition to determining the maximum amount of electrical work that can be extracted from a reaction, the Gibbs free energy is also useful in determining the spontaneity of a reaction. Obviously, if ΔG is zero, then no electrical work can be extracted from a reaction. Worse yet, if ΔG is greater than zero, then work must be input for a reaction to occur. Therefore, the sign of ΔG indicates whether or not a reaction is spontaneous:

$\Delta G > 0$ Nonspontaneous (energetically unfavorable)

$\Delta G = 0$ Equilibrium

$\Delta G < 0$ Spontaneous (energetically favorable)

A spontaneous reaction is energetically favorable; it is a "downhill" process. Although spontaneous reactions are energetically favorable, spontaneity is no guarantee that a reaction will occur, nor does it indicate how fast a reaction will occur. Many spontaneous reactions do not occur because they are impeded by kinetic barriers. For example, at STP, the conversion of diamond to graphite is energetically favorable ($\Delta G < 0$). Fortunately for diamond lovers, kinetic barriers prevent this conversion from occurring. Fuel cells, too, are constrained by kinetics. The rate at which electricity can be produced from a fuel cell is limited by several kinetic phenomena. These phenomena are covered in Chapters 3–5. Before we get to kinetics, however, we need to understand how the electrical work capacity of a fuel cell is translated into a cell *voltage*.

2.3.4   Relationship between Gibbs Free Energy and Voltage

The potential of a system to perform electrical work is measured by voltage (also called *electrical potential*). The electrical work done by moving a charge Q, measured in coulombs, through an electrical potential difference E in volts is

$$W_{\text{elec}} = E Q \tag{2.52}$$

If the charge is assumed to be carried by electrons

$$Q = nF \tag{2.53}$$

where n is number of moles of electrons transferred and F is Faraday's constant. Combining Equations 2.51, 2.52, and 2.53 yields

$$\Delta \hat{g} = -nFE \tag{2.54}$$

Thus, the Gibbs free energy sets the magnitude of the reversible voltage for an electrochemical reaction. For example, in a hydrogen–oxygen fuel cell, the reaction

$$H_2 + \tfrac{1}{2}O_2 \rightleftharpoons H_2O \tag{2.55}$$

has a Gibbs free-energy change of -237 kJ/mol under standard-state conditions for liquid water product. The reversible voltage generated by a hydrogen–oxygen fuel cell under standard-state conditions is thus

$$
\begin{aligned}
E^0 &= -\frac{\Delta \hat{g}^0_{rxn}}{nF} \\
&= -\frac{-237{,}000 \, J/mol}{(2 \, mol \, e^-/mol \, reactant)(96{,}400 \, C/mol)} \\
&= 1.23 \, V
\end{aligned}
\tag{2.56}
$$

where E^0 is the standard-state reversible voltage and $\Delta \hat{g}^0_{rxn}$ is the standard-state free-energy change for the reaction.

At STP, thermodynamics dictates that the highest voltage attainable from a H_2–O_2 fuel cell is 1.23 V. If we need 10 V, forget about it. In other words, the chemistry of the fuel cell sets the reversible cell voltage. By picking a different fuel cell chemistry, we could establish a different reversible cell voltage. However, most feasible fuel cell reactions have reversible cell voltages in the range of 0.8–1.5 V. To get 10 V from fuel cells, we usually have to stack several cells together in series.

2.3.5 Standard Electrode Potentials: Computing Reversible Voltages

Although we learned how to calculate cell voltage using Equation 2.54, the cell potentials of many reactions have already been calculated for us in *standard electrode potential* tables. It is often easier to determine reversible voltages using these electrode potential tables. Standard electrode potential tables compare the standard-state reversible voltages of various electrochemical half reactions relative to the hydrogen reduction reaction. In these tables, the standard-state potential of the hydrogen reduction reaction is defined as zero, thus making it easy to compare other reactions.

To illustrate the concept of electrode potentials, a brief list is presented in Table 2.1. A more complete set of electrode potentials is provided in Appendix C.

To find the standard-state voltage produced by a complete electrochemical system, we simply sum all the potentials in the circuit:

$$
E^0_{cell} = \sum E^0_{half \, reactions}
\tag{2.57}
$$

TABLE 2.1. Selected List of Standard Electrode Potentials

Electrode Reaction	E^0 (V)
$Fe^{2+} + 2e^- \rightleftharpoons Fe$	−0.440
$CO_2 + 2H^+ + 2e^- \rightleftharpoons CHOOH_{(aq)}$	−0.196
$2H^+ + 2e^- \rightleftharpoons H_2$	0.000
$CO_2 + 6H^+ + 6e^- \rightleftharpoons CH_3OH + H_2O$	0.03
$O_2 + 4H^+ + 4e^- \rightleftharpoons 2H_2O$	1.229

THE QUANTITY *NF*

When studying fuel cells or other electrochemical systems, we will frequently encounter expressions containing the quantity nF. This quantity is our bridge from the world of thermodynamics (where we talk about moles of chemical species) to the world of electrochemistry (where we talk about current and voltage). In fact, the quantity nF expresses one of the most fundamental aspects of electrochemistry: the quantized transfer of electrons, in the form of an electrical current, between reacting chemical species. In any electrochemical reaction, there exists an integer correspondence between the moles of chemical species reacting and the moles of electrons transferred. For example, in the H_2–O_2 fuel cell reaction, 2 mol of electrons is transferred for every mole of H_2 gas reacted. In this case, $n = 2$. To convert this molar quantity of electrons to a quantity of charge, we must multiply n by Avogadro's number ($N_A = 6.022 \times 10^{23}$ electrons/mol) to get the number of electrons and then multiply by the charge per electron ($q = 1.68 \times 10^{-19}$ C/electron) to get the total charge. Thus we have

$$Q = nN_Aq = nF \tag{2.58}$$

What we call Faraday's constant is really the quantity $N_Aq[F = (6.022 \times 10^{23}$ electrons/mol) $\times (1.68 \times 10^{-19}$ C/electron) $\simeq 96{,}400$ C/mol]. Interestingly, the fact that Faraday's constant is a large number has important technological repercussions. Because F is large, a little chemistry produces a lot of electricity. This relationship is one of the factors that makes fuel cells technologically feasible.

Students are often confused whether they should base the number of moles of electrons transferred (n) in a reaction on a per-mole reactant basis, per-mole product basis, and so on. The answer is that it does not matter as long as you are consistent. For example, consider the reaction

$$A + 2B \rightarrow C + 2e^- \qquad \Delta G_{rxn} \tag{2.59}$$

In this reaction, $n = 2$ on a per-mole of A reacted basis, or per-mole of C produced, or per 2 mol of B reacted. If n is desired on a per-mole of B reacted basis instead, then the reaction stoichiometry must be adjusted as

$$\tfrac{1}{2}A + B \rightarrow \tfrac{1}{2}C + e^- \qquad \tfrac{1}{2}\Delta G_{rxn} \tag{2.60}$$

Now, on a per-mole of B reacted basis, n is 1. Also $n = 1$ per $\tfrac{1}{2}$ mol of A reacted or per $\tfrac{1}{2}$ mol of C produced. However, keep in mind that ΔG for reaction 2.60 is now $\tfrac{1}{2}\Delta G$ of the original reaction. As long as n and ΔG are kept consistent with the reaction stoichiometry, you should not suffer any confusion.

For example, the standard-state potential of the hydrogen–oxygen fuel cell is determined by

$$
\begin{array}{ll}
H_2 \rightarrow 2H^+ + 2e^- & E^0 = -0.000 \\
+\frac{1}{2}(O_2 + 4H^+ + 4e^- \rightarrow 2H_2O) & E^0 = +1.229 \\
\hline
= H_2 + \frac{1}{2}O_2 \rightarrow H_2O & E^0_{cell} = +1.229
\end{array}
$$

Note that we multiply the O_2 reaction by $\frac{1}{2}$ to get the correct stoichiometry. However, do not multiply the E^0 values by $\frac{1}{2}$. The E^0 values are independent of reaction amounts. Note also that in this calculation we reverse the direction of the hydrogen reaction (in a hydrogen–oxygen fuel cell, hydrogen is oxidized, not reduced). When we reverse the direction of a reaction, we reverse the sign of its potential. For the hydrogen reaction, this makes no difference, since $+0.000$ V $= -0.000$ V. However, the standard-state potential of the iron *oxidation* reaction, for example,

$$
Fe \rightleftharpoons Fe^{2+} + 2e^- \tag{2.61}
$$

would be $+0.440$V.

A complete electrochemical reaction generally consists of two half reactions, a reduction reaction *and* an oxidation reaction. However, electrode potential tables list all reactions as reduction reactions. For a set of coupled half reactions, how do we know which reaction will spontaneously proceed as the reduction reaction and which reaction will proceed as the oxidation reaction? The answer is found by comparing the *size* of the electrode potentials for the reactions. Because electrode potentials really represent free energies, increasing potential indicates increasing "reaction strength." For a matched pair of electrochemical half reactions, the reaction with the larger electrode potential will occur as written, while the reaction with the smaller electrode potential will occur opposite as written. For example, consider the Fe^{2+}–H^+ reaction couple from the list above. Because the hydrogen reduction reaction has a larger electrode potential compared to the iron reduction reaction (0 V > -0.440 V), the hydrogen reduction reaction will occur as written. The iron reaction will proceed in the opposite direction as written:

$$
\begin{array}{ll}
2H^+ + 2e^- \rightarrow H_2 & E^0 = 0.000 \\
Fe \rightarrow Fe^{2+} + 2e^- & E^0 = +0.440 \\
\hline
Fe + 2H^+ \rightarrow Fe^{2+} + H_2 & E^0 = +0.440
\end{array}
$$

Thus, thermodynamics predicts that in this system iron will be spontaneously oxidized to Fe^{2+} and hydrogen gas will be evolved, with a net cell potential of $+0.440$ V. This is the thermodynamically spontaneous reaction direction under standard-state conditions. Any thermodynamically spontaneous electrochemical reaction will have a positive cell potential. of course, the reaction could be made to occur in the reverse direction if an external voltage greater than 0.440 V is applied to the cell. In this case, a power supply would be doing work to the cell in order to overcome the thermodynamics of the system.

Example 2.3. A direct methanol fuel cell uses methanol (CH_3OH) as fuel instead of hydrogen:

$$CH_3OH + \tfrac{3}{2}O_2 \rightarrow CO_2 + 2H_2O \qquad (2.62)$$

Calculate the standard-state reversible potential for a direct methanol fuel cell.

Solution: We break this overall reaction into two electrochemical half reactions:

$$
\begin{array}{ll}
CH_3OH + H_2O \rightleftharpoons CO_2 + 6H^+ + 6e^- & E^0 = -0.03 \\
\tfrac{3}{2}(O_2 + 4H^+ + 4e^- \rightleftharpoons 2H_2O) & E^0 = +1.229 \\
\hline
CH_3OH + \tfrac{3}{2}O_2 \rightarrow CO_2 + 2H_2O & E^0 = +1.199
\end{array}
$$

Thus, the net cell potential for a methanol fuel cell is $+1.199$ V—almost the same as for a H_2–O_2 fuel cell. Note that although we multiplied the oxygen reduction reaction by $\tfrac{3}{2}$ to get a balanced reaction, we *did not* multiply the E^0 value by $\tfrac{3}{2}$. The E^0 values are *independent* of reaction amounts.

2.4 PREDICTING REVERSIBLE VOLTAGE OF A FUEL CELL UNDER NON-STANDARD-STATE CONDITIONS

Standard-state reversible fuel cell voltages (E^0 values) are only useful under standard-state conditions (room temperature, atmospheric pressure, unit activities of all species). Fuel cells are frequently operated under conditions that vary greatly from the standard state. For example, high-temperature fuel cells operate at 700–1000°C, automotive fuel cells often operate under 3–5 atm of pressure, and almost all fuel cells cope with variations in the concentration (and therefore activity) of reactant species.

In the following sections, we systematically define how reversible fuel cell voltages are affected by departures from the standard state. First, the influence of temperature on the reversible fuel cell voltage will be explored, then the influence of pressure. Finally, contributions from species activity (concentration) will be delineated, which will result in the formulation of the Nernst equation. In the end, we will have thermodynamic tools to predict the reversible voltage of a fuel cell under any arbitrary set of conditions.

2.4.1 Reversible Voltage Variation with Temperature

To understand how the reversible voltage varies with temperature, we need to go back to our original differential expression for the Gibbs free energy:

$$dG = -S\,dT + V\,dp \qquad (2.63)$$

From which we can write

$$\left(\frac{dG}{dT}\right)_p = -S \qquad (2.64)$$

For molar reaction quantities, this becomes:

$$\left(\frac{d(\Delta \hat{g})}{dT}\right)_p = -\Delta \hat{s} \qquad (2.65)$$

We have previously shown that the Gibbs free energy is related to the reversible cell voltage by

$$\Delta \hat{g} = -nFE \qquad (2.66)$$

Combining Equations 2.65 and 2.66 allows us to express how the reversible cell voltage varies as a function of temperature:

$$\left(\frac{dE}{dT}\right)_p = \frac{\Delta \hat{s}}{nF} \qquad (2.67)$$

We define E_T as the reversible cell voltage at an arbitrary temperature T. At constant pressure, E_T can be calculated by

$$E_T = E^0 + \frac{\Delta \hat{s}}{nF}(T - T_0) \qquad (2.68)$$

Generally, we assume $\Delta \hat{s}$ to be independent of temperature. If a more accurate value of E_T is required, it may be calculated by integrating the heat capacity related temperature dependence of $\Delta \hat{s}$.

As Equation 2.68 indicates, if $\Delta \hat{s}$ for a chemical reaction is positive, then E_T will increase with temperature. If $\Delta \hat{s}$ is negative, then E_T will decrease with temperature. For most fuel cell reactions $\Delta \hat{s}$ is negative; therefore reversible fuel cell voltages tend to *decrease* with increasing temperature.

For example, consider our familiar H_2–O_2 fuel cell. As can be calculated from the data in Appendix B, $\Delta \hat{s}_{rxn} = -44.43$ J/(mol $\cdot$ K) (for $H_2O_{(g)}$ as product). The variation of cell voltage with temperature is approximated as

$$E_T = E^0 + \frac{-44.43 \text{ J/(mol} \cdot \text{K)}}{(2)(96,400)}(T - T_0)$$
$$= E^0 - (2.304 \times 10^{-4} \text{ V/K})(T - T_0) \qquad (2.69)$$

Thus, for every 100 degrees increase in cell temperature, there is an approximate 23-mV decrease in cell voltage. A H_2–O_2 SOFC operating at 1000 K would have a reversible voltage of around 1.07 V. The temperature variation for the electrochemical oxidation of a number of different fuels is given in Figure 2.4. (After Broers [3]).

Since most reversible fuel cell voltages decrease with increasing temperature, should we operate a fuel cell at the lowest temperature possible? The answer is NO! As you will learn in Chapters 3 and 4, kinetic losses tend to decrease with increasing temperature. Therefore,

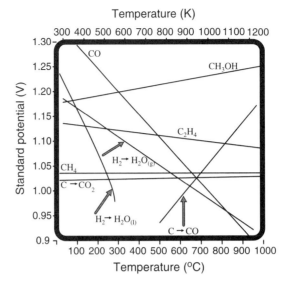

Figure 2.4. Reversible voltage (E_T) versus temperature for electrochemical oxidation of a variety of fuels [3].

real fuel cell performance typically *increases* with increasing temperature even though the thermodynamically reversible voltage decreases.

2.4.2 Reversible Voltage Variation with Pressure

Like temperature effects, the pressure effects on cell voltage may also be calculated starting from the differential expression for the Gibbs free energy:

$$dG = -S\,dT + V\,dp \qquad (2.70)$$

This time, we note

$$\left(\frac{dG}{dp}\right)_T = V \qquad (2.71)$$

Written for molar reaction quantities, this becomes

$$\left(\frac{d(\Delta\hat{g})}{dT}\right)_T = \Delta\hat{v} \qquad (2.72)$$

We have previously shown that the Gibbs free energy is related to the reversible cell voltage by

$$\Delta\hat{g} = -nFE \qquad (2.73)$$

Substituting this equation into Equation 2.72 allows us to express how the reversible cell voltage varies as a function of pressure:

$$\left(\frac{dE}{dp}\right)_T = -\frac{\Delta\hat{v}}{nF} \tag{2.74}$$

In other words, the variation of the reversible cell voltage with pressure is related to the volume change of the reaction. If the volume change of the reaction is negative (if fewer moles of gas are generated by the reaction than consumed, for instance), then the cell voltage will increase with increasing pressure. This is an example of *Le Chatelier's principle*; increasing the pressure of the system favors the reaction direction which relieves the stress on the system.

Usually, only gas species produce an appreciable volume change. Assuming the ideal gas law applies, we can write Equation 2.74 as

$$\left(\frac{dE}{dp}\right)_T = -\frac{\Delta n_g RT}{nFp} \tag{2.75}$$

where Δn_g represents the change in the total number of moles of gas upon reaction. If n_p is the number of product moles of gas and n_r is the number of reactant moles of gas, then $\Delta n_g = n_p - n_r$.

Pressure, like temperature, turns out to have a minimal effect on reversible voltage. As you will see in a forthcoming example, pressurizing a H_2–O_2 fuel cell to 3 atm H_2 and 5 atm O_2 increases the reversible voltage by only 15 mV.

2.4.3 Reversible Voltage Variation with Concentration: Nernst Equation

To understand how the reversible voltage varies with concentration, we need to introduce the concept of *chemical potential*. Chemical potential measures how the Gibbs free energy of a system changes as the chemistry of the system changes. Each chemical species in a system is assigned a chemical potential. Formally

$$\mu_i^\alpha = \left(\frac{\partial G}{\partial n_i}\right)_{T,P,n_{j\neq i}} \tag{2.76}$$

where μ_i^α is the chemical potential of species i in phase α and $(\partial G/\partial n_i)_{T,P,n_{j\neq i}}$ expresses how much the Gibbs free energy of the system changes for an infinitesimal increase in the quantity of species i (while temperature, pressure, and the quantities of all other species in the system are held constant). When we change the amounts (concentrations) of chemical species in a fuel cell, we are changing the free energy of the system. This change in free energy in turn changes the reversible voltage of the fuel cell. Understanding chemical potential is key to understanding how changes in concentration affect the reversible voltage.

Chemical potential is related to concentration through *activity a*:

$$\mu_i = \mu_i^0 + RT \ln a_i \tag{2.77}$$

where μ_i^0 is the reference chemical potential of species i at standard-state conditions and a_i is the activity of species i. The activity of a species depends on its chemical nature:

- *For an ideal gas, $a_i = p_i/p^0$*, where p_i is the partial pressure of the gas and p^0 is the standard-state pressure (1 atm). For example, the activity of oxygen in air at 1 atm is approximately 0.21. The activity of oxygen in air pressurized to 2 atm would be 0.42. Since we accept $p^0 = 1$ atm, we are often lazy and write $a_i = p_i$, recognizing that p_i is a *unitless* gas partial pressure.
- *For a nonideal gas, $a_i = \gamma_i(p_i/p^0)$*, where γ_i is an activity coefficient describing the departure from ideality ($0 < \gamma_i < 1$).
- *For a dilute (ideal) solution, $a_i = c_i/c^0$*, where c_i is the molar concentration of the species and c^0 is the standard-state concentration (1 M = 1 mol/L). For example, the activity of Na^+ ions in 0.1 M NaCl is 0.10.
- *For nonideal solutions, $a_i = \gamma_i(c_i/c^0)$*. Again, we use γ_i to describe departures from ideality ($0 < \gamma_i < 1$)
- *For pure components, $a_i = 1$*. For example, the activity of gold in a chunk of pure gold is 1. The activity of platinum in a platinum electrode is 1. The activity of liquid water is usually taken as 1.

Combining Equations 2.76 and 2.77, it is possible to calculate changes in the Gibbs free energy for a system of i chemical species by

$$dG = \sum_i \mu_i \, dn_i = \sum_i (\mu_i^0 + RT \ln a_i) \, dn_i \tag{2.78}$$

Consider an arbitrary chemical reaction placed on a molar basis for species A in the form

$$1A + bB \rightleftharpoons mM + nN \tag{2.79}$$

where A and B are reactants, M and N are products, and 1, b, m, and n represent the number of moles of A, B, M, and N, respectively. On a molar basis for species A, $\Delta \hat{g}$ for this reaction may be calculated from the chemical potentials of the various species participating in the reaction (assuming a single phase):

$$\Delta \hat{g} = (m\mu_M^0 + n\mu_N^0) - (\mu_A^0 + b\mu_B^0) + RT \ln \frac{a_M^m a_N^n}{a_A^1 a_B^b} \tag{2.80}$$

Recognizing that the lumped standard-state chemical potential terms represent the standard-state molar free-energy change for the reaction, $\Delta \hat{g}^0$, the equation can be simplified to a final form:

$$\Delta \hat{g} = \Delta \hat{g}^0 + RT \ln \frac{a_M^m a_N^n}{a_A^1 a_B^b} \tag{2.81}$$

This equation, called the van't Hoff isotherm, tells how the Gibbs free energy of a system changes as a function of the activities (read concentrations or gas pressures) of the reactant and product species.

From previous thermodynamic explorations (Section 2.3.4), we know that the Gibbs free energy and the reversible cell voltage are related:

$$\Delta \hat{g} = -nFE \tag{2.82}$$

Combining Equations 2.81 and 2.82 allows us to see how the reversible cell voltage varies as a function of chemical activity:

$$E = E^0 - \frac{RT}{nF} \ln \frac{a_M^m a_N^n}{a_A^1 a_B^b} \tag{2.83}$$

For a system with an arbitrary number of product and reactant species, this equation takes the general form

$$E = E^0 - \frac{RT}{nF} \ln \frac{\prod a_{products}^{v_i}}{\prod a_{reactants}^{v_i}} \tag{2.84}$$

Always take care to raise the activity of each species by its corresponding stoichiometric coefficient (v_i). For example, if a reaction involves $2Na^+$, the activity of Na^+ must be raised to the power of 2 (e.g., $a_{Na^+}^2$).

This important result is known as the Nernst equation. The Nernst equation outlines how reversible electrochemical cell voltages vary as a function of species concentration, gas pressure, and so on. This equation is the centerpiece of fuel cell thermodynamics. Remember it forever.

As an example of the utility of this equation, we will apply it to the familiar hydrogen–oxygen fuel cell reaction:

$$H_2 + \tfrac{1}{2}O_2 \rightleftharpoons H_2O \tag{2.85}$$

We write the Nernst equation for this reaction as

$$E = E^0 - \frac{RT}{2F} \ln \frac{a_{H_2O}}{a_{H_2} a_{O_2}^{1/2}} \tag{2.86}$$

Following our activity guidelines, we replace the activities of hydrogen and oxygen gases by their unitless partial pressures ($a_{H_2} = p_{H_2}$, $a_{O_2} = p_{O_2}$). If the fuel cell is operated below 100°C, so that liquid water is produced, we set the activity of water to unity ($a_{H_2O} = 1$). This yields

$$E = E^0 - \frac{RT}{2F} \ln \frac{1}{p_{H_2} p_{O_2}^{1/2}} \tag{2.87}$$

From this equation, it is apparent that pressurizing the fuel cell in order to increase the reactant gas partial pressures will increase the reversible voltage. However, because the pressure terms appear within a natural logarithm, the voltage improvements are slight. For example, if we operate a room temperature H_2–O_2 fuel cell on 3 atm pure H_2 and 5 atm

PRESSURE, TEMPERATURE, AND NERNST EQUATION

The Nernst equation accounts for the same pressure effects that were previously discussed in Section 2.4.2. Either Equation 2.84 or Equation 2.74 can be used to determine how the reversible voltage varies with pressure. If you use one, do not also use the other. The Nernst equation allows you to calculate voltage effects directly in terms of reactant and product pressures, while Equation 2.74 requires the volume change for the reaction (which you will have to express in terms of reactant gas pressures using the ideal gas law). The Nernst equation is generally more convenient.

Although temperature enters into the Nernst equation as a variable, the Nernst equation *does not* fully account for how the reversible voltage varies with temperature. At an arbitrary temperature $T \neq T_0$, the Nernst equation must be modified as:

$$E = E_T - \frac{RT}{nF} \ln \frac{\prod a_{\text{products}}^{v_i}}{\prod a_{\text{reactants}}^{v_i}} \tag{2.88}$$

where E_T is given from Equation 2.68 as

$$E_T = E^0 + \frac{\Delta \hat{s}}{nF}(T - T_0) \tag{2.89}$$

In summary, to properly account for both temperature and pressure changes, make sure to use Equation 2.88 or Equations 2.68 and 2.74.

air, thermodynamics predicts a cell voltage of 1.244 V:

$$E = 1.229 - \frac{(8.314)(298.15)}{(2)(96400)} \ln \frac{1}{(3)(5*0.21)^{1/2}}$$

$$= 1.244 \text{ V} \tag{2.90}$$

This is not much of an increase for all the extra work of pressurizing the fuel cell stack! From a thermodynamic perspective it is not worth the trouble; however, as you will learn in Chapters 3 and 5, there may be kinetic reasons to pressurize a fuel cell.

In contrast, what does the Nernst equation indicate about low pressure operation? Perhaps we are worried that almost all fuel cells operate on air instead of pure oxygen. Air is only about 21% oxygen, so at 1 atm, the partial pressure of oxygen in air is only 0.21. How much does this affect the reversible voltage of a room temperature H_2–O_2 fuel cell?

$$E = 1.229 - \frac{(8.314)(298.15)}{(2)(96400)} \ln \frac{1}{(1)(0.21)^{1/2}}$$

$$= 1.219 \text{ V} \tag{2.91}$$

Operation in air drops the reversible voltage by only 10 mV. Again, kinetic factors can introduce more deleterious penalties for air operation. However, as far as thermodynamics is concerned, air operation is not a problem.

2.4.4 Concentration Cells

The curious phenomenon of the concentration cell highlights some of the most fascinating implications of the Nernst equation. In a concentration cell, the *same* chemical species is present at both electrodes but at different concentrations. Amazingly, such a cell will develop a voltage because the concentration (activity) of the chemical species is different at one electrode versus the other electrode. For example, a *salt water battery* consisting of salt water at one electrode and freshwater at the other will produce a voltage because the concentration of salt differs at the two electrodes.

As a second example, consider the *hydrogen concentration cell* shown in Figure 2.5, which consists of a pressurized hydrogen fuel compartment and an evacuated ultra-low-pressure vacuum compartment separated by a composite platinum–electrolyte–platinum membrane structure. This "hydrogen fuel cell" contains no oxygen to react with the hydrogen, yet it will still produce a significant voltage. Thus, you could even use this fuel cell in outer space, where oxygen is unavailable. The thermodynamic voltage produced by the cell is related to the concentration of hydrogen in the fuel compartment relative to the vacuum compartment. For example, if the hydrogen fuel compartment is pressurized to 100 atm H_2 and the vacuum compartment is evacuated to 10^{-8} atm (presumably what remains will be mostly H_2), then this device will exhibit a voltage as determined by the Nernst equation:

$$E = 0 - \frac{(8.314)(298.15)}{(2)(96400)} \ln \frac{10^{-8}}{100}$$

$$= 0.296 \text{ V} \tag{2.92}$$

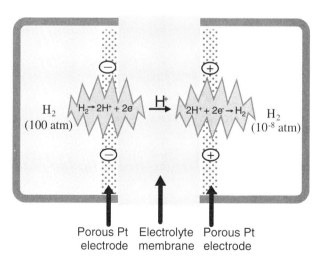

Figure 2.5. Hydrogen concentration cell. A high-pressure hydrogen compartment and a low pressure hydrogen compartment are separated by a platinum–electrolyte–platinum membrane structure. This device will develop a voltage due to the difference in the chemical potential of hydrogen between the two compartments.

At room temperature, we can extract almost 0.3 V just by exploiting a difference in hydrogen concentration. How is this possible? A voltage develops because the chemical potential of the hydrogen on one side of the membrane is dramatically different from the chemical potential of the hydrogen on the other side of the membrane. Driven by the chemical potential gradient, some of the hydrogen in the fuel compartment decomposes on the platinum catalyst electrode to protons and electrons. The protons flow through the electrolyte to the vacuum compartment, where they react with electrons in the second platinum catalyst electrode to reproduce hydrogen gas. If the two platinum electrodes are not connected, then very quickly excess electrons will accumulate on the fuel side while electrons will be depleted on the vacuum side, setting up an electrical potential gradient. This electrical potential gradient retards further movement of hydrogen from the fuel compartment to the vacuum compartment. Equilibrium is established when this electrical potential gradient builds up sufficiently to exactly balance the chemical potential gradient. (This is very similar to the "built-in voltage" that occurs at semiconductor p–n junctions.) The chemical potential difference created by the vastly different hydrogen concentrations at the two electrodes is offset by the development of an electrical potential which is equal but opposite in magnitude. The concept of chemical and electrical potentials offsetting one another to maintain thermodynamic equilibrium is summarized by a quantity called the *electrochemical potential*:

$$\tilde{\mu}_i = \mu_i + z_i F \phi_i \qquad (2.93)$$

where $\tilde{\mu}_i$ is the electrochemical potential of species i, μ_i is the chemical potential of species i, z_i is the charge number on the species (e.g., $z_{e^-} = -1$, $z_{Cu^{2+}} = +2$), F is Faraday's constant, and ϕ_i is the electrical potential experienced by species i. At equilibrium, the net change in the electrochemical potential for the species taking part in the system must be zero; in other words, the chemical and electrical potentials offset one another. For a reaction

$$\left(\sum_i \nu_i \mu_i \right)_{\text{products}} - \left(\sum_i \nu_i \mu_i \right)_{\text{reactants}} = -z_i F \Delta \phi_i \quad \text{(at equilibrium)} \qquad (2.94)$$

Compare this to Equation 2.54. Do you see how these two equations are really expressing the same thing? Following procedures analogous to Equations 2.77–2.81, we can rederive the Nernst equation from the basis of the electrochemical potential:

$$\tilde{\mu}_i = \mu_i^0 + RT \ln a_i + z_i F \phi_i = 0 \qquad (2.95)$$

The trick to rederiving the Nernst equation is to write out the change in electrochemical potential for the reactants being converted into products while also including the change in electrochemical potential for the electrons as they move from the anode to the cathode. Solving for the difference in the electrical potential for the electrons at the cathode versus the anode ($\Delta \phi_{e^-}$) gives the cell potential E. If n moles of electrons move from the anode to the cathode per-mole of chemical reaction, then

$$\Delta\phi_{e^-} = E = -\frac{\Delta\hat{g}^0}{nF} - \frac{RT}{nF}\ln\frac{\prod a_{products}^{v_i}}{\prod a_{reactants}^{v_i}} \qquad (2.96)$$

which gives

$$E = E^0 - \frac{RT}{nF}\ln\frac{\prod a_{products}^{v_i}}{\prod a_{reactants}^{v_i}} \qquad (2.97)$$

The details of this derivation are left as a homework problem at the end of this chapter.

Based on this discussion of concentration cells, you should see that it is possible to think of an H_2–O_2 fuel cell as simply a hydrogen concentration cell. Oxygen is used at the cathode merely as a convenient way to chemically "tie up" hydrogen. The O_2 gas keeps the cathode concentration of hydrogen to extremely low effective levels, allowing a significant thermodynamic voltage to be produced.

2.4.5 Summary

Let us briefly summarize the effects of nonstandard-state conditions on reversible electrochemical cell voltages. In the past few pages, we have used classical thermodynamics to predict how changes in temperature, pressure, and chemical composition affect the reversible voltages of fuel cells. (Incidentally, these relations are equally applicable to any other electrochemical system, not just fuel cells.)

- The variation of the reversible cell voltage with temperature is

$$\left(\frac{dE}{dT}\right)_p = \frac{\Delta\hat{s}}{nF} \qquad (2.98)$$

- The variation of the reversible cell voltage with pressure is

$$\left(\frac{dE}{dp}\right)_T = -\frac{\Delta n_g RT}{nFp} = -\frac{\Delta\hat{v}}{nF} \qquad (2.99)$$

- The variation of the reversible cell voltage with chemical activity (chemical composition, concentration, etc.) is given by the Nernst equation:

$$E = E^0 - \frac{RT}{nF}\ln\frac{\prod a_{products}^{v_i}}{\prod a_{reactants}^{v_i}} \qquad (2.100)$$

The Nernst equation accounts for the pressure effects on reversible cell voltage (it supersedes Equation 2.99) but does not fully account for the temperature effects. When $T \neq T_0$, E^0 in the Nernst equation should be replaced by E_T.

These equations give us the ability to predict the reversible voltage of a fuel cell under an arbitrary set of conditions.

2.5 FUEL CELL EFFICIENCY

For any energy conversion device, efficiency is of great importance. Central to a discussion of efficiency are the concepts of "ideal" (or reversible) efficiency and "real" (or practical) efficiency. Although you might be tempted to think that the ideal efficiency of a fuel cell should be 100%, this is not true. Just as thermodynamics tells us that the electrical work available from a fuel cell is limited by ΔG, the ideal efficiency of a fuel cell is also limited by ΔG. The story for real fuel cell efficiency is even worse. A real fuel cell must always be less efficient than an ideal fuel cell because real fuel cells incur nonideal irreversible losses during operation. A discussion of real fuel cell efficiency motivates forthcoming chapters, where these nonthermodynamic losses are discussed.

2.5.1 Ideal Reversible Fuel Cell Efficiency

We define the *efficiency* ε of a conversion process as the amount of useful energy that can be extracted from the process relative to the total energy evolved by that process:

$$\varepsilon = \frac{\text{useful energy}}{\text{total energy}} \tag{2.101}$$

If we wish to extract work from a chemical reaction, the efficiency is

$$\varepsilon = \frac{\text{work}}{\Delta \hat{h}} \tag{2.102}$$

For a fuel cell, recall that the maximum amount of energy available to do work is given by the Gibbs free energy. Thus, the reversible efficiency of a fuel cell can be written as

$$\varepsilon_{\text{thermo, fc}} = \frac{\Delta \hat{g}}{\Delta \hat{h}} \tag{2.103}$$

At room temperature and pressure, the H_2–O_2 fuel cell has $\Delta \hat{g}^0 = -237.3$ kJ/mol and $\Delta \hat{h}^0_{\text{HHV}} = -286$ kJ/mol. This yields a 83% reversible HHV efficiency for the H_2–O_2 fuel cell at STP:

$$\varepsilon_{\text{thermo, fc}} = \frac{-237.3}{-286} = 0.83 \tag{2.104}$$

In contrast to a fuel cell, the maximum theoretical efficiency of a conventional heat/expansion engine is described by the Carnot cycle. This efficiency may be derived from classical thermodynamics. We do not repeat the derivation here, but we provide the result:

$$\varepsilon_{\text{Carnot}} = \frac{T_H - T_L}{T_H} \tag{2.105}$$

HIGHER HEATING VALUE EFFICIENCY

To convert water from the liquid to the vapor state requires heat input. The quantity of heat required is called the latent heat of vaporization. Due to this latent heat of vaporization, the $\Delta \hat{h}_{rxn}$ for a hydrogen–oxygen fuel cell is significantly different depending on whether vapor or liquid water product is assumed. When liquid water is produced, $\Delta \hat{h}_{rxn}^0 = -286$ kJ/mol; when water vapor is produced, $\Delta \hat{h}_{rxn}^0 = -241$ kJ/mol. Basically, the difference between these two numbers tells us that more total heat is recoverable if the product water can be condensed to the liquid form. The extra heat recovered by going to liquid water is precisely the latent heat of vaporization. Because condensation to liquid water results in more heat recovery, the $\Delta \hat{h}_{rxn}^0$ involving liquid water is called the higher heating value (HHV) while the $\Delta \hat{h}_{rxn}^0$ involving water vapor is called the lower heating value (LHV).

Which of these values should be used in computing a fuel cell's efficiency? The most equitable calculations of fuel cell efficiency use the HHV. Using the HHV instead of the LHV is appropriate because it acknowledges the true total heat that could theoretically be recovered from the hydrogen combustion reaction. Use of the LHV will result in higher, but perhaps misleading, efficiency numbers.

All calculations and examples in this book will make use of the HHV. Thus, we should rewrite Equation 2.103 to explicitly reflect this fact:

$$\varepsilon_{thermo,\ fc} = \frac{\Delta \hat{g}}{\Delta \hat{h}_{HHV}} \tag{2.106}$$

In these efficiency calculations, it is important to note that $\Delta \hat{g}$ should still be calculated by properly accounting for phase transitions. Thus, for a hydrogen–oxygen fuel cell above 100°C, the calculation of $\Delta \hat{g}$ should use formation enthalpies and entropies for water vapor. Below 100°C, the calculation of $\Delta \hat{g}$ should use the formation enthalpies and entropies for liquid water. You should recognize that calculating $\Delta \hat{g}$ based on water vapor above 100°C while simultaneously using $\Delta \hat{h}_{HHV}$ (based on liquid water) for efficiency calculations does not represent a contradiction. What this calculation says is that, in a fuel cell operating above 100°C, we are losing the ability to convert the latent heat of vaporization of the product water into useful work.

In this expression, T_H is the maximum temperature of the heat engine and T_L is the rejection temperature of the heat engine. For a heat engine that operates at 400°C (673 K) and rejects heat at 50°C (323 K), the reversible efficiency is 52%.

From the Carnot equation, it is apparent that the reversible efficiency of a heat engine *improves* as the operating temperature increases. In contrast, the reversible efficiency of a fuel cell tends to *decrease* as the operating temperature increases.

As an example, the reversible HHV efficiency of a H_2–O_2 fuel cell is compared to the reversible efficiency of a heat engine as a function of temperature in Figure 2.6. Fuel cells hold a significant thermodynamic efficiency advantage at low temperature but lose this

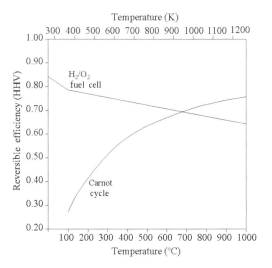

Figure 2.6. Reversible HHV efficiency of H_2–O_2 fuel cell compared to reversible efficiency of heat engine (Carnot cycle, rejection temperature 273.15 K). Fuel cells hold a significant thermodynamic efficiency advantage at low temperature but lose this advantage at higher temperatures. The kink in the fuel cell efficiency curve at 100°C arises from the entropy difference between liquid water and water vapor.

advantage at higher temperatures. Note the kink in the fuel cell efficiency curve at 100°C. This change in slope arises from the entropy difference between liquid water and water vapor.

2.5.2 Real (Practical) Fuel Cell Efficiency

As mentioned previously, the real efficiency of a fuel cell must always be less than the reversible thermodynamic efficiency. The two major reasons are as follows:

1. Voltage losses
2. Fuel utilization losses

The real efficiency of a fuel cell, $\varepsilon_{\text{real}}$, may be calculated as

$$\varepsilon_{\text{real}} = (\varepsilon_{\text{thermo}}) \times (\varepsilon_{\text{voltage}}) \times (\varepsilon_{\text{fuel}}) \tag{2.107}$$

where $\varepsilon_{\text{thermo}}$ is the reversible thermodynamic efficiency of the fuel cell, $\varepsilon_{\text{voltage}}$ is the voltage efficiency of the fuel cell, and $\varepsilon_{\text{fuel}}$ is the fuel utilization efficiency of the fuel cell. Each of these terms are briefly discussed:

- The *reversible thermodynamic efficiency* $\varepsilon_{\text{thermo}}$ was described in the previous section. It reflects how, even under ideal conditions, not all the enthalpy contained in the fuel can be exploited to perform useful work.

- The *voltage efficiency of the fuel cell* $\varepsilon_{\text{voltage}}$ incorporates the losses due to irreversible kinetic effects in the fuel cell. Recall from Section 1.7 that these losses are captured in the operational i–V curve of the fuel cell. The voltage efficiency of a fuel cell is the ratio of the real operating voltage of the fuel cell (V) to the thermodynamically reversible voltage of the fuel cell (E):

$$\varepsilon_{\text{voltage}} = \frac{V}{E} \tag{2.108}$$

Note that the operating voltage of a fuel cell depends on the current (i) drawn from the fuel cell, as given by the i–V curve. Therefore, $\varepsilon_{\text{voltage}}$ will change depending on the current drawn from the cell. The higher the current load, the lower the voltage efficiency. Therefore, fuel cells are *most efficient at low load*. This is in direct contrast to combustion engines, which are generally most efficient at maximum load.

- The *fuel utilization efficiency* $\varepsilon_{\text{fuel}}$ accounts for the fact that not all of the fuel provided to a fuel cell will participate in the electrochemical reaction. Some of the fuel may undergo side reactions that do not produce electric power. Some of the fuel will simply flow through the fuel cell without ever reacting. The fuel utilization efficiency, then, is the ratio of the fuel used by the cell to generate electric current versus the total fuel provided to the fuel cell. If i is the current generated by the fuel cell and v_{fuel} is the rate at which fuel is supplied to the fuel cell (mol/sec), then

$$\varepsilon_{\text{fuel}} = \frac{i/nF}{v_{\text{fuel}}} \tag{2.109}$$

If an overabundance of fuel is supplied to a fuel cell, it will be wasted, as reflected in $\varepsilon_{\text{fuel}}$. Generally, the supply of fuel to a fuel cell is adjusted according to the current so that the fuel cell is always supplied with just a bit more fuel than it needs at any load. Fuel cells operated in this manner are described by a *stoichiometric factor*. For example, a fuel cell supplied with 1.5 times more fuel than would be required for 100% fuel utilization is operating at 1.5 times stoichiometric. (The stoichiometric factor λ for this fuel cell is 1.5.) For fuel cells operating under a stoichiometric condition, fuel utilization is independent of current, and we can write the fuel utilization efficiency as

$$\varepsilon_{\text{fuel}} = \frac{1}{\lambda} \tag{2.110}$$

Combining effects of thermodynamics, irreversible kinetic losses, and fuel utilization losses, we can write the practical efficiency of a real fuel cell as

$$\varepsilon_{\text{real}} = \left(\frac{\Delta \hat{g}}{\Delta \hat{h}_{\text{HHV}}} \right) \left(\frac{V}{E} \right) \left(\frac{i/nF}{v_{\text{fuel}}} \right) \tag{2.111}$$

For a fuel cell operating under a fixed stoichiometric condition, this equation simplifies to

$$\varepsilon_{\text{real}} = \left(\frac{\Delta\hat{g}}{\Delta\hat{h}_{\text{HHV}}}\right)\left(\frac{V}{E}\right)\left(\frac{1}{\lambda}\right) \tag{2.112}$$

In the next few chapters, we explore the underlying causes for the kinetic losses in real fuel cells. In other words, we learn why V is different than E. These losses are mainly associated with the reaction, conduction, and mass transport steps taking place inside a fuel cell. These three steps are discussed in the next three chapters.

CHAPTER SUMMARY

The purpose of this chapter is to understand the theoretical limits to fuel cell performance by applying the principles of thermodynamics. The main points introduced in this chapter include the following:

- Thermodynamics provides the theoretical limits or ideal case for fuel cell performance.
- The heat potential of a fuel is given by the fuel's heat of combustion or, more generally, the enthalpy of reaction.
- Not all of the heat potential of a fuel can be utilized to perform useful work. The work potential of the fuel is given by the Gibbs free energy, ΔG.
- Electrical energy can only be extracted from a spontaneous ("downhill") chemical reaction. The magnitude of ΔG gives the amount of energy that is available ("free") to do electrical work. Thus, the sign of ΔG indicates whether or not electrical work can be done, and the size of ΔG indicates how much electrical work can be done.
- The reversible voltage of a fuel cell, E, is related to the molar Gibbs free energy by $\Delta\hat{g} = -nFE$.
- ΔG scales with reaction amount whereas $\Delta\hat{g}$ and E do not scale with reaction amount.
- E varies with temperature as $dE/dT = \Delta\hat{s}/nF$. For fuel cells, $\Delta\hat{s}$ is generally negative, therefore reversible fuel cell voltages tend to decrease with increasing temperature. E varies with pressure as: $dE/dp = -\Delta n_g RT/(nFp) = -\Delta\hat{v}/nF$
- The Nernst equation describes how E varies with reactant/product activities:

$$E = E^0 - \frac{RT}{nF}\ln\frac{\prod a_{\text{products}}^{v_i}}{\prod a_{\text{reactants}}^{v_i}}$$

- The Nernst equation intrinsically includes the pressure effects on reversible cell voltage but does not fully account for the temperature effects.
- Ideal HHV fuel cell efficiency $\varepsilon_{\text{thermo}} = \Delta\hat{g}/\Delta\hat{h}_{\text{HHV}}$.

- Thermodynamic fuel cell efficiency generally decreases as temperature increases. Contrast this to heat engines, for which thermodynamic efficiency generally increases as temperature increases.
- Real fuel cell efficiency is always less than the ideal thermodynamic efficiency. Major reasons are irreversible kinetic losses and fuel utilization losses. Total overall efficiency is given by the product of individual efficiencies.

CHAPTER EXERCISES

Review Questions

2.1 If an isothermal reaction involving gases exhibits a large negative volume change, will the entropy change for the same reaction likely be negative or positive? Why?

2.2 (a) If $\Delta\hat{h}$ for a reaction is negative and $\Delta\hat{s}$ is positive, can you say anything about the spontaneity of the reaction? (b) What if $\Delta\hat{h}$ is negative and $\Delta\hat{s}$ is negative? (c) What if $\Delta\hat{h}$ is positive and $\Delta\hat{s}$ is negative? (d) What if $\Delta\hat{h}$ is positive and $\Delta\hat{s}$ is positive?

2.3 Reaction A has $\Delta\hat{g}_{rxn} = -100\,kJ/mol$. Reaction B has $\Delta\hat{g}_{rxn} = -200\,kJ/mol$. Can you say anything about the relative speeds (reaction rates) for these two reactions?

2.4 Why does ΔG for a reaction scale with reaction quantity but E does not? For example, ΔG^0_{rxn} for the combustion of 1 mol of hydrogen is $1 \times -237\,kJ/mol = -237\,kJ$, while ΔG^0_{rxn} for the combustion of 2 mol of hydrogen is $2 \times -237\,kJ/mol = -474\,kJ$. In both cases, however, the reversible cell voltage produced by the reaction, E^0, is 1.23 V.

2.5 In general, will increasing the concentration (activity) of reactants increase or decrease the reversible cell voltage of an electrochemical system?

2.6 Derive the Nernst equation starting from Equation 2.95 for a general chemical reaction of the from

$$1A + bB \rightleftharpoons mM + nN \tag{2.113}$$

2.7 Can the thermodynamic efficiency of a fuel cell, as defined by $\varepsilon = \Delta\hat{g}/\Delta\hat{h}$, ever be greater than unity? Explain why or why not. Consider all fuel cell chemistries, not just H_2–O_2 fuel cells.

Calculations

2.8 In Example 2.2, we assumed that $\Delta\hat{h}_{rxn}$ and $\Delta\hat{s}_{rxn}$ were independent of temperature. We are now interested in determining how much of an error this assumption introduced into our solution. Rework Example 2.2 assuming constant-heat-capacity values for all species involved in the reaction. Heat capacity values are provided in the table below:

Chemical Species	c_p(J/mol · K)
CO	29.2
CO_2	37.2
H_2	28.8
$H_2O_{(g)}$	33.6

Note that a more accurate calculation is made by using temperature-dependent heat capacity equations. These equations generally use polynomial series to reflect how the heat capacity changes with temperature. Such calculations are tedious and are now mostly done via computer programs.

2.9 (a) If a fuel cell has a reversible voltage of E_1 at $p = p_1$ and $T = T_1$, write an expression for the temperature T_2 that would be required to maintain the fuel cell voltage at E_1 if the cell pressure is adjusted to p_2. (b) For a H_2–O_2 fuel cell operating at room temperature and atmospheric pressure (on pure oxygen), what temperature would be required to maintain the original reversible voltage if the operating pressure is reduced by one order of magnitude?

2.10 In Section 2.4.4, it was mentioned that you could think of a hydrogen–oxygen fuel cell as simply a hydrogen concentration cell, where oxygen is used to chemically "tie up" hydrogen at the cathode. Oxygen's ability to chemically tie up hydrogen is measured by the Gibbs free energy of the hydrogen–oxygen reaction. At STP (assuming air at the cathode), what is the effective hydrogen pressure that oxygen is able to chemically maintain at the cathode of a hydrogen–oxygen(air) fuel cell?

2.11 A typical H_2–O_2 PEMFC might operate at a voltage of 0.75 V and $\lambda = 1.10$. At STP, what is the efficiency of such a fuel cell (use HHV and assume pure oxygen at the cathode)?

CHAPTER 3

FUEL CELL REACTION KINETICS

Having learned what is "ideally" possible with fuel cells in the previous chapter, our journey now enters the realm of the practical, beginning in this chapter with a discussion of fuel cell reaction kinetics. Fuel cell reaction kinetics discusses the nuts and bolts of *how* fuel cell reactions occur.

At the most fundamental level, a fuel cell reaction (or any electrochemical reaction) involves the transfer of electrons between an electrode surface and a chemical species adjacent to the electrode surface. In fuel cells, we harness *thermodynamically favorable* electron transfer processes to extract electrical energy (in the form of an electron current) from chemical energy. Previously, we learned how to distinguish thermodynamically favorable electrochemical reactions. Here, we study the *kinetics* of electrochemical reactions. In other words, we study the mechanisms by which electron transfer processes occur. Because each electrochemical reaction event results in the transfer of one or more electrons, the current produced by a fuel cell (number of electrons per time) depends on the rate of the electrochemical reaction (number of reactions per time). Increasing the rate of the electrochemical reaction is therefore crucial to improving fuel cell performance. Catalysis, electrode design, and other methods to increase the rate of the electrochemical reaction will be introduced.

3.1 INTRODUCTION TO ELECTRODE KINETICS

This section discusses a few basic concepts about electrochemical systems that tend to cause confusion. Crystallize these basic concepts in your mind and you will be on your way to understanding electrochemistry.

3.1.1 Electrochemical Reactions Are Different from Chemical Reactions

All electrochemical reactions involve the transfer of charge (electrons) between an electrode and a chemical species. This distinguishes electrochemical reactions from chemical reactions. In chemical reactions, charge transfer occurs directly between two chemical species without the liberation of free electrons.

3.1.2 Electrochemical Processes Are Heterogeneous

Because electrochemistry deals with the transfer of charge between an electrode and a chemical species, electrochemical processes are necessarily *heterogeneous*. Electrochemical reactions, like the HOR,

$$H_2 \rightleftharpoons 2H^+ + 2e^- \tag{3.1}$$

can only take place at the *interface* between an electrode and an electrolyte. In Figure 3.1, it is obvious that hydrogen gas and protons cannot exist inside the metal electrode, while free electrons cannot exist within the electrolyte. Therefore the reaction between hydrogen, protons, and electrons must occur where the electrode and electrolyte intersect.

3.1.3 Current Is a Rate

Because electrons are generated or consumed by electrochemical reactions, the current i evolved by an electrochemical reaction is a direct measure of the rate of the electrochemical reaction. The unit of current is the ampere; an ampere is a coulomb per second (C/s). From Faraday's law

$$i = \frac{dQ}{dt} \tag{3.2}$$

where Q is the charge (C) and t is time. Thus, current expresses the rate of charge transfer. If each electrochemical reaction event results in the transfer of n electrons, then

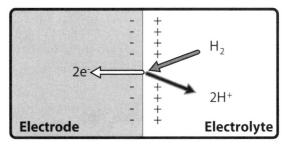

Figure 3.1. Electrochemical reactions are heterogeneous. As this schematic shows, the HOR is a surface-limited reaction. It can take place only at the interface between an electrode and an electrolyte.

$$i = nF\frac{dN}{dt} \tag{3.3}$$

where dN/dt is the rate of the electrochemical reaction (mol/s) and F is Faraday's constant. (Faraday's constant is necessary to convert mol of electrons to charge in coulombs.)

Example 3.1. Assuming 100% fuel utilization, how much current can a fuel cell produce if provisioned with 5 sccm H_2 gas at STP? (1 sccm = 1 standard cubic centimeter per minute.) Assume sufficient oxidant is also supplied.

Solution: In this problem, we are provided with a volumetric flow rate of H_2 gas. To get current, we need to convert volumetric flow rate into molar flow rate and then convert molar flow rate into current. Treating H_2 as an ideal gas, the molar flow rate is related to the volumetric flow rate via the ideal gas law:

$$\frac{dN}{dt} = \frac{P(dV/dt)}{RT} \tag{3.4}$$

where dN/dt is the molar flow rate and dV/dt is the volumetric flow rate. At STP

$$\frac{dN}{dt} = \frac{(1\ \text{atm})(0.005\ \text{L/min})}{[0.082\ \text{L}\cdot\text{atm/(mol}\cdot\text{K)}](298.15\ \text{K})} = 2.05 \times 10^{-4}\ \text{mol}\ H_2/\text{min} \tag{3.5}$$

Since two mol of electrons are transferred for every mole of H_2 gas reacted, $n = 2$. Inserting n and dN/dt into Equation 3.3 and converting from minutes to seconds gives

$$i = nF\frac{dN}{dt} = (2)(96{,}400\ \text{C/mol})(2.05 \times 10^{-4}\ \text{mol}\ H_2/\text{min})(1\ \text{min}/60\ \text{s})$$

$$= 0.657\ \text{A} \tag{3.6}$$

Thus a flow rate of 5 sccm H_2 is sufficient to sustain a 0.657 A current, assuming 100% fuel utilization.

3.1.4 Charge Is an Amount

If we integrate a rate, we obtain an amount. Integrating Faraday's law (Equation 3.2) gives

$$\int_0^t i\,dt = Q = nFN \tag{3.7}$$

The total amount of electricity produced, as measured by the accumulated charge Q in coulombs, is proportional to the number of moles of material processed in the electrochemical reaction.

Example 3.2. A fuel cell operates for 1 h at 2 A current load and then operates for 2 more h at 5 A current load. Calculate the total number of moles of H_2 consumed by the fuel cell over the course of this operation. To what mass of H_2 does this correspond? Assume 100% fuel utilization.

Solution: From the time–current profile that we are given, we can calculate the total amount of electricity produced by this fuel cell (as measured by the accumulated charge). Then, using Equation 3.7, we can calculate the total number of moles of H_2 processed by the reaction.

The total amount of electricity produced is calculated by integrating the current load profile over the operation time. For this particular example, the calculation is easy:

$$Q_{tot} = i_1 t_1 + i_2 t_2 = (2 \text{ A})(3600 \text{ s}) + (5 \text{ A})(7200 \text{ s}) = 43,200 \text{ C} \qquad (3.8)$$

Since 2 mol of electrons is transferred for every mole of H_2 reacted, $n = 2$. Thus, the total number of moles of H_2 processed by this fuel cell is

$$N_{H_2} = \frac{Q_{tot}}{nF} = \frac{43,200 \text{ C}}{(2)(96,400 \text{ C/mol})} = 0.224 \text{ mol } H_2 \qquad (3.9)$$

Since the molar mass of H_2 is approximately 2 g/mol, this corresponds to about 0.448 g of H_2.

3.1.5 Current Density Is More Fundamental Than Current

Because electrochemical reactions only occur at interfaces, the current produced is usually directly proportional to the area of the interface. Doubling the interfacial area available for reaction should double the rate. Therefore current density (current per unit area) is more fundamental than current; it allows the reactivity of different surfaces to be compared on a per-unit area basis. Current density j is usually expressed in units of amperes per square centimeter (A/cm^2):

$$j = \frac{i}{A} \qquad (3.10)$$

where A is the area. In a similar fashion to current density, electrochemical reaction rates can also be expressed on a per unit-area basis. We give per unit-area reaction rates the symbol υ:

$$\upsilon = \frac{1}{A}\frac{dN}{dt} = \frac{i}{nFA} = \frac{j}{nF} \qquad (3.11)$$

3.1.6 Potential Controls Electron Energy

Potential (voltage) is a measure of electron energy. According to band theory, the electron energy in a metal is measured by the *Fermi level*. By controlling the electrode potential, we control the electron energy in an electrochemical system (Fermi level), thereby influencing the direction of a reaction. For example, consider a general electrochemical reaction occurring at an electrode between the oxidized (Ox) and reduced (Re) forms of a chemical species:

$$Ox + e^- \rightleftharpoons Re \qquad (3.12)$$

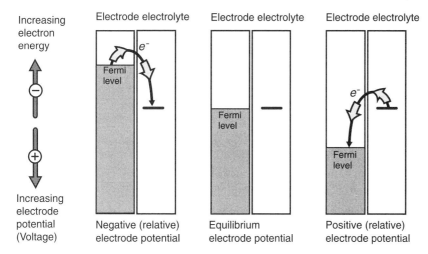

Figure 3.2. Electrode potential can be manipulated to trigger reduction (left) or oxidation (right). The thermodynamic equilibrium electrode potential (middle) corresponds to the situation where the oxidation and reduction processes are balanced.

If the potential of the electrode is made relatively more negative than the equilibrium potential, the reaction will be biased toward the formation of Re. (Consider that a more negative electrode makes the electrode less "hospitable" to electrons, forcing electrons out of the electrode and onto the electroactive species.) On the other hand, if the electrode potential is made relatively more positive than the equilibrium potential, the reaction will be biased toward the formation of Ox. (A more positive electrode "attracts" electron to the electrode, "pulling" them off of the electroactive species.) Figure 3.2 illustrates this concept schematically.

Using potential to control reactions is key to electrochemistry. Later in this chapter, we develop this principle more fully to understand how rate (and therefore the current produced by an electrochemical reaction) is related to cell voltage.

3.1.7 Reaction Rates Are Finite

It should be obvious that the rate of an electrochemical reaction, or any reaction for that matter, is finite. This means that the current produced by an electrochemical reaction is limited. Reaction rates are finite even if they are energetically "downhill" because an energy barrier (called an *activation energy*) impedes the conversion of reactants into products. As illustrated in Figure 3.3, in order for reactants to be converted into products, they must first make it over this activation "hill." The *probability* that reactant species can make it over this barrier determines the rate at which the reaction occurs. In the next section, we discuss *why* electrochemical reactions have activation barriers.

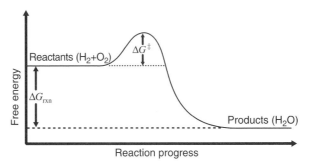

Figure 3.3. An activation barrier ($\Delta G^{\ddagger}$) impedes the conversion of reactants to products. Because of this barrier, the rate at which reactants are converted into products (the reaction rate) is limited.

3.2   WHY CHARGE TRANSFER REACTIONS HAVE AN ACTIVATION ENERGY

Even reactions as elementary as the HOR actually consist of a series of even simpler basic steps. For example, the overall reaction $H_2 \rightleftharpoons 2H^+ + 2e^-$ might occur by the following series of basic steps:

1. Mass transport of H_2 gas to the electrode:

$$(H_{2(bulk)} \rightarrow H_{2(near\ electrode)})$$

2. Absorption of H_2 onto the electrode surface:

$$(H_{2(near\ electrode)} + M \rightarrow M \cdots H_2)$$

3. Separation of the H_2 molecule into two individually bound (*chemisorbed*) hydrogen atoms on the electrode surface:

$$(M \cdots H_2 + M \rightarrow 2M \cdots H)$$

4. Transfer of electrons from the chemisorbed hydrogen atoms to the electrode, releasing H^+ ions into the electrolyte:

$$2 \times \left[M \cdots H \rightarrow (M + e^-) + H^+_{(near\ electrode)} \right]$$

5. Mass transport of the H^+ ions away from the electrode:

$$2 \times \left[H^+_{(near\ electrode)} \rightarrow H^+_{(bulk\ electrolyte)} \right]$$

Just as an army can only march as fast as its slowest member, the overall reaction rate will be limited by the slowest step in the series. Suppose that the overall reaction above is limited by the electron transfer step between chemisorbed hydrogen and the metal electrode surface (step 4 above). This step can be represented as

$$M \cdots H \rightleftharpoons (M + e^-) + H^+ \qquad (3.13)$$

In this equation, $M \cdots H$ represents a hydrogen atom chemisorbed on the metal surface and $(M + e^-)$ represents a liberated metal surface site and a free electron in the metal. This reaction is depicted physically in Figure 3.4, while Figure 3.5 illustrates the energetics. First consider curve 1 of Figure 3.5. This curve depicts the free-energy of the chemisorbed

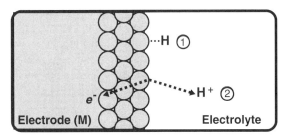

Figure 3.4. Schematic of chemisorbed hydrogen charge transfer reaction. The reactant state, a chemisorbed hydrogen atom $(M \cdots H)$, is shown at 1. Completion of the charge transfer reaction, as shown at 2, liberates a free electron into the metal and a free proton into the electrolyte $((M + e^-) + H^+)$.

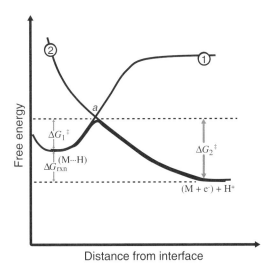

Figure 3.5. Schematic of energetics of chemisorbed hydrogen charge transfer reaction. Curve 1 shows the free energy of the reactant state ($[M \cdots H]$) as a function of the distance of separation between the H atom and the metal surface. Curve 2 shows the free energy of the product state ($[(M + e^-) + H^+]$) as a function of the distance of separation between the H^+ ion and the metal surface. The dark line denotes the "easiest" (minimum) energy path for the conversion of $[M \cdots H]$ to $[(M + e^-) + H^+]$. The activated state is represented by a.

atomic hydrogen, H, which increases with distance from the metal electrode surface. We know that atomic hydrogen is not very stable; stability improves with chemisorption of the atomic hydrogen to the metal electrode surface. Chemisorption to the metal surface allows the hydrogen to partially satisfy its bonding requirements, lowering its free energy. Separating the atomic hydrogen from the metal surface destroys this bond, thus increasing the free energy.

Now consider curve 2, which depicts the free-energy of a H^+ ion in the electrolyte. This curve shows that energy is required to bring the H^+ ion toward the surface, working against the repulsive forces between the charged ion and the metal surface. This energy increases dramatically as the H^+ ion is brought closer and closer to the surface because it is energetically unfavorable for the H^+ ion to exist within the metal phase. The free energy of the H^+ ion is lowest when it is deep within the electrolyte, far from the metal surface.

The "easiest" (minimum) energy path for the conversion of chemisorbed hydrogen to H^+ and $(M + e^-)$ is given by the dark solid line in Figure 3.5. Note that this energy path necessarily involves overcoming a *free-energy maximum*. This maximum occurs because any deviation from the energetically stable reactant and product states involves an increase in free energy (as detailed by curves 1 and 2). The point marked a on the diagram is called the *activated state*. Species in the activated state have overcome the free-energy barrier; they can be converted into either products or reactants without further impediment.

3.3   ACTIVATION ENERGY DETERMINES REACTION RATE

Only species in the activated state can undergo the transition from reactant to product. Therefore, the rate of conversion of reactants to products depends on the probability that a reactant species will find itself in the activated state. While it is beyond the scope of this book to treat theoretically, statistical mechanics arguments hold that the probability of finding a species in the activated state is exponentially dependent on the size of the activation barrier:

$$P_{act} = e^{-\Delta G_1^{\ddagger}/(RT)} \tag{3.14}$$

where P_{act} is the probability of finding a reactant species in the activated state, $\Delta G_1^{\ddagger}$ is the size of the energy barrier between the reactant and activated states, R is the gas constant, and T is the temperature (K). Starting from this probability, we can describe a reaction rate as a statistical process involving the number of reactant species available to participate in the reaction (per unit reaction area), the probability of finding those reactant species in the activated state, and the frequency at which those activated species decay to form products:

$$
\begin{aligned}
\upsilon_1 &= c_R^* \times f_1 \times P_{act} \\
&= c_R^* f_1 e^{-\Delta G_1^{\ddagger}/(RT)}
\end{aligned}
\tag{3.15}
$$

where υ_1 is the reaction rate in the forward direction (reactants $\rightarrow$ products), c_R^* is the reactant surface concentration (mol/cm^2), and f_1 is the decay rate to products. The decay

MORE ON THE DECAY RATE (OPTIONAL)

As was mentioned above, the decay rate to products is given by the lifetime of the activated species and the likelihood that it will convert to a product instead of back to a reactant:

$$f_1 = \frac{P_{a \to p}}{\tau_a} \tag{3.16}$$

Here, $P_{a \to p}$ is the probability that the activated state will decay to the product state and τ_a is the lifetime of the activated state. Both decay rates to products (f_1) and decay rates to reactants (f_2) can be computed. In general, the decay rates are determined by the curvature of the free-energy surface in the vicinity of the activated state.

For simplicity, it is often assumed that there is an equal likelihood of conversion to the reactant (r) or product (p) states ($P_{a \to p} = P_{a \to r} = \frac{1}{2}$). In addition, τ_a can often be approximated as $h/2kT$, where k is Boltzmann's constant and h is Planck's constant. In these cases, the decay rate to products and reactants are equal, reducing to

$$f_1 = f_2 = \frac{kT}{h} \tag{3.17}$$

Combining this simplified decay rate expression with our reaction rate equation (Equation 3.15) yields the following reduced expression for reaction rate:

$$\upsilon_1 = c_R^* \frac{kT}{h} e^{-\Delta G_1^{\ddagger}/(RT)} \tag{3.18}$$

rate to products is given by the lifetime of the activated species and the likelihood that it will convert to a product instead of back to a reactant. (A species in the activated state can "fall" either way.) More details on the decay rate are presented in a discussion box.

3.4 CALCULATING NET RATE OF A REACTION

When evaluating the overall rate of a reaction, we must consider the rates for both the forward and reverse directions of the reaction. The net rate is given by the difference in rates between the forward and reverse reactions. For example, the chemisorbed hydrogen reaction (Equation 3.13) can be split into forward and reverse reactions:

$$\text{Forward reaction:} \quad M \cdots H \to (M + e^-) + H^+ \tag{3.19}$$

$$\text{Reverse reaction:} \quad M \cdots H \leftarrow (M + e^-) + H^+ \tag{3.20}$$

with corresponding reaction rates given by υ_1 for the forward reaction and υ_2 for the reverse reaction. The net reaction rate υ is defined as

$$\upsilon = \upsilon_1 - \upsilon_2 \tag{3.21}$$

In general, the rates for the forward and reverse reactions may not be equal. In our example of the chemisorbed hydrogen reaction, the free-energy diagram in Figure 3.5 shows that the activation barrier for the forward reaction is much smaller than the activation barrier for the reverse reaction ($\Delta G_1^\ddagger < \Delta G_2^\ddagger$). In this situation, it stands to reason that the forward reaction rate should be much greater than the reverse reaction rate.

Using our reaction rate formula (Equation 3.15), the net reaction rate υ may be written as

$$\upsilon = c_R^* f_1 e^{-\Delta G_1^\ddagger/(RT)} - c_P^* f_2 e^{-\Delta G_2^\ddagger/(RT)} \tag{3.22}$$

where c_R^* is the reactant surface concentration, c_P^* is the product surface concentration, $\Delta G_1^\ddagger$ is the activation barrier for the forward reaction, and $\Delta G_2^\ddagger$ is the activation barrier for the reverse reaction. From the figure it is obvious that $\Delta G_2^\ddagger$ is related to $\Delta G_1^\ddagger$ and ΔG_{rxn}. In calculating the relationship between these activation energies, it is imperative to be careful with signs: ΔG quantities are always calculated as *final state − initial state*. For $\Delta G_1^\ddagger$ and $\Delta G_2^\ddagger$, the final state is the activated state; thus, activation barriers are always positive. If signs are properly accounted for, then

$$\Delta G_{\text{rxn}} = \Delta G_1^\ddagger - \Delta G_2^\ddagger \tag{3.23}$$

Equation 3.22 can be expressed in terms of only the forward activation barrier $\Delta G_1^\ddagger$:

$$\upsilon = c_R^* f_1 e^{-\Delta G_1^\ddagger/(RT)} - c_P^* f_2 e^{-(\Delta G_1^\ddagger - \Delta G_{\text{rxn}})/(RT)} \tag{3.24}$$

Thus, Equation 3.24 states that the net rate of a reaction is given by the difference between the forward and reverse reaction rates, both of which are exponentially dependent on an activation barrier, $\Delta G_1^\ddagger$.

3.5 RATE OF REACTION AT EQUILIBRIUM: EXCHANGE CURRENT DENSITY

For fuel cells, we are interested in the *current* produced by an electrochemical reaction. Therefore we want to recast these reaction rate expressions in terms of current density. Recall from Section 3.1.3 that current density j and reaction rate υ are related by $j = nF\upsilon$. Therefore the forward current density can be expressed as

$$j_1 = nF c_R^* f_1 e^{-\Delta G_1^\ddagger/(RT)} \tag{3.25}$$

and the reverse current density is given by

$$j_2 = nF c_P^* f_2 e^{(\Delta G_1^\ddagger - \Delta G_{\text{rxn}})/(RT)} \tag{3.26}$$

At thermodynamic equilibrium, we recognize that the forward and reverse current densities must balance so that there is no net current density ($j = 0$). In other words

$$j_1 = j_2 = j_0 \qquad \text{(at equilibrium)} \tag{3.27}$$

We call j_0 the *exchange current density* for the reaction. Although at equilibrium the net reaction rate is zero, both forward and reverse reactions are taking place at a rate which is characterized by j_0—this is called *dynamic* equilibrium.

3.6 POTENTIAL OF A REACTION AT EQUILIBRIUM: GALVANI POTENTIAL

Another way to understand the equilibrium state of a reaction is presented in Figure 3.6 which revisits our chemisorbed hydrogen system. Figure 3.6*a* is a simplified version of

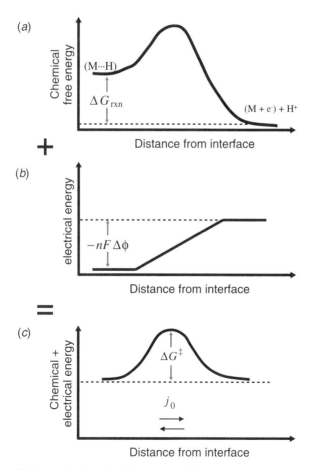

Figure 3.6. At equilibrium, the chemical free-energy difference (*a*) across a reaction interface is balanced by an electrical potential difference (*b*), resulting in a zero net reaction rate (*c*).

Figure 3.5, showing the chemical free-energy path for the chemisorbed hydrogen reaction. The lower free energy of the product state ($[M + e^-] + H^+$) compared to the reactant state ($M \cdots H$) leads to unequal activation barriers for the forward- versus reverse-reaction directions. Therefore, as we have previously discussed, we expect the forward reaction rate to proceed faster than the reverse reaction rate. However, these unequal rates quickly result in a buildup of charge, with e^- accumulating in the metal electrode and H^+ accumulating in the electrolyte. The charge accumulation continues until the resultant potential difference ($\Delta\phi$) across the reaction interface [as shown in Figure 3.6b] exactly counterbalances the chemical free-energy difference between the reactant and product states. This balance expresses the thermodynamic statement of electrochemical equilibrium that we developed in Equation 2.94. The combined effect of the chemical and electrical potentials is shown in Figure 3.6c, where the net force balance leads to equal rates for the forward and reverse reactions. As we have previously seen, the speed of this equilibrium reaction rate is captured in the exchange current density j_0.

Recall that before the buildup of the interfacial potential ($\Delta\phi$), the forward rate was much faster than the reverse rate. The buildup of an interfacial potential effectively equalizes the situation by increasing the forward activation barrier from $\Delta G_1^{\ddagger}$ to $\Delta G^{\ddagger}$ while decreasing the reverse activation barrier from $\Delta G_2^{\ddagger}$ to $\Delta G^{\ddagger}$. We can write the forward and reverse current densities at equilibrium as

$$j_1 = nFc_R^* f_1 e^{-\Delta G^{\ddagger}/(RT)} \tag{3.28}$$

$$j_2 = nFc_P^* f_2 e^{-(\Delta G^{\ddagger} - \Delta G_{\text{rxn}} + nF\Delta\phi)/(RT)} \tag{3.29}$$

While we have discussed Figure 3.6 in terms of the hydrogen reaction, it could just as easily represent the situation for the oxygen reaction at a fuel cell cathode. As in the hydrogen reaction, a difference in chemical free energy between the reactant and product states at the cathode will lead to an electrical potential difference. At equilibrium, the two force contributions balance, leading to a dynamic equilibrium with zero net reaction.

As shown in Figure 3.7, the sum of the interfacial electrical potential differences at the anode and cathode yield the overall thermodynamic equilibrium voltage for the fuel cell.

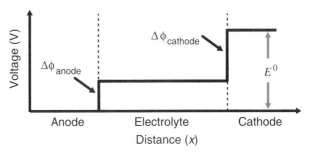

Figure 3.7. One hypothetical possibility for the shape of the fuel cell voltage profile, since scientists can determine E^0, but not $\Delta\phi_{\text{anode}}$ or $\Delta\phi_{\text{cathode}}$. The Galvani potentials at the anode and cathode of a fuel cell must sum to give the overall thermodynamic cell voltage E^0.

The anode ($\Delta\phi_{\text{anode}}$) and cathode ($\Delta\phi_{\text{cathode}}$) interfacial potentials shown in Figure 3.7 are called *Galvani potentials*. For reasons we will not discuss, the exact magnitude of these Galvani potentials are as-yet unknowable. While scientists know that the anode and cathode Galvani potentials must sum to give the net thermodynamic voltage of the fuel cell as a whole ($E^0 = \Delta\phi_{\text{anode}} + \Delta\phi_{\text{cathode}}$), they are unable to determine *how much* of this potential may be attributed to the anode interface versus the cathode interface. Thus, Figure 3.7 illustrates only one possible view of the fuel cell voltage profile. As a homework problem, you will sketch other possible voltage profiles.

3.7 POTENTIAL AND RATE: BUTLER–VOLMER EQUATION

A distinguishing feature of electrochemical reactions is the ability to *manipulate* the size of the activation barrier by varying the cell potential. Charged species are involved as either reactants or products in all electrochemical reactions. The free energy of a charged species is sensitive to voltage. Therefore, changing the cell voltage changes the free energy of the charged species taking part in a reaction, thus affecting the size of the activation barrier.

Figure 3.8 illustrates this idea. If we neglect to benefit from the full Galvani potential across a reaction interface, we can bias the system energetics such that the forward reaction rate is favored. By sacrificing part of the thermodynamically available cell voltage, we can produce a net current from our fuel cell. The Galvani potentials at the anode and the cathode must both be reduced (though not necessarily in equal amounts) to extract a net current from a fuel cell. Figure 3.9 shows how reducing the anode and cathode Galvani potentials results in a smaller net fuel cell voltage.

As shown in Figure 3.8c, decreasing the Galvani potential by η reduces the forward activation barrier ($\Delta G_1^{\ddagger} < \Delta G^{\ddagger}$) and increases the reverse activation barrier ($\Delta G_2^{\ddagger} > \Delta G^{\ddagger}$). A careful inspection of the figure shows that the forward activation barrier is decreased by $\alpha n F\eta$ while the reverse activation barrier is increased by $(1 - \alpha)n F\eta$.

The value of α depends on the symmetry of the activation barrier. Called the transfer coefficient, α expresses how the change in the electrical potential across the reaction interface changes the sizes of the forward versus reverse activation barrier. The value of α is always between 0 and 1. For "symmetric" reactions, $\alpha = 0.5$. For most electrochemical reactions, α ranges from about 0.2 to 0.5.

At equilibrium, the current densities for the forward and reverse reactions are both given by j_0. Away from equilibrium, we can write the new forward and reverse current densities by starting from j_0 and taking into account the changes in the forward and reverse activation barriers:

$$j_1 = j_0 e^{(\alpha n F\eta/(RT))} \tag{3.30}$$

$$j_2 = j_0 e^{-(1-\alpha)n F\eta/(RT)} \tag{3.31}$$

The net current ($j_1 - j_2$) is then

$$j = j_0\left(e^{\alpha n F\eta/(RT)} - e^{-(1-\alpha)n F\eta/(RT)}\right) \tag{3.32}$$

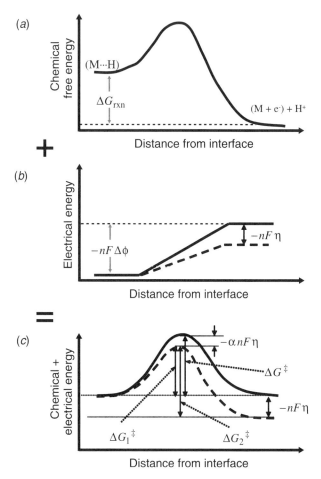

Figure 3.8. If the Galvani potential across a reaction interface is reduced, the free energy of the forward reaction will be favored over the reverse reaction. While the chemical energy (*a*) of the reaction system is the same as before, changing the electrical potential (*b*) upsets the balance between the forward and reverse activation barriers (*c*). In this diagram, reducing the Galvani potential by η reduces the forward activation barrier ($\Delta G_1^{\ddagger} < \Delta G^{\ddagger}$) and increases the reverse activation barrier ($\Delta G_2^{\ddagger} > \Delta G^{\ddagger}$).

Although it may not be obvious, this equation assumes that the concentration of reactant and product species at the electrode are unaffected by the presence of a net reaction rate. (Remember that j_0 depends on c_R^* and c_P^*; see Equations 3.25 and 3.26.) In reality, however, a net reaction rate will likely affect the surface concentrations of the reactant and product species. For example, if the forward reaction rate increases dramatically while the reverse reaction rate decreases dramatically, the reactant species surface concentration will tend to become depleted. In this case we can explicitly reflect the concentration dependence

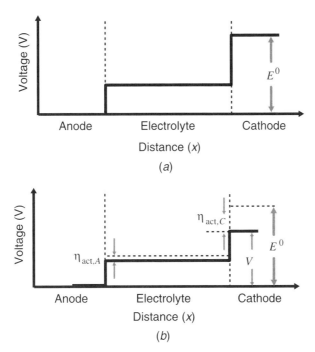

Figure 3.9. Extracting a net current from a fuel cell requires sacrificing a portion of both the anode and cathode Galvani potentials. In this figure, the anode Galvani potential is lowered by $\eta_{act,A}$ while the cathode Galvani potential is lowered by $\eta_{act,C}$. As the figure indicates, $\eta_{act,A}$ and $\eta_{act,C}$ are not necessarily equal. For a typical H_2–O_2 fuel cell, $\eta_{act,C}$ is generally much larger than $\eta_{act,A}$.

of the exchange current density in our equation as follows:

$$j = j_0^0 \left(\frac{c_R^*}{c_R^{0*}} e^{\alpha n F \eta/(RT)} - \frac{c_P^*}{c_P^{0*}} e^{-(1-\alpha)n F \eta/(RT)} \right) \qquad (3.33)$$

where η is the voltage loss, n is the number of electrons transferred in the electrochemical reaction, c_R^* and c_P^* are the actual surface concentrations of the rate-limiting species in the reaction, and j_0^0 is measured at the reference reactant and product concentration values c_R^{0*} and c_P^{0*}. Effectively, j_0^0 represents the exchange current density at a "standard concentration."

Equation 3.32 (or 3.33), known as the Butler–Volmer equation, is considered the cornerstone of electrochemical kinetics. It is used as the primary departure point for all attempts to describe how current and voltage are related in electrochemical systems. Remember it forever. The Butler–Volmer equation basically states that the current produced by an electrochemical reaction increases exponentially with *activation overvoltage*. Activation overvoltage is the label given to η, recognizing that η represents voltage which is sacrificed (lost) to overcome the activation barrier associated with the electrochemical reaction. Thus,

THE ACTIVATION OVERVOLTAGE, η_{act}

To clarify that η represents a voltage loss due to activation, it is typically given the subscript act, as in η_{act}. This distinguishes it from other voltage losses that you will read about in the upcoming chapters (which are also given the symbol η). From now on, we refer to the activation loss appearing in the Butler–Volmer equation as η_{act}, the activation overvoltage.

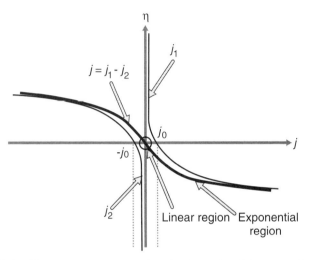

Figure 3.10. Relationship between η and j as given by the Butler–Volmer equation. The fine solid lines show the individual contributions from the forward (j_1) and reverse (j_2) current density terms while the dark solid line shows the net current density (j) given by the complete Butler–Volmer equation. Note that the Butler–Volmer curve is distinctly linear at low current density and distinctly exponential at high current density. In these regions, simplifications of the Butler–Volmer equation (as developed in Section 3.9) may be used.

the Butler–Volmer equation tells us that if we want more electricity (current) from our fuel cell, we must pay a price in terms of lost voltage. Figure 3.10 shows the functional form of the Butler–Volmer equation. Two distinct regions are indicated where simplifications of Equation 3.32 lead to easier kinetic treatment. These simplifications will be discussed in Section 3.9.

While we derived the Butler–Volmer equation using a specific reaction example, the Butler–Volmer equation applies to all single-step electrochemical reactions (or to multistep electrochemical reactions where the rate-determining step is intrinsically much slower than the other steps). For multistep reactions where several steps have approximately the same intrinsic rate, modifications to the Butler–Volmer equation are required. (While important, this treatment is beyond the scope of this book.) Even for these complex multistep

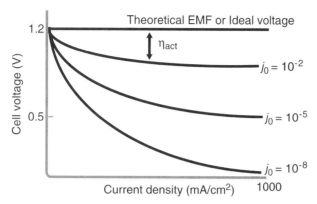

Figure 3.11. Effect of activation overvoltage on fuel cell performance. Reaction kinetics typically inflicts an exponential loss on a fuel cell's i–V curve as determined by the Butler–Volmer equation. The magnitude of this loss is influenced by the size of j_0. (Curves calculated for various j_0 values with $\alpha = 0.5$, $n = 2$, and $T = 298.15$ K.)

reactions, however, Butler–Volmer kinetics often proves to be an excellent first approximation.

For simple electrochemical systems, variations between reactions can be treated in terms of variations in kinetic parameters such as α and j_0 using the Butler–Volmer equation. As far as fuel cell performance is concerned, reaction kinetics induces a characteristic, exponentially shaped loss on a fuel cell's i–V curve, as shown in Figure 3.11. This curve was calculated by starting with E_{thermo} and then subtracting η_{act}. The functional dependence of η_{act} on j was given by the Butler–Volmer equation 3.32. The magnitude of the activation loss (in other words, the size of η_{act}) depends on the reaction kinetic parameters. The loss especially depends on the size of j_0, as shown in Figure 3.11. Having a high j_0 is absolutely critical to good fuel cell performance. As we will now discuss, there are several effective ways to increase j_0.

Example 3.3. If a fuel cell reaction exhibits $\alpha = 0.5$ and $n = 2$ at room temperature, what activation overvoltage is required to increase the forward current density by one order of magnitude and decrease the reverse current density by one order of magnitude?

Solution: Since $\alpha = 0.5$, the reaction is symmetric. We can look at either the forward or reverse term in the Butler–Volmer equation to calculate the overvoltage necessary to cause an order-of-magnitude change in current density. Using the forward term,

$$\frac{10 j_1}{j_1} = \frac{j_0 \left(e^{\alpha n F \eta_{\text{act2}}/(RT)} \right)}{j_0 \left(e^{\alpha n F \eta_{\text{act1}}/(RT)} \right)}$$

$$10 = e^{\alpha n F \Delta \eta_{\text{act}}/(RT)} \tag{3.34}$$

where we have defined $\Delta \eta_{\text{act}}$ as the change in activation overvoltage ($\eta_{\text{act2}} - \eta_{\text{act1}}$) necessary to increase the forward current density 10-fold. Solving for $\Delta \eta_{\text{act}}$ gives

$$\Delta\eta_{act} = \frac{RT}{\alpha n F} \ln 10 = \frac{(8.314)(298.15)}{(0.5)(2)(96,400)} \ln 10 = 0.059 \text{ V} \qquad (3.35)$$

Thus an activation overvoltage of approximately 60 mV is required to increase the forward current density by one order of magnitude and decrease the reverse current density by one order of magnitude for this reaction. If the exchange current density for this reaction was 10^{-6} A/cm^2, increasing the net current density to 1 A/cm^2 (a typical fuel cell operating current density) would require an activation overvoltage of 6×60 mV $= 0.36$ V.

3.8 EXCHANGE CURRENTS AND ELECTROCATALYSIS: HOW TO IMPROVE KINETIC PERFORMANCE

Improving kinetic performance stems from increasing j_0. To understand how we can increase j_0, recall how j_0 is defined. Remember that j_0 represents the "rate of exchange" between the reactant and product states at equilibrium. We can define j_0 from either the forward- or reverse-reaction direction. Taking the forward reaction for simplicity (see Equation 3.25) and including the concentration effects,

$$j_0 = n F c_R^* f_1 e^{-\Delta G_1^{\ddagger}/(RT)} \qquad (3.36)$$

By including reactant concentration effects in j_0, we must then use Equation 3.32 for the Butler–Volmer equation. Examining Equation 3.36, it is clear that we cannot change n, F, f_1 (not significantly), or R. Therefore, we have only three ways to increase j_0. In fact, there are four major ways to increase j_0, although the fourth method is not apparent from our equation:

1. Increase the reactant concentration c_R^*.
2. Decrease the activation barrier $\Delta G_1^{\ddagger}$.
3. Increase the temperature T.
4. Increase the number of possible reaction sites (i.e., increase the reaction interface roughness).

Each of these is discussed below.

3.8.1 Increase Reactant Concentration

In the last chapter, we noted that the thermodynamic benefit to increasing reactant concentration is minor, due to the logarithmic form of the Nernst equation. In contrast, the kinetic benefit to increasing reactant concentration is significant, with a linear rather than logarithmic impact. By operating fuel cells at higher pressure, we can increase the concentrations of the reactant gas species, improving the kinetics commensurately. Unfortunately, the kinetic penalty due to *decreasing* reactant concentration is likewise significant.

In real fuel cells, kinetic reactant concentration effects generally work against us for several reasons. First, most fuel cells use air instead of pure oxygen at the cathode. This leads to an approximate $5\times$ reduction in the oxygen kinetics compared to pure oxygen operation. Second, as will be discussed in Chapter 5, reactant concentrations tend to decrease at fuel cell electrodes during high-current-density operation (due to mass transport limitations). Essentially, the reactants are being consumed at the electrodes faster than they can be replenished, causing the local reactant concentrations to diminish. This depletion effect leads to further kinetic penalties. This interaction between kinetics and mass transport is the heart of the concentration loss effect described in Chapter 5.

3.8.2 Decrease Activation Barrier

As is apparent from Equation 3.36, decreasing the size of the activation barrier $\Delta G_1^{\ddagger}$ will increase j_0. A decrease in $\Delta G_1^{\ddagger}$ represents the catalytic influence of the surface of the electrode: A catalytic electrode is one which significantly lowers the activation barrier for the reaction. Because $\Delta G_1^{\ddagger}$ appears as an exponent, even small decreases in the activation barrier can cause large effects. Using a highly catalytic electrode therefore provides a way to dramatically increase j_0.

How does a catalytic electrode lower the activation barrier? *By changing the free energy surface of the reaction.* If you recall Figure 3.5, the size of the activation barrier for the hydrogen charge transfer reaction is related to the shape of the $[M \cdots H]$ and $[(M + e^-) + H^+]$ free-energy curves. Thus, the free-energy curves shown in Figure 3.5 will depend on the nature of the electrode metal, M. Different free-energy curves and therefore different activation barriers arise depending on the chemical nature of the $M \cdots H$ bond.

For the case of the hydrogen charge transfer reaction, an intermediate-strength bond provides the greatest catalytic effect. Why is an intermediate-strength bond most effective? If the $[M \cdots H]$ bond is too weak, then it is difficult for hydrogen to bond to the electrode surface in the first place, and it is furthermore difficult to transfer charge from the hydrogen to the electrode. On the other hand, if the $[M \cdots H]$ bond is too strong, the hydrogen bonds too well to the electrode surface. We then find it difficult to liberate free protons (H^+), and the electrode surface becomes clogged with unreactive $[M \cdots H]$ pairs. The optimal compromise between bonding and reactivity occurs for intermediate-strength $[M \cdots H]$ bonds. This peak in catalytic activity coincides with platinum-group metals and their neighbors, such as Pt, Pd, Ir, and Rh.

CHOICE OF CATALYST ALSO AFFECTS α

Note that the value of α will also be affected by the choice of catalyst. Recall that α is based on the symmetry of the free-energy curve in the vicinity of the activated state. Therefore, changes in the electrode free-energy curve can also be expected to change α. The Butler–Volmer equation predicts that increasing α will result in a higher net current density. Therefore, catalysts with a high α should be desired over catalysts with a low α. Generally, α changes only slightly with choice of catalyst, so it is often overlooked compared to other catalytic effects.

3.8.3 Increase Temperature

Equation 3.36 shows that increasing the temperature of reaction will also increase j_0. By increasing the reaction temperature, we are increasing the thermal energy available in the system; all particles in the system now move about and vibrate with increased intensity. This higher level of thermal activity increases the likelihood that a given reactant will possess sufficient energy to reach the activated state, thus increasing the rate of reaction. Like changing the activation barrier, changing the temperature has an exponential effect on j_0.

In reality, the complete story about temperature is a little more complicated than described here. At high overvoltage levels, increasing the temperature can actually decrease the current density. This effect is explained for the interested reader in a future dialogue box.

3.8.4 Increase Reaction Sites

Although not evident from Equation 3.36, the fourth method for increasing j_0 is to increase the number of available reaction sites per unit area. It is helpful to remember that j_0 represents a current density, or a reaction current *per unit area*. Current densities are generally based on the *plane*, or projected geometric area of an electrode. If an electrode surface is extremely rough, the true electrode surface area can be orders of magnitude larger than the geometric electrode area. As far as the kinetics are concerned, a highly rough electrode surface provides many more sites for reaction compared to a smooth electrode surface. Therefore the effective j_0 of a rough electrode surface will be greater than the j_0 of a smooth electrode surface simply because of the greater surface area. This relationship can be summarized by the equation

$$j_0 = j_0' \frac{A}{A'} \tag{3.37}$$

where j_0' represents the intrinsic exchange current density of a perfectly smooth electrode surface. The ratio A/A' expresses the surface area enhancement of a real electrode (area A) compared to an ideally smooth electrode (area A'). This definition has the benefit that j_0' can be considered an intrinsic property of an electrode for a specific electrochemical reaction. For example, the standard state j_0' for the HOR on platinum in sulfuric acid is widely considered to be around 10^{-3} A/cm^2. A platinum catalyst electrode with an effective surface area 1000 times greater than smooth platinum would therefore show an effective j_0 for the HOR of approximately 1 A/cm^2.

3.9 SIMPLIFIED ACTIVATION KINETICS: TAFEL EQUATION

When dealing with fuel cell reaction kinetics, the Butler–Volmer equation often proves unnecessarily complicated. In this section, we simplify the kinetics with two useful approximations. These approximations apply when the activation overvoltage (η_{act}) in the Butler–Volmer equation is either very small or very large:

- **When η_{act} Is Very Small.** For small η_{act} (less than about 15 mV at room temperature), a Taylor series expansion of the exponential terms can be performed with powers higher than 1 neglected ($e^x \approx 1 + x$ for small x). This treatment produces

$$j = j_0 \frac{nF\eta_{act}}{RT} \qquad (3.38)$$

which indicates that current and overvoltage are linearly related for small deviations from equilibrium and are independent of α. Theoretically, j_0 values can therefore be obtained from measurements of j versus η_{act} at low values of η_{act} (i.e., low current densities). As previously stated, j_0 is critical to fuel cell performance, so the ability to measure it would prove extremely useful. Unfortunately, experimental sources of error such as impurity currents, ohmic losses, and mass transport effects make these measurements difficult. Instead, j_0 values are usually extracted from high overvoltage measurements (see below).

- **When η_{act} Is Large.** When η_{act} is large (greater than 50–100 mV at room temperature), the second exponential term in the Butler–Volmer equation becomes negligible. In other words, the forward-reaction direction dominates, corresponding to a completely *irreversible* reaction process. The Butler–Volmer equation simplifies to

$$j = j_0 e^{\alpha nF\eta_{act}/(RT)} \qquad (3.39)$$

Solving this equation for η_{act} yields

$$\eta_{act} = -\frac{RT}{\alpha nF} \ln j_0 + \frac{RT}{\alpha nF} \ln j \qquad (3.40)$$

A plot of η_{act} versus $\ln j$ should be a straight line. Determination of j_0 and α is possible by fitting the line of η_{act} versus $\ln j$ or $\log j$. For good results, the fit should persist for at least one order of magnitude in current, preferably more. If this equation is generalized in the form

$$\eta_{act} = a + b \log j \qquad (3.41)$$

it is known as the *Tafel equation*, and b is called the *Tafel slope*. Like its relative, the Butler–Volmer equation, this equation is also quite important to electrochemical kinetics. Actually, the Tafel equation predates the Butler–Volmer equation. It was first developed as an empirical law based on electrochemical observations. Much later, kinetic theory provided an explanation for the Tafel equation based on basic principles!

For fuel cells, we are primarily interested in situations where large amounts of net current are produced. This situation corresponds to the case of an irreversible reaction process in which the forward-reaction direction dominates. Therefore, the second simplification of the Butler–Volmer equation (the Tafel equation) proves more useful in most discussions.

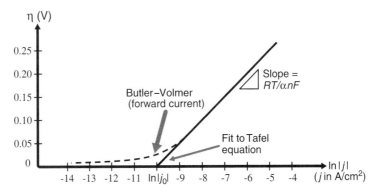

Figure 3.12. The j–η representation of a hypothetical electrochemical reaction. At high overvoltages, a linear fit of the kinetics to the Tafel approximation allows determination of j_0 and α. The Tafel approximation deviates from Butler–Volmer kinetics at low overvoltages.

An example of a Tafel plot showing the linear η–$\ln j$ behavior of a typical electrochemical reaction is shown in Figure 3.12. At high overvoltages, the linear Tafel equation applies very well to the curve. However, at low overvoltages, the Tafel approximation deviates from Butler–Volmer kinetics. From the slope and intercept of a linear fit to this plot, you should be able to calculate j_0 and α. (Note that most Tafel plots give η_{act} vs. $\log j$. Be aware of the conversion necessary to switch from $\log j$ to $\ln j$.)

> *Example 3.4.* Calculate j_0 and α for the hypothetical reaction in Figure 3.12. Assume the kinetic response depicted in the figure is for an electrochemical reaction at room temperature with $n = 2$.
>
> *Solution:* Using the linear Tafel fit of the data in Figure 3.12, we can extract both j_0 and α. From the figure, the j-axis intercept of the Tafel line gives $\ln j_0 = -10$. Therefore
>
> $$j_0 = e^{-10} = 4.54 \times 10^{-5} \text{A/ cm}^2 \qquad (3.42)$$
>
> Approximating the Tafel slope of this figure gives
>
> $$\text{Slope} \approx \frac{0.25 - 0.10}{-5 - (-8)} = 0.05 \qquad (3.43)$$
>
> From the Tafel equation, this slope is equal to $RT/\alpha n F$. Solving for α gives
>
> $$\alpha = \frac{RT}{\text{slope} \times n F} = \frac{(8.314)(298.15)}{(0.05)(2)(96{,}400)} = 0.257 \qquad (3.44)$$
>
> Thus α for this reaction is fairly small at 0.257 and j_0 is moderate at 4.54×10^{-5} A/cm^2. These kinetic parameters signify a moderate-to-slow electrochemical reaction.

MORE ON TEMPERATURE EFFECTS (OPTIONAL)

At high overvoltage levels, increasing the temperature can actually decrease the current density. How is this possible? While increasing temperature increases j_0, it has the opposite effect on the activation overvoltage. At high enough overvoltage levels, this "bad" temperature effect actually outweighs the "good" temperature effect. Since this reversal only occurs at high overvoltage levels, we can use the Tafel approximation of the Butler–Volmer equation to further discuss the situation:

$$j = j_0 e^{\alpha n F \eta_{\text{act}}/(RT)} \qquad (3.45)$$

If we then incorporate the temperature effect of j_0 and lump all the nontemperature-dependent constants into a constant, A, we get

$$j = A e^{-\Delta G_1^\ddagger/(RT)} e^{\alpha n F \eta_{\text{act}}/(RT)} \qquad (3.46)$$

From this equation, it is apparent that the current density j will increase with increasing temperature when $\alpha n F \eta_{\text{act}} < \Delta G_1^\ddagger$, but the current density will decrease with increasing temperature when $\alpha n F \eta_{\text{act}} > \Delta G_1^\ddagger$. In other words, for activation overvoltage levels greater than $\Delta G_1^\ddagger/\alpha n F$, increasing the temperature is no longer helpful; instead, it causes the current density to decrease.

This subtle temperature effect is seldom seen experimentally. Other positive effects of increasing the temperature (such as improvements in ion conductivity and mass transport) usually outweigh this reaction kinetics effect. Nonetheless, the phenomenon provides an interesting side note that highlights the complexity of electrochemical reaction kinetics.

3.10 DIFFERENT FUEL CELL REACTIONS PRODUCE DIFFERENT KINETICS

As was previously mentioned, the Butler–Volmer equation applies in general to all simple electrochemical reactions. Variations between reactions can be treated in terms of variations in the kinetic parameters α, j_0, ... Sluggish reaction kinetics (low α and j_0 values) result in severe performance penalties, while fast reaction kinetics (high α and j_0 values) result in minor performance penalties. As an example, consider the basic H_2–O_2 fuel cell. In a H_2–O_2 fuel cell, the HOR kinetics are extremely fast, while the ORR kinetics are extremely slow. Therefore the bulk of the activation overvoltage loss occurs at the cathode, where the ORR takes place. The difference between the anode and cathode activation losses in a typical low-temperature H_2–O_2 fuel cell is illustrated in Figure 3.13.

The ORR is sluggish because it is complicated. Completion of the ORR requires many individual steps and significant molecular reorganization. In comparison, the HOR is relatively straightforward. The contrast between H_2 and O_2 kinetics is highlighted in Tables 3.1 and 3.2, which present lists of j_0' values for the HOR and ORR at a variety of metal sur-

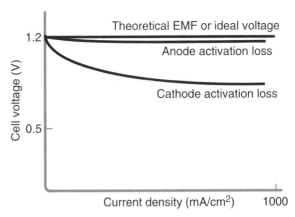

Figure 3.13. Relative contributions to activation loss from H_2–O_2 fuel cell anode versus cathode. The bulk of the activation overvoltage loss occurs at the cathode due to the sluggishness of the oxygen reduction kinetics.

TABLE 3.1. Standard-State ($T \approx 300$ K, 1 atm) Exchange Current Densities for Hydrogen Oxidation Reaction on Various Metal Surfaces

Surface	Electrolyte	j_0' (A/cm^2)
Pt	Acid	10^{-3}
Pt	Alkaline	10^{-4}
Pd	Acid	10^{-4}
Rh	Alkaline	10^{-4}
Ir	Acid	10^{-4}
Ni	Alkaline	10^{-4}
Ni	Acid	10^{-5}
Ag	Acid	10^{-5}
W	Acid	10^{-5}
Au	Acid	10^{-6}
Fe	Acid	10^{-6}
Mo	Acid	10^{-7}
Ta	Acid	10^{-7}
Sn	Acid	10^{-8}
Al	Acid	10^{-10}
Cd	Acid	10^{-12}
Hg	Acid	10^{-12}

Note: Rounded to nearest decade. Values are normalized per *real unit surface area* of metal [4, 5].

TABLE 3.2. Standard-State ($T \approx 300$ K, 1 atm) Exchange Current Densities for Oxygen Reduction Reaction on Various Surfaces

Surface	Electrolyte	j_0' (A/cm^2)
Metal Surfaces in Acid Electrolyte		
Pt	Acid	10^{-9}
Pd	Acid	10^{-10}
Ir	Acid	10^{-11}
Rh	Acid	10^{-11}
Au	Acid	10^{-11}
Pt Alloys in PEMFC		
Pt–C	Nafion	3×10^{-9}
PtMn–C	Nafion	6×10^{-9}
PtCr–C	Nafion	9×10^{-9}
PtFe–C	Nafion	7×10^{-9}
PtCo–C	Nafion	6×10^{-9}
PtNi–C	Nafion	5×10^{-9}

Note: Values are normalized per *real unit surface area* of metal. The exchange current density for the ORR is orders of magnitude smaller than for the HOR, although the same group of metals shows the highest activity for both reactions. Pt alloys may show a slight performance enhancement over pure Pt in a PEMFC environment [4, 6].

faces. Although Pt surfaces are most active for both reactions, the j_0' values for the ORR are still at least six orders of magnitude lower than for the HOR. Furthermore, most fuel cells run on air instead of pure oxygen. Although we saw in the previous chapter that air operation does not cause a significant thermodynamic penalty, it does cause a significant kinetic penalty. Because the oxygen concentration shows up in either the Butler–Volmer equation or j_0 (depending on which version of the Butler–Volmer equation you choose), operation in air (approximately $\frac{1}{5}$ oxygen) causes an additional 5× kinetic penalty compared to operation on pure oxygen.

Because the HOR is straightforward and kinetically fast, there is a significant kinetic advantage to using hydrogen fuel. When more complex hydrocarbon fuels are used, the anode kinetics become just as complicated and sluggish as the cathode kinetics, if not more so. Furthermore, fuels that involve carbon tend to generate undesirable intermediates that "poison" the fuel cell. The most serious of these for low-temperature fuel cells is CO. Carbon monoxide permanently absorbs onto platinum, clogging up reaction sites. The

CO passivated Pt surface is thus "poisoned," and the desired electrochemical reactions no longer occur.

Many of these kinetic problems are resolved in high-temperature fuel cells. For SOFCs, CO can act as a fuel rather than a poison. Furthermore, high temperature improves the oxygen kinetics, dramatically reducing the oxygen activation losses. The reactivity of hydrocarbon fuels also improves.

Not only do fuel cell reaction kinetics change depending on the type of fuel and temperature used, but they also change depending on the type of electrolyte used. For example, the hydrogen reduction reaction in a polymer electrolyte membrane (acidic) fuel cell, where H^+ is the charge carrier, occurs as

$$H_2 \rightarrow 2H^+ + 2e^- \tag{3.47}$$

Compare this to the hydrogen reduction reaction in an alkaline fuel cell, where OH^- is the charge carrier:

$$H_2 + 2OH^- \rightarrow 2H_2O + 2e^- \tag{3.48}$$

Compare this, yet again, to the hydrogen reduction reaction in a SOFC, where O^{2-} is the charge carrier:

$$H_2 + O^{2-} \rightarrow H_2O + 2e^- \tag{3.49}$$

The differences in reaction chemistry and temperature for these fuel cell types means that different catalysts are used. For low-temperature acidic fuel cells (PEMFCs and PAFCs) a Pt-based catalyst is used. For AFCs, nickel-based catalysts are used. For SOFCs, nickel-based or ceramic-based catalysts are used. For the interested reader, Sections 8.2–8.6 cover some of the specifics about catalyst materials for various fuel cell types.

3.11 CATALYST–ELECTRODE DESIGN

As we have seen, activation losses are minimized by maximizing the exchange current density. Since the exchange current density is a strong function of the catalyst material and the total reaction surface area, catalyst–electrode design focuses on these two parameters to achieve optimal performance.

To maximize reaction surface area, highly porous, nanostructured electrodes are fabricated to achieve intimate contact between gas-phase pores, the electrically conductive electrode, and the ion-conductive electrolyte. This nanostructuring is a deliberate attempt to maximize the total number of reaction sites in the fuel cell. In the fuel cell literature, these reaction sites are often called *triple phase zones* or *triple phase boundaries* (TPBs). This name refers to the fact that the fuel cell reactions can only occur where the three important phases—electrolyte, gas, and electrically connected catalyst regions—are in contact. The TPB is where all the action occurs! A simplified schematic of the TPBs is shown in Figure 3.14.

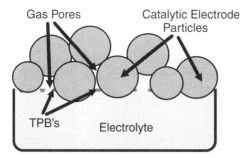

Figure 3.14. Simplified schematic of electrode–electolyte interface in a fuel cell, illustrating TPB reaction zones where catalytically active electrode particles, electrolyte phase, and gas pores intersect.

The second parameter, optimal catalyst material, is a function of the fuel cell chemistry and operating temperature, as previously discussed. The major requirements for an effective catalyst are as follows:

- High mechanical strength
- High electrical conductivity
- Low corrosion
- High porosity
- Ease of manufacturability
- High catalytic activity (high j_0)

For PEMFC, platinum is currently the best known catalyst. For higher temperature fuel cells, nickel- or ceramic-based catalysts are often used. As mentioned earlier, technology-specific catalyst selections are discussed in detail in Sections 8.2–8.6. Designing new catalysts is an area of intense research. In the next section, quantum mechanical approaches to catalyst simulation and design are briefly discussed.

Regardless of the type of catalyst, catalyst layer thickness is another variable that requires careful attention. In practice, the thickness of most fuel cell catalyst layers is between $\sim$ 10 and 50 μ m. While a thin layer is preferred for better gas diffusion and catalyst utilization, a thick layer incorporates higher catalyst loading and presents more TPBs. Thus, catalyst layer optimization requires a delicate balance between mass transport and catalytic activity concerns.

Usually, the catalyst layer is reinforced by a thicker porous electrode support layer. In a PEMFC, this electrode support layer is called the gas diffusion layer (GDL). The GDL protects the often delicate catalyst structure, provides mechanical strength, allows easy gas access to the catalyst, and enhances electrical conductivity. Electrode supports typically range in thickness from 100 to 400 μ m. As with the catalyst layer, a thinner electrode support generally provides better gas access but may also present increased electrical resistance or decreased mechanical strength.

The specifics of catalyst–electrode design vary by fuel cell type. Chapter 8 provides details for each of the main fuel cell types.

3.12   QUANTUM MECHANICS: FRAMEWORK FOR UNDERSTANDING CATALYSIS IN FUEL CELLS

Understanding the role of the catalyst in a fuel cell is crucial for designing next-generation fuel cell systems. As discussed in the previous section, virtually all PEMFC today rely on the availability of platinum or platinum alloys as catalytic materials. Unfortunately, platinum is scarce and expensive. This is fueling the drive toward novel catalyst design.

Most catalysts to date have been discovered with a trial-and-error approach. Considering the vast space of materials combinations, however, it is quite likely that better catalysts are waiting to be discovered. Unfortunately, finding optimal catalysts by trial and error is too time consuming and expensive. Fortunately, a cost-effective systematic approach involving simulation followed by experimental verification has recently become possible. For fuel cells, this simulation approach may soon help identify novel material systems with equivalent or possibly better catalytic performance when compared to platinum. Modern quantum mechanical simulation tools will play a key role in this search. A rudimentary understanding of their capability will be important for the next-generation of fuel cell scientists and engineers. In this section, we provide a glimpse into how quantum mechanics might contribute to the quest for new catalysts.

How exactly does a fuel cell catalyst work? Up to now, we have discussed catalysis from a continuum viewpoint. However, quantum-mechanics-based simulations can give us further insight. For example, consider the fuel cell anode from a quantum perspective. Hydrogen gas enters the fuel cell anode as a molecular species. As shown in Figure 3.15*a*, the hydrogen molecule consists of two hydrogen atoms strongly held together by an electron bond. The three-dimensional (3D) surface drawn around the hydrogen molecule in Figure 3.15*a* is a physical representation of the electron density in the molecule. In effect, the electron density distribution defines the spatial "extent" and "shape" of the molecule. Figure 3.15 was calculated using a quantum mechanical simulation technique known as density functional theory (DFT). Specifically, a commercially available tool called Gaussian[1] was used, which is capable of determining the electron density and the minimum energy of a quantum system. It is only in the last decade that commercially available quantum tools like Gaussian have become widely available. They rely on the mathematical framework of quantum mechanics, the details of which are presented for the interested student in Appendix D.

In Figure 3.15*b*, we watch as the hydrogen molecule begins to interact with a platinum catalyst cluster. As the hydrogen molecule gets closer and closer (Figure 3.15*b* through *d*) bonds between the hydrogen molecule and the platinum atoms are formed. The new emerging bonds between platinum and hydrogen lead to weakening of the hydrogen–hydrogen bond and ultimately to complete separation. Thus, the platinum catalyst facilitates the separation of the hydrogen molecule into hydrogen atoms. In the absence of the platinum clus-

---

[1]Gaussian is a computational tool predicting energies, molecular structures, and vibrational frequencies of molecular systems by Gaussian Inc.

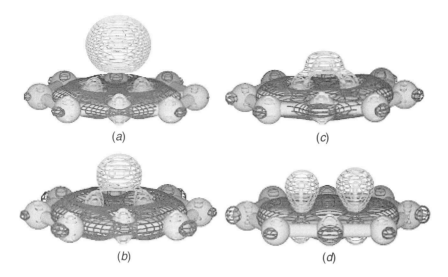

Figure 3.15. Evolution of electron orbitals as a hydrogen molecule approaches a cluster of platinum atoms. (*a*) Platinum and hydrogen molecule are not yet interacting. (*b*), (*c*) Atomic orbitals begin overlapping and forming bonds. (*d*) Complete separation of hydrogen atoms occurs almost simultaneously with reaching the lowest energy configuration.

ter, this reaction would not occur spontaneously; instead, significant energy input would be required to induce separation.

Each separated hydrogen atom in Figure 3.15*d* is sharing its electron with the platinum cluster. In the next reaction step, the hydrogen atoms must be removed from the platinum surface (as hydrogen ions) while leaving their electrons behind. The electrons can then be collected from the electrode and generate useful current. In most PEMFC environments, it is believed that the hydrogen ions are removed from the platinum surface by binding to water molecules, forming hydronium ions (H_3O^+). Figure 3.16 illustrates this reaction sequence.

Once a hydronium ion is formed, it may depart from the platinum surface. The formation of hydronium and its subsequent detachment from the catalyst surface may require overcoming a small energy barrier. This energy can be provided by the random motion of surrounding water molecules or by the thermal vibration of the platinum surface. Once the hydronium ion has departed, the platinum surface is available to participate in another reaction. A fresh hydrogen molecule can bind to the platinum surface and will be subject to the same set of reactions.

Figure 3.17 illustrates the situation at the fuel cell cathode. Figure 3.17*a* shows the *p* electron of an oxygen molecule approaching a platinum surface. Figure 3.17*b* indicates the bond formation of oxygen on the surface of the platinum cluster. As this figure indicates, splitting O_2 on the surface of a platinum substrate does not occur as readily as for H_2. The oxygen–oxygen bond is weakened but not destroyed after binding to platinum. The remaining bond strength is still 2.3 eV. In contrast, the bond strength of O_2 without a

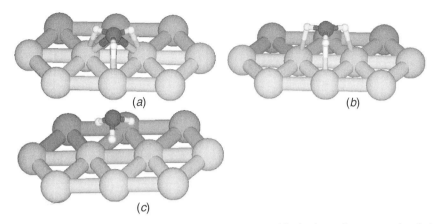

Figure 3.16. Formation of hydronium. Water attaches to a positively charged proton on the platinum surface, forming a hydronium ion. The hydronium ion then desorbs from the surface. For simplicity only atomic nuclei (no electron orbitals) are shown.

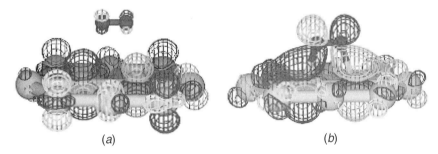

Figure 3.17. (*a*) Oxygen molecule approaching a platinum catalyst surface. (*b*) Even after having reached lowest energy configuration via hybrid orbital formation, the oxygen molecule is not completely separated into individual oxygen atoms.

platinum catalyst surface is 8.8 eV. Thus, significant energy is still required to complete the fuel cell reaction between this absorbed oxygen species and protons (hydronium ions) to form water. This quantum mechanical picture provides an explanation for why the oxygen reaction occurs more slowly than the hydrogen reaction.

It is important to realize that the picture painted in these figures is necessarily simplified. Various details, including the influence of voltage, platinum surface structure, and the involvement of additional water molecules, are ignored. For example, more sophisticated simulations of the cathode show that interactions of OH^- groups with partially broken oxygen molecules and protons further reduces the energy required for complete oxygen breakup.[2] This mechanism is believed to occur in many low-temperature PEMFCs.

---

[2] Also, the spin states of the electrons in platinum influence the energy required to break the oxygen bonds. See Appendix D for further explanations.

In spite of these issues, the simple model discussed above provides good qualitative insight into how a catalyst works. It also shows the promise that next-generation quantum tools might provide in exploring alternative catalyst materials and nanostructured surfaces.

CHAPTER SUMMARY

The purpose of this chapter is to explain how fuel cell reaction processes lead to performance losses. The study of reaction processes is called reaction kinetics, and the voltage loss caused by kinetic limitations is known as an activation loss.

- Electrochemical reactions involve the transfer of electrons and occur at surfaces.
- Because electrochemical reactions involve electron transfer, the current generated is a measure of the reaction rate.
- Because electrochemical reactions occur at surfaces, the rate (current) is proportional to the reaction surface area.
- Current density is more fundamental than current. We use current density (current per unit area) to normalize the effects of system size.
- An activation barrier impedes the conversion of reactants to products (and vice versa).
- A portion of the fuel cell voltage is sacrificed to lower the activation barrier, thus increasing the rate at which reactants are converted into products and the current density generated by the reaction.
- The sacrificed (lost) voltage is known as activation overvoltage η_{act}.
- The relationship between the current density output and the activation overvoltage is exponential. It is described by the Butler–Volmer equation $j = j_0(e^{\alpha n F \eta_{act}/(RT)} - e^{-(1-\alpha) n F \eta_{act}/(RT)})$.
- The exchange current density j_0 measures the equilibrium rate at which reactant and product species are exchanged in the absence of an activation overvoltage. A high j_0 indicates a facile reaction, while a low j_0 indicates a sluggish reaction.
- Activation overvoltage losses are minimized by maximizing j_0. There are four major ways to increase j_0: (1) increase reactant concentration, (2) increase reaction temperature, (3) decrease the activation barrier (by employing a catalyst), and (4) increase the number of reaction sites (by fabricating high-surface-area electrodes and 3D structured reaction interfaces).
- Fuel cells are usually operated at relatively high current densities (high activation overvoltages). At high activation overvoltage, fuel cell kinetics can be approximated by a simplified version of the Butler–Volmer equation, $j = j_0 e^{\alpha n F \eta_{act}/(RT)}$. In a generalized logarithmic form, this is known as the Tafel equation $\eta_{act} = a + b \log j$, where b is the Tafel slope.
- For a H_2–O_2 fuel cell, the hydrogen (anode) kinetics are generally facile and produce only a small activation loss. In contrast, the oxygen kinetics are sluggish and lead to a significant activation loss (at low temperature).

- The details of fuel cell reaction kinetics are dependent on the fuel, electrolyte chemistry, and operation temperature. For low-T fuel cells, Pt is commonly used as a catalyst. High-T fuel cells employ nickel- or ceramic-based catalysts.
- The main requirements for an effective fuel cell catalyst are (1) activity, (2) conductivity, and (3) stability (specifically thermal, mechanical, and chemical stability in the fuel cell environment).
- To increase j_0, fuel cell catalyst–electrodes are designed to maximize the number of reaction sites per unit area. Increasing the number of reaction sites means maximizing triple-phase boundary regions, where the electrolyte, reactant, and catalytically active electrode phases meet. The best catalyst–electrodes are carefully optimized, porous, high-surface area structures.

CHAPTER EXERCISES

Review Questions

3.1   This problem is composed of three parts:

    (a)   For the reaction

$$\tfrac{1}{2}O_2 + 2H^+ + 2e^- \rightleftharpoons H_2O$$

the standard electrode potential is $+1.23$ V. Under standard-state conditions, if the electrode potential is reduced to 1.0 V, will this bias the reaction in the forward or reverse direction?

    (b)   For the reaction

$$H_2 \rightleftharpoons 2H^+ + 2e^-$$

the standard electrode potential is 0.0 V. Under standard-state conditions, if the electrode potential is increased to 0.10 V, will this bias the reaction in the forward or reverse direction?

    (c)   Considering your answers to parts (a) and (b), in a H_2–O_2 fuel cell, if we increase the overall rate of the fuel cell reaction,

$$H_2 + \tfrac{1}{2}O_2 \rightleftharpoons H_2O$$

which is made up of the half reactions

$$H_2 \rightleftharpoons 2H^+ + 2e^-$$
$$\tfrac{1}{2}O_2 + 2H^+ + 2e^- \rightleftharpoons H_2O$$

what happens to the potential difference (voltage output) for the reaction?

3.2   Figure 3.7 presented one possible case for the voltage profile of a fuel cell. Draw two other possible voltage profiles that yield the same overall cell voltage but show vastly different individual Galvani potentials. Is it possible for one of the Galvani potentials to be negative yet still have the overall cell voltage be positive?

3.3 What is α? Assuming that the Galvani potential varies linearly across a reaction interface, sketch free-energy curves that result in situations where $\alpha > 0.5, \alpha = 0.5$, and $\alpha < 0.5$.

3.4 What does the exchange current density represent?

3.5 (a) In the Tafel equation, how is the Tafel slope b related to α?

   (b) How is the intercept a related to the exchange current density? (Remember that the Tafel equation is defined using log instead of ln.)

3.6 For a SOFC (where the charge carrier in the electrolyte is O^{2-}), CO is considered a fuel rather than a poison. Write an electrochemical half reaction showing how CO can be utilized as a fuel in the SOFC.

3.7 List the major requirement for an effective fuel cell catalyst material. List the major requirements for an effective fuel cell catalyst–electrode structure.

Calculations

3.8 Consider two electrochemical reactions. Reaction A results in the transfer of 2 mol of electrons per mole of reactant and generates a current of 5 A on an electrode 2 cm^2 in area. Reaction B results in the transfer of 3 mol of electrons per mole of reactant and generates a current of 15 A on an electrode 5 cm^2 in area. What are the net reaction rates for reactions A and B (in mol of reactant per square centimeter per second)? Which reaction has the higher net reaction rate?

3.9 Any sound theory of electrochemical kinetics must collapse to the thermodynamic prediction under equilibrium conditions. Prove that the Butler–Volmer kinetic model collapses to the thermodynamic prediction (the Nernst equation) at equilibrium.

3.10 This problem has several parts:

   (a) If a portable electronic device draws 1 A current at a voltage of 2.5 V, what is the power requirement for the device?

   (b) You have designed a fuel cell that delivers 1 A at 0.5 V. How many of your fuel cells are required to supply the above portable electronic device with its necessary voltage and current requirements?

   (c) You would like the portable electronic device to have an operating lifetime of 100 h. Assuming 100% fuel utilization, what is the minimum amount of H_2 fuel (in grams) required?

   (d) If this H_2 fuel is stored as a compressed gas at 500 atm, what volume would it occupy (assume ideal gas, room temperature)? If it is stored as a metal hydride at 5 wt % hydrogen, what volume would it occupy? (Assume the metal hydride has a density of 10 g/cm^3.)

3.11 Everything else being equal, write a general expression showing how the exchange current density for a reaction changes as a function of temperature [e.g., write an expression for $j_0(T)$ at an arbitrary temperature T as a function of $j_0(T_0)$ at a reference temperature T_0.] If a reaction has $j_0 = 10^{-8}$ A/cm^2 at 300 K and $j_0 = 10^{-4}$ A/cm^2

at 600 K, what is $\Delta G_1^{\ddagger}$ for the reaction? Assume that the preexponent portion of j_0 is temperature independent.

3.12 **(a)** Everything else being equal, write a general expression showing how the exchange current density varies as a function of reactant concentration.

(b) Use this result and your answer from problem 3.11 to answer the following question: For a reaction with $\Delta G_1^{\ddagger} = 20$ kJ/mol, what temperature change (starting from 300 K) has the same effect on j_0 as increasing the reactant concentration by one order of magnitude? Assume that the preexponent portion of j_0 is temperature independent.

3.13 All else being equal, at a given activation overvoltage, which effect produces a greater increase in the net current density for a reaction: doubling the temperature (in degrees Kelvin) or halving the activation barrier? Defend your answer with an equation. Assume that the preexponent portion of j_0 is temperature independent.

3.14 Estimate the thermal energy required to separate molecular oxygen with and without a platinum catalyst. Convert this energy into temperature (degrees centigrade) and comment on the role of platinum as catalyst in a PEMFC.

CHAPTER 4

FUEL CELL CHARGE TRANSPORT

The previous chapter on reaction kinetics detailed one of the most pivotal steps in the electrochemical generation of electricity: the production and consumption of charge via electrochemical half reactions. In this chapter, we address an equally important step in the electrochemical generation of electricity: charge transport. Charge transport "completes the circuit" in an electrochemical system, moving charges from the electrode where they are produced to the electrode where they are consumed.

There are two major types of charged species: electrons and ions. Since both electrons and ions are involved in electrochemical reactions, both types of charge must be transported. The transport of electrons versus ions is fundamentally different, primarily due to the large difference in mass between the two. In most fuel cells, ion charge transport is far more difficult than electron charge transport; therefore we are mainly concerned with ionic conductivity.

As we will discover, resistance to charge transport results in a voltage loss for fuel cells. Because this voltage loss obeys Ohm's law, it is called an *ohmic*, or *IR*, loss. Ohmic fuel cell losses are minimized by making electrolytes as thin as possible and employing high-conductivity materials. The search for high-ionic-conductivity materials will lead to a discussion of the fundamental mechanisms of ionic charge transport and a review of the most important electrolyte material classes.

4.1 CHARGES MOVE IN RESPONSE TO FORCES

The rate at which charges move through a material is quantified in terms of *flux* (denoted with the symbol J). Flux measures how much of a given quantity flows through a material *per unit area per unit time*. Figure 4.1 illustrates the concept of flux: Imagine water flowing

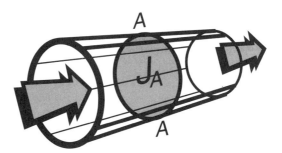

Figure 4.1. Schematic of flux. Imagine water flowing down this tube at a volumetric flow rate of 10 L/s. Dividing this flow rate by the cross-sectional area of the tube (A) gives the flux J_A of water moving down the tube. Generally, flux is measured in molar rather than volumetric quantities, so in this example the liters of water should be converted to moles.

down this tube at a volumetric flow rate of 10 L/s. If we divide this flow rate by the cross-sectional area of the tube (A), we get the volumetric flux J_A of water moving down the tube. In other words, J_A gives the per-unit-area flow rate of water through the tube. Be careful! Remember that flux and flow rate are not the same thing. By computing a flux, you normalize the flow rate by a cross-sectional area.

The most common type of flux is a molar flux (typical units are mol/cm$^2 \cdot$ s). *Charge flux* is a special type of flux that measures the *amount of charge* that flows through a material per unit area per unit time. Typical units for charge flux are C/cm$^2 \cdot$ s = A/cm^2. From these units, you may recognize that charge flux is the same thing as current density. To denote that charge flux represents a current density and carries different units than molar flux, we give it the symbol j. The quantity $z_i F$ is required to convert from molar flux J to charge flux j, where z_i is the charge number for the carrier (e.g., z_i is $+1$ for Na$^+$, -2 for O^{2-}, etc.) and F is Faraday's constant:

$$j = z_i F J \tag{4.1}$$

In all materials, a force must be acting on the charge carriers (i.e., the mobile electrons or ions in the material) for charge transport to occur. If there is no force acting on the charge carriers, there is no reason for them to move! The governing equation for transport can be generalized (in one dimension) as

$$J_i = \sum_k M_{ik} F_k \tag{4.2}$$

where J_i represents a flux of species i, F_k represent the k different forces acting on i, and the M_{ik}'s are the coupling coefficients between force and flux. The coupling coefficients reflect the relative ability of a species to respond to a given force with movement as well as the effective strength of the driving force itself. The coupling coefficients are therefore a property both of the species that is moving and the material through which it is moving. This general equation is valid for any type of transport (charge, heat, mass, etc.). In fuel cells, there are three major driving forces that give rise to charge transport: electrical driving forces (as represented by an electrical potential gradient dV/dx), chemical driving forces (as represented by a chemical potential gradient $d\mu/dx$), and mechanical driving forces (as represented by a pressure gradient dP/dx).

As an example of how these forces give rise to charge transport in a fuel cell, consider our familiar hydrogen–oxygen PEMFC (Figure 4.2). As hydrogen reacts in this fuel cell, protons and electrons accumulate at the anode, while protons and electrons are being consumed at the cathode. The accumulation/depletion of electrons at the two electrodes creates a voltage gradient which drives the transport of electrons from the anode to the cathode. In the electrolyte, accumulation/depletion of protons creates both a voltage gradient and a concentration gradient. These coupled gradients then drive the transport of protons from the anode to the cathode.

In the metal electrodes, only a voltage gradient drives electron charge transport. However, in the electrolyte, both a concentration (chemical potential) gradient and a voltage (electrical potential) gradient drive ion transport. How do we know which of these two driving forces is more important? In almost all situations, the electrical driving force dominates fuel cell ion transport. In other words, the electrical effect of the accumulated/depleted protons is far more important for charge transport than the chemical concentration effect of the accumulated/depleted protons. The underlying reasons why electrical driving forces dominate fuel cell charge transport are explained for the interested reader in an optional section near the end of this chapter (Section 4.7).

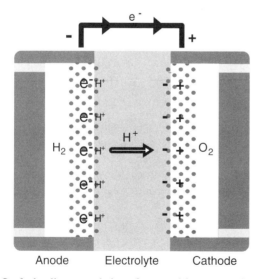

Figure 4.2. In a H_2–O_2 fuel cell, accumulation of protons/electrons at the anode and depletion of protons/electrons at the cathode lead to voltage gradients which drive charge transport. The electrons move from the negatively charged anode electrode to the positively charged cathode electrode. The protons move from the (relatively) positively charged anode side of the electrolyte to the (relatively) negatively charged cathode side of the electrolyte. The relative charge in the electrolyte at the anode versus the cathode arises due to differences in the concentration of protons. This concentration difference can also contribute to proton transport between the anode and cathode.

TABLE 4.1. Summary of Transport Processes Relevant to Charge Transport

Transport Process	Driving Force	Coupling Coefficient	Equation
Conduction	Electrical potential gradient, $\dfrac{dV}{dx}$	Conductivity σ	$J = \dfrac{\sigma}{|z_i|F}\dfrac{dV}{dx}$
Diffusion	Concentration gradient, $\dfrac{dc}{dx}$	Diffusivity D	$J = -D\dfrac{dc}{dx}$
Convection	Pressure gradient, $\dfrac{dp}{dx}$	Viscosity μ	$J = \dfrac{Gc}{\mu}\dfrac{dp}{dx}$

Note: The transport equation for convection in this table is based on Poiseuille's law, where G is a geometric constant and c is the concentration of the transported species. Convection flux is often calculated simply as $J = vc_i$, where v is the transport velocity.

For the case where charge transport is dominated by electrical driving forces, Equation 4.2 can be rewritten as

$$j = \sigma \frac{dV}{dx} \tag{4.3}$$

where j represents the charge flux (not molar flux), dV/dx is the electric field providing the driving force for charge transport, and σ is the conductivity, which measures the propensity of a material to permit charge flow in response to an electric field. This important application of Equation 4.2 simplifies the terms of fuel cell charge transport. In certain rare situations, both the concentration effects and electric potential effects may become important; in these cases, the charge transport equations become considerably more difficult.

In comparing Equation 4.3 to Equation 4.2, it is apparent that conductivity σ is nothing more than the name of the coupling coefficient that describes how flux and electrical driving forces are related. The relevant coupling coefficient that describes transport due to a chemical potential (concentration) gradient is called *diffusivity*. For transport due to a pressure gradient, the relevant coupling coefficient is called *viscosity*. These transport processes are summarized in Table 4.1 using molar flux quantities.

4.2 CHARGE TRANSPORT RESULTS IN A VOLTAGE LOSS

Unfortunately, charge transport is not a frictionless process. It occurs at a cost. For fuel cells, the penalty for charge transport is a loss in cell voltage. Why does charge transport result in a voltage loss? The answer is because fuel cell conductors are not perfect—they have an intrinsic resistance to charge flow.

Consider the uniform conductor pictured in Figure 4.3. This conductor has a constant cross-sectional area A and length L. Applying this example conductor geometry to our charge transport equation (4.3) produces

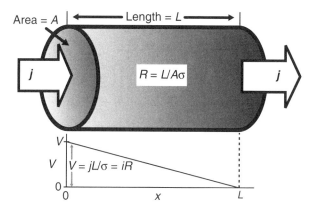

Figure 4.3. Illustration of charge transport along a uniform conductor of cross-sectional area A, length L, and conductivity σ. A voltage gradient dV/dx drives the transport of charge down the conductor. From the charge transport equation $j = \sigma(dV/dx)$ and the conductor geometry, we can derive Ohm's law: $V = iR$. The resistance of the conductor is dependent on the conductor's geometry and conductivity: $R = L/\sigma A$.

$$j = \sigma \frac{V}{L} \qquad (4.4)$$

Solving for V yields

$$V = j\left(\frac{L}{\sigma}\right) \qquad (4.5)$$

You might recognize that this equation is similar to Ohm's law: $V = iR$. In fact, since charge flux (current density) and current are related by $i = jA$, we can rewrite Equation 4.5 as

$$V = i\left(\frac{L}{A\sigma}\right) = iR \qquad (4.6)$$

where we identify the quantity $L/A\sigma$ as the resistance R of our conductor. The voltage V in this equation represents the voltage which must applied in order to transport charge at a rate given by i. Thus, this voltage represents a loss; it is the voltage which was expended, or sacrificed, in order to accomplish charge transport. This voltage loss arises due to our conductor's intrinsic resistance to charge transport, as embodied by $1/\sigma$.

Because this voltage loss obey's Ohm's law, we call it an "ohmic" loss. Like the activation overvoltage loss (η_{act}) introduced in the previous chapter, we give this voltage loss the symbol η. Specifically, we label it η_{ohmic} to distinguish it from η_{act}. Rewriting Equation 4.6 to reflect our nomenclature and explicitly including both the electronic (R_{elec}) and ionic (R_{ionic}) contributions to fuel cell resistance gives

$$\eta_{ohmic} = iR_{ohmic} = i(R_{elec} + R_{ionic}) \qquad (4.7)$$

Because ionic charge transport tends to be more difficult than electronic charge transport, the ionic contribution to R_{ohmic} tends to dominate.

The direction of the voltage gradient in an operating fuel cell electrolyte can often seem nonintuitive. As Figure 4.4c illustrates, although overall fuel cell voltage increases from the anode to the cathode, the cell voltage must *decrease* between the anode side of the electrolyte and the cathode side of the electrolyte to provide a driving force for charge transport.

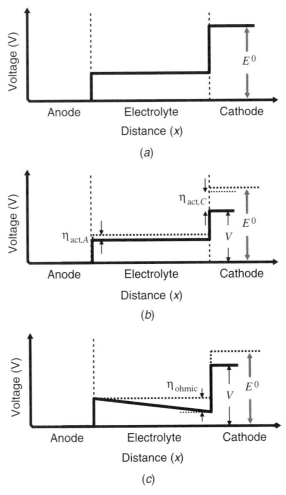

Figure 4.4. (*a*) Hypothetical voltage profile of a fuel cell at thermodynamic equilibrium (recall Figure 3.7). The thermodynamic voltage of the fuel cell is given by E^0. (*b*) Effect of anode and cathode activation losses on the fuel cell voltage profile (recall Figure 3.9). (*c*) Effect of ohmic losses on fuel cell voltage profile. Although the overall fuel cell voltage increases from the anode to the cathode, the cell voltage must decrease between the anode side of the electrolyte and the cathode side of the electrolyte to provide a driving force for charge transport.

Example 4.1. A 10-cm^2 PEMFC employs an electrolyte membrane with a conductivity of 0.10 $\Omega^{-1} \cdot$ cm^{-1}. For the fuel R_{elec} has been determined to be 0.005 Ω. Assuming the only other contribution to cell resistance comes from the electrolyte membrane, determine the ohmic voltage loss (η_{ohmic}) for the fuel cell at a current density of 1 A/cm^2 in the following cases: (a) the electrolyte membrane is 100 μm thick; (b) the electrolyte membrane is 50 μm thick.

Solution: We need to calculate R_{ionic} based on the electrolyte dimensions and then use Equation 4.7 to calculate η_{ohmic}. Since the fuel cell has an area of 10 cm^2, the current i of the fuel cell is 10 A:

$$i = jA = 1 \text{ A/cm}^2 \times 10 \text{ cm}^2 = 10 \text{ A} \qquad (4.8)$$

From Equation 4.6 we can calculate R_{ionic} for the the two cases (a), (b) given in this problem:

$$\text{Case (a):} \quad R_{ionic} = \frac{L}{\sigma A} = \frac{0.01 \text{ cm}}{(0.10 \ \Omega^{-1} \cdot \text{cm}^{-1})(10 \text{ cm}^2)} = 0.01 \ \Omega$$

$$(4.9)$$

$$\text{Case (b):} \quad R_{ionic} = \frac{0.005 \text{ cm}}{(0.10 \ \Omega^{-1} \cdot \text{cm}^{-1})(10 \text{ cm}^2)} = 0.005 \ \Omega$$

Inserting these values into Equation 4.7 and using $i = 10$A gives the following values for η_{ohmic}:

Case (a): $\eta_{ohmic} = i(R_{elec} + R_{ionic}) = 10 \text{ A}(0.005 \ \Omega + 0.01 \ \Omega) = 0.15 \text{ V}$

$$(4.10)$$

Case (b): $\eta_{ohmic} = 10 \text{ A}(0.005 \ \Omega + 0.005 \ \Omega) = 0.10 \text{ V}$

With everything else equal, making the membrane thinner reduces the ohmic loss! However, note that the payoff does not scale directly with membrane thickness. Although the membrane thickness was cut in half in this example, the ohmic loss was only reduced by one-third. This occurs because not all of the fuel cell's resistance contributions come from the electrolyte.

4.3 CHARACTERISTICS OF FUEL CELL CHARGE TRANSPORT RESISTANCE

As Equation 4.7 implies, charge transport linearly decreases fuel cell operating voltage as current increases. Figure 4.5 illustrates this effect. Obviously, if fuel cell resistance is decreased, fuel cell performance will improve.

Fuel cell resistance exhibits several important properties. First, resistance is geometry dependent, as Equation 4.6 clearly implies. Fuel cell resistance scales with area: To normalize out this effect, area-specific resistances are used to compare fuel cells of different sizes. Fuel cell resistance also scales with thickness; for this reason fuel cell electrolytes are generally made as thin as possible. Additionally, fuel cell resistances are additive;

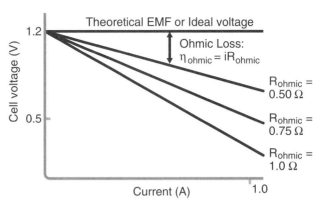

Figure 4.5. Effect of ohmic loss on fuel cell performance. Charge transport resistance contributes a linear decrease in fuel cell operating voltage as determined by Ohm's law (Equation 4.7). The magnitude of this loss is determined by the size of R_{ohmic}. (Curves calculated for R_{ohmic} equal 0.50 Ω, 0.75 Ω, and 1.0 Ω, respectively.)

resistance losses occurring at different locations within a fuel cell can be summed together in series. An investigation of the various contributions to fuel cell resistance reveals that the ionic (electrolyte) component to fuel cell resistance usually dominates. Thus, performance improvements may be won by the development of better ion conductors. Each of these important points will now be addressed.

4.3.1 Resistance Scales with Area

Since fuel cells are generally compared on a per-unit-area basis using current density instead of current, it is generally necessary to use area-normalized fuel cell resistances when discussing ohmic losses. Area-normalized resistance, also known as area-specific resistance (ASR), carries units of $\Omega \cdot cm^2$. By using ASR, ohmic losses can be calculated from current density as

$$\eta_{ohmic} = j\,(ASR_{ohmic}) \tag{4.11}$$

where ASR_{ohmic} is the ASR of the fuel cell. Area-specific resistance accounts for the fact that fuel cell resistance scales with area, thus allowing fuel cells of different sizes to be compared. It is calculated by multiplying a fuel cell's ohmic resistance R_{ohmic} by its area:

$$ASR_{ohmic} = A_{fuel\ cell}\,R_{ohmic} \tag{4.12}$$

Be careful, you must *multiply* resistance by area to get ASR, not divide! This calculation will probably seem nonintuitive at first. Because a large fuel cell has so much more area to flow current through compared to a small fuel cell, its resistance is far lower. However, on a per-unit-area basis, their resistances should be about the same; therefore the resistance of the large fuel cell must be multiplied by its area. This concept may be more understandable if you recall the original definition of resistance in Equation 4.6:

$$R = \frac{L}{A\sigma} \tag{4.13}$$

Since resistance is inversely proportional to area, multiplication by area is necessary to get area-independent resistances. This point is reinforced by Example 4.2.

> *Example 4.2.* Consider the two fuel cells illustrated in Figure 4.6. At a current density of 1 A/cm^2, calculate the ohmic voltage losses for both fuel cells. Which fuel cell incurs the larger ohmic voltage loss?
>
> *Solution:* There are two ways to solve this problem. To calculate voltage loss based on current density, we can either convert the resistances of the fuel cells to ASRs and then use Equation 4.11 (solution 1) or we can convert the current densities into currents and use Equation 4.6 (solution 2).
>
> *Solution 1:* Calculating the ASRs for the two fuel cells gives
>
> $$ASR_1 = R_1 A_1 = (0.1\ \Omega)(1\ cm^2) = 0.1\ \Omega \cdot cm^2$$
> $$ASR_2 = R_2 A_2 = (0.02\ \Omega)(10\ cm^2) = 0.2\ \Omega \cdot cm^2 \tag{4.14}$$
>
> Then, the ohmic voltage losses for the two cells can be calculated using Equation 4.11:
>
> $$\eta_{1,ohmic} = j(ASR_1) = (1\ A/cm^2)(0.1\ \Omega\ cm^2) = 0.1\ V$$
> $$\eta_{2,ohmic} = j(ASR_2) = (1\ A/cm^2)(0.2\ \Omega\ cm^2) = 0.2\ V \tag{4.15}$$
>
> *Solution 2:* Converting current densities for the two fuel cells into currents gives
>
> $$i_1 = jA_1 = (1\ A/cm^2)(1\ cm^2) = 1A$$
> $$i_2 = jA_2 = (1\ A/cm^2)(10\ cm^2) = 10A \tag{4.16}$$

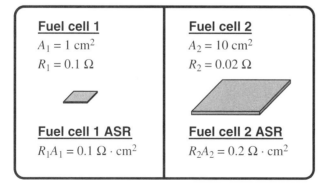

Figure 4.6. The importance of ASR is illustrated by these two fuel cells. Fuel cell 2 has lower total resistance than fuel cell 1 but yields a larger ohmic loss for a given current density. Fuel cell resistance is best compared using ASR rather than R.

Then, the ohmic voltage losses for the two cells can be calculated using Equation 4.6:

$$\eta_{1,\text{ohmic}} = i_1(R_1) = (1 \text{ A})(0.1 \text{ }\Omega) = 0.1 \text{ V}$$
$$\eta_{2,\text{ohmic}} = i_1(R_2) = (10 \text{ A})(0.02 \text{ }\Omega) = 0.2 \text{ V}$$

(4.17)

In both solutions, the same answer is obtained; cell 2 incurs a greater voltage loss. Although the total resistance of cell 2 is lower than cell 1 (0.02 Ω versus 0.1 Ω), the ASR of cell 2 is higher than that of cell 1. Thus, on an area-normalized basis, cell 2 is actually more "resistive" than cell 1 and leads to poorer fuel cell performance.

4.3.2 Resistance Scales with Thickness

Referring again to Equation 4.6, it is apparent that resistance scales not only with the cross-sectional area of the conductor but also with the length (thickness) of the conductor. If we normalize by using ASR, then:

$$\text{ASR} = \frac{L}{\sigma}$$

(4.18)

The shorter the conductor length L, the lower the resistance. It is intuitive that a shorter path results in less resistance.

Ionic conductivity is orders-of-magnitude lower than the electronic conductivity of metals, so minimizing the resistance of the fuel cell electrolyte is essential. Hence, we want the shortest path possible for ions between the anode and the cathode. Fuel cell electrolytes, therefore, are designed to be as thin as possible. Although reducing electrolyte thickness improves fuel cell performance, there are several practical issues that limit how thin the electrolyte can be made. The most important limitations are as follows:

- **Mechanical Integrity.** For solid electrolytes, the membrane cannot be made so thin that it risks breaking or develops pinholes. Membrane failure can result in catastrophic mixing of the fuel and oxidant!
- **Nonuniformities.** Even mechanically sound, pinhole-free electrolytes may fail if the thickness varies considerably across the fuel cell. Thin electrolyte areas may become "hot spots" that are subject to rapid deterioration and failure.
- **Shorting.** Extremely thin electrolytes (solid or liquid) risk electrical shorting, especially when the electrolyte thickness is on the same order of magnitude as the electrode roughness.
- **Fuel Crossover.** As the electrolyte thickness is reduced, the crossover of reactants may increase. This leads to an undesirable parasitic loss which can eventually become so large that further thickness decreases are counterproductive.
- **Contact Resistance.** Part of the electrolyte resistance is associated with the interface between the electrolyte and the electrode. This "contact" resistance is independent of electrolyte thickness.
- **Dielectric Breakdown.** The ultimate physical limit to solid-electrolyte thickness is given by the electrolyte's dielectric breakdown properties. This limit is reached when

the electrolyte is made so thin that the electric field across the membrane exceeds the dielectric breakdown field for the material.

For most solid-electrolyte materials, the ultimate limit on thickness, as predicted by the dielectric breakdown field, is on the order of several nanometers. However, the other practical limitations listed above currently limit achievable thickness to about 10–100 μm, depending on the electrolyte.

4.3.3 Fuel Cell Resistances Are Additive

As Figure 4.7 conceptually illustrates, the total ohmic resistance presented by a fuel cell is actually a combination of resistances coming from different components of the device. Depending on how much precision is needed, it is possible to assign individual resistances to the electrical interconnections, anode electrode, cathode electrode, anode catalyst layer, cathode catalyst layer, electrolyte, and so on. It is also possible to ascribe *contact resistances* associated with the interfaces between the various layers in the fuel cell (e.g., a flow structure/electrode contact resistance). Because the current produced by the fuel cell must flow serially through all of these regions, the total fuel cell resistance is simply the sum of all the individual resistance contributions. Unfortunately, it is experimentally very difficult to distinguish between all the various sources of resistance loss.

You might think that it should be a relatively easy experimental task to measure the resistance of each component in a fuel cell (e.g., the electrodes, the flow structures, the interconnections, the membrane) before assembling them together into a device. However, such measurements never completely reflect the true total resistance of a fuel cell device.

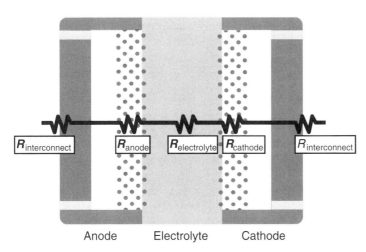

Figure 4.7. The total ohmic resistance presented by a fuel cell is actually a combination of resistances, each attributed to different components of the fuel cell. In this diagram, fuel cell resistance is divided into interconnect, anode, electrolyte, and cathode components. Since current flows serially through all components, total fuel cell resistance is given by the series sum of the individual resistance components.

Variations in contact resistances, assembly processes, and operating conditions make total fuel cell resistance difficult to predict. These factors make fuel cell characterization extremely challenging, as discussed in Chapter 7, and emphasize the necessity of in situ fuel cell characterization. Despite the experimental difficulties involved in pinpointing all the sources of fuel cell resistance loss, the electrolyte yields the biggest resistance loss for most fuel cell devices.

 ### 4.3.4   Ionic (Electrolyte) Resistance Usually Dominates

The best electrolytes employed in fuel cells have ionic conductivities of around $0.10 \ \Omega^{-1} \cdot \text{cm}^{-1}$. Even at a thickness of 50 μm (very thin), this produces an ASR of $0.05 \ \Omega \cdot \text{cm}^2$. In contrast, a 50-$\mu$m-thick porous carbon cloth electrode would have an ASR of less than $5 \times 10^{-6} \ \Omega \cdot \text{cm}^2$. This example illustrates how electrolyte resistance usually dominates fuel cells.

Well-designed fuel cells have a total ASR in the range of 0.05–$0.10 \ \Omega \cdot \text{cm}^2$, and electrolyte resistance accounts for most of the total. If electrolyte thickness cannot be reduced, decreasing ohmic loss depends on finding high-σ ionic conductors. Unfortunately, developing satisfactory ionic conductors is challenging. The three most widely used electrolyte classes, discussed in Sections 4.5.1–4.5.3, are the aqueous, polymer, and ceramic electrolytes. The conductivity mechanisms and materials properties of these three electrolyte classes are quite different. Before we get to that discussion, however, it is helpful to develop a clear physical picture of conductivity in general terms.

4.4   PHYSICAL MEANING OF CONDUCTIVITY

Conductivity quantifies the ability of a material to permit the flow of charge when driven by an electric field. In other words, conductivity is a measure of how well a material accommodates charge transport. A material's conductivity is influenced by two major factors: *how many* carriers are available to transport charge and *the mobility* of those carriers within the material. The following equation defines σ in those terms:

$$\sigma_i = (|z_i|F)c_i u_i \tag{4.19}$$

where c_i represents the *molar* concentration of charge carriers (how many moles of carrier is available per unit volume) and u_i is the mobility of the charge carriers within the material. The quantity $|z_i|F$ is necessary to convert charge carrier concentration from units of moles to units of coulombs. Here, z_i is the charge number for the carrier (e.g., $z_i = +2$ for Cu^{2+}, $z_i = -1$ for e^-, etc.), the absolute-value function ensures that conductivity is always a positive number, and F is Faraday's constant.

A material's conductivity is therefore determined by *carrier concentration* c_i and *carrier mobility* u_i. These properties are in turn set by the structure and conduction mechanisms within the material. Up to this point, the charge transport equations we have learned apply equally well to both electronic and ionic conduction. Now, however, their paths will diverge. Because electronic and ionic conduction mechanisms are vastly different, electronic and ionic conductivities are also quite different.

CONDUCTIVITY AND MOBILITY

The difference between conductivity and mobility can be understood by an analogy. Pretend that we are studying the transport of people (in cars) down an interstate highway. Mobility describes how fast the cars are driving down the highway. Conductivity, however, would also include information about how many cars are on the highway and how many people each car can hold. This analogy is not perfect but may help keep the two terms straight.

4.4.1 Electronic versus Ionic Conductors

Differences in the fundamental nature of electrons versus ions lead to differences in the mechanisms for electronic versus ionic conduction. Figure 4.8 schematically contrasts a typical electronic conductor (a metal) and a typical ionic conductor (a solid electrolyte).

Figure 4.8a illustrates the free-electron model of a metallic electron conductor. In this model, the valence electrons associated with the atoms of the metal become detached from the atomic lattice and are free to move about the metal. Meanwhile, the metal ions remain

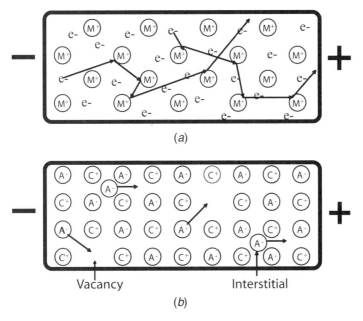

Figure 4.8. Illustration of charge transport mechanisms. (*a*) Electron transport in a free-electron metal. Valence electrons detach from immobile metal atom cores and move freely in response to an applied field. Their velocity is limited by scattering from the lattice. (*b*) Charge transport in this crystalline ionic conductor is accomplished by mobile anions which "hop" from position to position within the lattice. The hopping process only occurs where lattice defects such as vacancies or interstitials are present.

intact and immobile. The free valence electrons constitute a "sea" of mobile charges which are able to move in response to an applied field.

By contrast, Figure 4.8b illustrates the hopping model of a solid-state ionic conductor. The crystalline lattice of this ion conductor consists of both positive and negative ions, all of which are fixed to specific crystallographic positions. Occasionally, defects such as missing atoms ("vacancies") or extra atoms ("interstitials") will occur in the material. Charge transport is accomplished by the site-to-site "hopping" of these defects through the material.

The structural differences between the two kinds of conductors lead to dramatic differences in carrier concentrations. In a metal, free electrons are populous, while carriers in a crystalline solid electrolyte are rare. The differences in the charge transport mechanisms as illustrated in Figure 4.8 also lead to dramatic differences in carrier mobility. Combined together, the differences in carrier concentration and carrier mobility lead to a very different picture for electron conductivity in a metal versus ion conductivity in a solid electrolyte. Let us briefly take a look at both.

4.4.2 Electron Conductivity in a Metal

For a simple electron conductor, such as a metal, the Drude model predicts that the mobility of free electrons in the metal will be limited by scattering (from phonons, lattice imperfections, impurities, etc.):

$$u = \frac{q\tau}{m} \tag{4.20}$$

where τ gives the mean free time between scattering events, m is the mass of the electron ($m = 9.11 \times 10^{-31}$ kg), and q is the elementary electron charge in coulombs ($q = 1.68 \times 10^{-19}$ C).

Inserting the results for electron mobility (Equation 4.20) into the expression for conductivity (Equation 4.19) gives

$$\sigma = \frac{|z_e|c_e q\tau}{m} \tag{4.21}$$

Carrier concentration in a metal may be calculated from the density of free electrons. In general, each metal atom will contribute approximately one free electron. Atomic packing densities are generally on the order of 10^{28} atoms/m^3, which yields molar carrier concentrations on the order of 10^4 mol/m^3.

Inserting typical numbers into Equation 4.21 allows us to calculate ballpark electronic conductivity values. The charge number on an electron is, of course, 1 ($|z_e| = 1$). Typical scattering times (in relatively pure metals) are 10^{-12}–10^{-14} s. Using $c_e \approx 10^4$ mol/m^3 yields typical electron conductivities for metals in the range of 10^6–10^8 $\Omega^{-1} \cdot$ m^{-1} (10^4–10^6 $\Omega^{-1} \cdot$ cm^{-1}).

4.4.3 Ion Conductivity in a Crystalline Solid Electrolyte

The conduction hopping process illustrated in Figure 4.8b for a solid ion conductor leads to a very different expression for mobility as compared to a metallic electron conductor. Ion mobility for the material in Figure 4.8*b* is dependent on the rate at which ions can hop from position to position within the lattice. This hopping rate, like the reaction rates studied in the previous chapter, is exponentially activated. The effectiveness of the hopping process is characterized by the material's diffusivity D:

$$D = D_0 e^{-\Delta G_{act}/(RT)} \tag{4.22}$$

where D_0 is a constant reflecting the attempt frequency of the hopping process, ΔG_{act} is the activation barrier for the hopping process, R is the gas constant, and T is the temperature (K). The overall mobility of ions in the solid electrolyte is then given by

$$u = \frac{|z_i|FD}{RT} \tag{4.23}$$

where $|z_i|$ is the charge number on the ion, F is Faraday's constant, R is the gas constant, and T is the temperature (K).

Inserting the expression for ion mobility (Equation 4.23) into our equation for conductivity (Equation 4.19) gives

$$\sigma = \frac{c(z_i F)^2 D}{RT} \tag{4.24}$$

Carrier concentration in a crystalline electrolyte is controlled by the density of the mobile defect species. Most crystalline electrolytes conduct via a vacancy mechanism. These vacancies are intentionally introduced into the lattice by doping. Maximum effective vacancy doping levels are around 8–10%, leading to carrier concentrations of 10^2–10^3 mol/m^3.

Typical ion diffusivities are 10^{-8} m^2/s for liquid electrolytes, 10^{-8} m^2/s for polymer electrolytes, and 10^{-11} m^2/s for ceramic electrolytes at 700–1000°C. Typical ion carrier concentrations are 10^3–10^4 mol/m^3 for liquid electrolytes, 10^2–10^3 mol/m^3 for polymer electrolytes, and 10^2–10^3 mol/m^3 for ceramic electrolytes at 700–1000°C. Inserting these values into Equation 4.24 yields ionic conductivity values of 10^{-4}–10^2 $\Omega^{-1} \cdot$ m^{-1} (10^{-6}–10^0 $\Omega^{-1} \cdot$ cm^{-1}).

Note that solid-electrolyte ionic conductivity values are well below electronic conductivity values for metals. As has been previously stated, ionic charge transport tends to be far more difficult than electronic charge transport. Therefore, much of the focus in fuel cell research is placed on finding better electrolytes.

4.5   REVIEW OF FUEL CELL ELECTROLYTE CLASSES

The search for better electrolytes has led to the development of three major candidate materials classes for fuel cells: aqueous, polymer, and ceramic electrolytes. Regardless of the class, however, any fuel cell electrolyte must meet the following requirements:

- High ionic conductivity
- Low electronic conductivity
- High stability (in both oxidizing and reducing environments)

- Low fuel crossover
- Reasonable mechanical strength (if solid)
- Ease of manufacturability

Other than the high-conductivity requirement, the electrolyte stability requirement is often the hardest to fulfill. It is difficult to find an electrolyte that is stable in both the highly reducing environment of the anode and the highly oxidizing environment of the cathode.

4.5.1 Ionic Conduction in Aqueous Electrolytes/Ionic Liquids

In this section we discuss ionic conduction in aqueous electrolytes and ionic liquids. An aqueous electrolyte is a water-based solution containing dissolved ions that can transport charge. An ionic liquid is a material which is *itself* simultaneously liquid and ionic. Sodium chloride dissolved in water is an example of an aqueous electrolyte. The NaCl separates into mobile Na^+ ions and mobile Cl^- ions which can transport charge by moving through the water solvent. Molten NaCl (when heated to high temperature) is an example of an ionic liquid. Pure H_3PO_4 at $50°C$ is another example of an ionic liquid. At room temperature, H_3PO_4 is a somewhat waxy, white crystalline solid. However, when heated above $42°C$, it becomes a viscous ionic liquid consisting of H^+ ions, PO_4^{3-} ions, and H_3PO_4 molecules.

Almost all aqueous/liquid electrolyte fuel cells use a matrix material to support or immobilize the electrolyte. The matrix generally accomplishes three tasks:

1. Provides mechanical strength to the electrolyte
2. Minimizes the distance between the electrodes while preventing shorts
3. Prevents crossover of reactant gases through the electrolyte

Reactant crossover, the last task on this list, is a particular problem for aqueous/liquid electrolytes (much more so than for solid electrolytes). In an unsupported liquid electrolyte, reactant gas crossover can be severe; in these situations, unbalanced-pressure or high-pressure operation is impossible. The use of a matrix material provides mechanical integrity and reduces gas crossover problems while still permitting thin (0.1–1.0-mm) electrolytes.

Alkaline fuel cells use concentrated aqueous KOH electrolytes, while phosphoric acid fuel cells use either concentrated aqueous H_3PO_4 electrolytes or pure H_3PO_4 (an ionic liquid). Molten carbonate fuel cells use molten $(K/Li)_2CO_3$ immobilized in a supporting matrix. The $(K/Li)_2CO_3$ material melts at around $450°C$ to become a liquid ("molten") electrolyte. (MCFCs must therefore obviously be operated above $450°C$.)

Ionic conductivity in aqueous/liquid environments can best be approached using a driving force/frictional force balance model. In liquids, an ion will accelerate under the force of an electric field until frictional drag exactly counteracts the electric field force. The balance between the electric field and frictional drag determines the terminal velocity of the ion.

The electric field force F_E is given by:

$$F_E = n_i q \frac{dV}{dx} \tag{4.25}$$

where n_i is the charge number of the ion and q is the fundamental electron charge (1.6×10^{-19} C). Although we do not show the derivation here, the frictional drag force F_D may be approximated from Stokes's law as

$$F_D = 6\pi \mu r v \qquad (4.26)$$

where μ is the viscosity of the liquid, r is the radius of the ion, and v is the velocity of the ion. Equating the two forces allows us to determine the mobility u_i, which is defined as the ratio between the applied electric field and the resulting ion velocity:

$$u_i = \frac{v}{dV/dx} = \frac{n_i q}{6\pi \mu r} \qquad (4.27)$$

Thus, mobility is determined by the ion size and the liquid viscosity. Intuitively, this expression makes sense: Bulky ions or highly viscous liquids should lead to lower mobilities, while nonviscous liquids and small ions should yield higher mobilities. The mobilities of a variety of ions in aqueous solution are given in Table 4.2. Note that in aqueous solutions the H^+ ion tends to be hydrated by one or more water molecules. This ionic species is therefore better thought of as H_3O^+ or $H \cdot (H_2O)_x{}^+$, where x represents the number of water molecules "hydrating" the proton.

Recall our expression for conductivity (Equation 4.19), which is repeated here for clarity:

$$\sigma_i = (|z_i|F)c_i u_i \qquad (4.28)$$

If the values of ion mobilities in Table 4.2 are inserted into this expression, the ionic conductivity of various aqueous electrolytes may be calculated. Unfortunately, these calculations are only accurate for dilute aqueous solutions when the ion concentration is low. At high ion concentration (or for ionic liquids) strong electrical interactions between the ions make conductivity far more difficult to calculate. In general, the conductivity of highly concentrated aqueous solutions or pure ionic liquids will be much lower than that predicted by Equation 4.28. For example, the conductivity of pure H_3PO_4 is experimentally determined to be 0.1–1.0 $\Omega^{-1} \cdot cm^{-1}$ (depending on the temperature), whereas Equation 4.28 predicts that the conductivity of pure H_3PO_4 should be approximately 18 $\Omega^{-1} \cdot cm^{-1}$.

TABLE 4.2. Selected Ionic Mobilities at Infinite Dilution in Aqueous Solutions at 25°C

Cation	Mobility, u (cm^2/V · s)	Anion	Mobility, u (cm^2/V · s)
$H^+(H_3O^+)$	3.63×10^{-3}	OH^-	2.05×10^{-3}
K^+	7.62×10^{-4}	Br^-	8.13×10^{-4}
Ag^+	6.40×10^{-4}	I^-	7.96×10^{-4}
Na^+	5.19×10^{-4}	Cl^-	7.91×10^{-4}
Li^+	4.01×10^{-4}	$HCO_3{}^-$	4.61×10^{-4}

Source: From Ref. [8].

Table 4.2 does offer some other useful insights. For example, it explains why KOH is the electrolyte of choice in alkaline fuel cells. Besides being extremely inexpensive, KOH exhibits the highest ionic conductivity of any of the hydroxide compounds. (Compare the u value for K^+ to other candidate hydroxide cations such as Na^+ or Li^+.) In alkaline fuel cells, fairly concentrated (30–65%) solutions of KOH are used, resulting in conductivities on the order of 0.1–0.5 $\Omega^{-1} \cdot cm^{-1}$. How much would the conductivity be reduced if a far more dilute electrolyte was used? To get an answer, refer to Example 4.3, where the approximate conductivity of a 0.1 M KOH electrolyte solution is calculated using Equation 4.28.

Example 4.3. Calculate the approximate conductivity of a 0.1 M aqueous solution of KOH.

Solution: We use Equation 4.28 as our guide. Assuming that 0.1 M KOH completely dissolves into K^+ ions and OH^- ions (it does), the concentration of K^+ and OH^- will also be 0.1 M. Converting these concentrations to units of moles per cubic centimeters gives

$$c_{K^+} = (0.1 \text{ mol/L})(1 \text{ L}/1000 \text{ cm}^3) = 1 \times 10^{-4} \text{ mol/cm}^3$$

$$c_{OH^-} = (0.1 \text{mol/L})(1 \text{ L}/1000 \text{ cm}^3) = 1 \times 10^{-4} \text{ mol/cm}^3 \tag{4.29}$$

The mobilities of K^+ and OH^- are given in Table 4.2. Inserting these numbers into Equation 4.28 yields

$$\sigma_{K^+} = (1)(96,400)(1 \times 10^{-4} \text{ mol/cm}^3)(7.62 \times 10^{-4} \text{ cm}^2/V \cdot s)$$

$$= 0.0073 \ \Omega^{-1} \cdot cm^{-1}$$

$$\sigma_{OH^-} = (1)(96,400)(1 \times 10^{-4} \text{ mol/cm}^3)(2.05 \times 10^{-3} \text{ cm}^2/V \cdot s) \tag{4.30}$$

$$= 0.0198 \ \Omega^{-1} \cdot cm^{-1}$$

The total ionic conductivity of the electrolyte is then given by the sum of the cation and anion conductivities:

$$\sigma_{\text{total}} = \sigma_{K^+} + \sigma_{OH^-} = 0.0073 + 0.0198 = 0.0271 \ \Omega^{-1} \cdot cm^{-1} \tag{4.31}$$

In reality, the conductivity of the 0.1 M KOH solution will likely be a little lower than this predicted value. Note that most of the conductivity is provided by the OH^- ion, rather than the K^+ ion. This is due to the higher mobility of the OH^- ion.

4.5.2 Ionic Conduction in Polymer Electrolytes

In general, ionic transport in polymer electrolytes follows the exponential relationship described by Equations 4.22 and 4.24. By combining these two equations, we can obtain (see problem 4.11)

$$\sigma T = \sigma_0 e^{-E_a/kT} \tag{4.32}$$

where σ_0 represents the conductivity at a reference state and E_a represents the activation energy (eV/mol) ($E_a = \Delta G_{act}/F$, where F is Faraday's constant). As this equation indicates, conductivity increases exponentially with increasing temperature. Most polymer and crystalline ion conductors obey this model quite well.

For a polymer to be a good ion conductor, at a minimum it should possess the following structural properties:

1. The presence of fixed charge sites
2. The presence of free volume ("open space")

The fixed charge sites should be of opposite charge compared to the moving ions, ensuring that the net charge balance across the polymer is maintained. The fixed charge sites provide temporary centers where the moving ions can be accepted or released. In a polymer structure, maximizing the concentration of these charge sites is critical to ensuring high conductivity. However, excessive addition of ionically charged side chains will significantly degrade the mechanical stability of the polymer, making it unsuitable for fuel cell use.

Free volume correlates with the spatial organization of the polymer. In general, a typical polymer structure is *not fully dense*. Small-pore structures (or free volumes) will almost always exist. Free volume improves the ability of ions to move across the polymer. Increasing the polymer free volume increases the range of small-scale structural vibrations and motions within the polymer. These motions can result in the *physical* transfer of ions from site to site across the polymer. (See Figure 4.9.)

Due to these free-volume effects, polymer membranes exhibit relatively high ionic conductivities compared to other solid-state ion-conducting materials (such as ceramics).

Polymer free volume also leads to another well-known transport mechanism, known as the *vehicle mechanism*. In the vehicle mechanism, ions are transported through free-volume spaces by hitching a ride on certain free species (the "vehicles") as these vehicles pass by. Water is a common vehicular species; as water molecules move through the free volumes in a polymer membrane, ions can go along for the ride. In this case, the conduction behavior of the ions in the polymer electrolyte is much like that in an aqueous electrolyte. Persulfonated

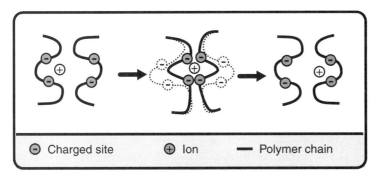

Figure 4.9. Schematic of ion transport between polymer chains. Polymer segments can move or vibrate in the free volume, thus inducing physical transfer of ions from one charged site to another.

polytetrafluoroethylene (PTFE)—more commonly known as Nafion—exhibits extremely high proton conductivity based on the vehicle mechanism. Since Nafion is the most popular and important electrolyte for PEMFC applications, we review its properties in the next section.

Ionic Transport in Nafion. Nafion has a backbone structure similar to polytetrafluoroethylene (Teflon). However, unlike Teflon, Nafion includes sulfonic acid ($SO_3^- H^+$) functional groups. The Teflon backbone provides mechanical strength while the sulfonic acid

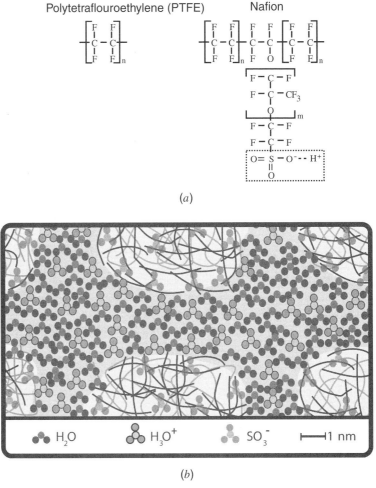

(a)

(b)

Figure 4.10. (*a*) Chemical structure of Nafion. Nafion has a PTFE backbone for mechanical stability with sulfonic groups to promote proton conduction. (*b*) Schematic microscopic view of proton conduction in Nafion. When hydrated, nanometer-sized pores swell and become largely interconnected. Protons bind with water molecules to form hydronium complexes. Sulfonic groups near the pore walls enable hydronium conduction.

$(SO_3^- H^+)$ chains provide charge sites for proton transport. Figure 4.10 illustrates the structure of Nafion.

It is believed that Nafion free volumes aggregate into interconnected nanometer-sized pores whose walls are lined by sulfonic acid $(SO_3^- H^+)$ groups. In the presence of water, the protons (H^+) in the pores form hydronium complexes (H_3O^+) and detach from the sulfonic acid side chains. When sufficient water exists in the pores, the hydronium ions can transport in the aqueous phase. Under these circumstances, ionic conduction in Nafion is similar to conduction in liquid electrolytes (Section 4.5.1). As a bonus, the hydrophobic nature of the Teflon backbone further accelerates water transport through the membrane, since the hydrophobic pore surfaces tend to repel water. Because of these factors, Nafion exhibits proton conductivity comparable to that of a liquid electrolyte. To maintain this extraordinary conductivity, Nafion must be fully hydrated with liquid water. Usually, hydration is achieved by humidifying the fuel and oxidant gases provisioned to the fuel cell. In the following paragraphs, we review the key properties of Nafion in more detail.[1]

Nafion Absorbs Significant Amounts of Water. The pore structure in Nafion can hold significant amounts of water. In fact, Nafion can accommodate so much water that its volume will increase up to 22% when fully hydrated. (Strongly polar liquids, like alcohols, can cause Nafion to swell up to 88%!) Since conductivity and water content are strongly related, determining water content is essential to determining the conductivity of the membrane. The water content λ in Nafion is defined as the ratio of the number of water molecules to the number of charged $(SO_3^- H^+)$ sites. Experimental results suggest that λ can vary from almost 0 (for completely dehydrated Nafion) to 22 (for full saturation, under certain conditions). For fuel cells, experimental measurements have related the water content in Nafion to the humidity condition of the fuel cell, as shown in Figure 4.11. Thus, if the humidity condition of the fuel cell is known, the water content in the membrane can be estimated. Humidity in Figure 4.11 is quantified by water vapor activity a_W (essentially relative humidity):

$$a_W = \frac{p_W}{p_{SAT}} \tag{4.33}$$

where p_W represents the actual partial pressure of water vapor in the system and p_{SAT} represents the saturation water vapor pressure for the system at the temperature of operation. The data in Figure 4.11 can be represented mathematically as

$$\lambda = \begin{cases} 0.0043 + 17.81a_W - 39.85a_W^2 + 36.0a_W^3 & \text{for} \quad 0 < a_W \leq 1 \\ \lambda = 14 + 1.4(a_W - 1) & \text{for} \quad 1 < a_W \leq 3 \end{cases} \tag{4.34}$$

Equation 4.34 does not consider the effects of temperature; however, it is reasonably accurate for PEMFCs operating near 80°C.

---

[1]The Nafion model reviewed here was suggested by Springer et al. [8]

WATER VAPOR SATURATION PRESSURE

When the partial pressure of water vapor (p_W) within a gas stream reaches the *water vapor saturation pressure* p_{SAT} for a given temperature, the water vapor will start to condense, generating water droplets. In other words, relative humidity is 100% when $p_W = p_{SAT}$, where p_{SAT} is a strong function of temperature:

$$\log_{10} p_{SAT} = -2.1794 + 0.02953T - 9.1837 \times 10^{-5}T^2 + 1.4454 \times 10^{-7}T^3$$
$$(4.35)$$

where p_{SAT} is given in bars (1 bar = 100,000 Pa) and T is the temperature in degrees Celsius. For example, if fully humidified air at 80°C and 3 atm is provided to a fuel cell, the water vapor pressure is [9]

$$p_{SAT} = 10^{-2.1794 + 0.02953 \times 80 - 9.1837 \times 10^{-5} \times 80^2 + 1.4454 \times 10^{-7} \times 80^3}$$

$$= 0.4669 \text{ bar} \qquad (4.36)$$

This gives the mole fraction of water in fully humidified air as 0.4669 bar/3 atm = 0.4669 bar/(3×1.0132501 bar) = 0.154 assuming an ideal gas.

Under these same conditions, if the air is instead only *partially humidified*, such that the water mole fraction is 0.1, then the water vapor activity (or relative humidity) would be (again assuming an ideal gas)

$$a = \frac{p_{H_2O}}{p_{SAT}} = \frac{x_{H_2O} \times p_{total}}{x_{H_2O,SAT} \times p_{total}} = \frac{0.1}{0.154} = 0.65 \qquad (4.37)$$

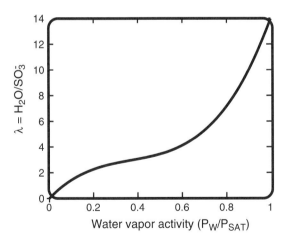

Figure 4.11. Water content versus water activity for Nafion 117 at 303 K (30°C) according to Equation 4.34. Water vapor activity is defined as the ratio of the actual water vapor pressure (p_W) for the system compared to the saturation water vapor pressure (p_{SAT}) for the system at the temperature of interest. (Reprinted with permission from Ref. [8], *Journal of the Electrochemical Society*, 138: 2334, 1991. Copyright 1991 by the Electrochemical Society.)

Nafion Conductivity Is Highly Dependent on Water Content. As previously mentioned, conductivity and water content are strongly related in Nafion. Conductivity and temperature are also strongly related. In general, the proton conductivity of Nafion increases linearly with increasing water content and exponentially with increasing temperature, as shown by the experimental data in Figures 4.12 and 4.13. In equation form, these experimentally determined relationships may be summarized as

$$\sigma(T, \lambda) = \sigma_{303K}(\lambda) \exp\left[1268\left(\frac{1}{303} - \frac{1}{T}\right)\right] \quad (4.38)$$

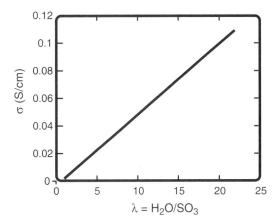

Figure 4.12. Ionic conductivity of Nafion versus water content λ according to Equations 4.38 and 4.39 at 303 K.

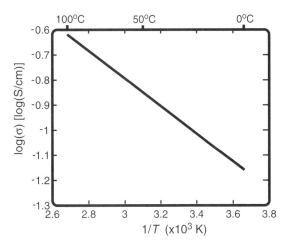

Figure 4.13. Ionic conductivity of Nafion versus temperature according to Equation 4.38 when $\lambda = 22$.

where

$$\sigma_{303K}(\lambda) = 0.005193\lambda - 0.00326 \tag{4.39}$$

where σ represents the conductivity (S/cm) of the membrane and and T (K) is the temperature.

Since the conductivity of Nafion can change locally depending on water content, the total resistance of a membrane is found by integrating the local resistance over the membrane thickness (t_m) as

$$R_m = \int_0^{t_m} \rho(z)\,dz = \int_0^{t_m} \frac{dz}{\sigma[\lambda(z)]} \tag{4.40}$$

Protons Drag Water with Them. Since conductivity in Nafion is dependent on water content, it is essential to know how water content varies across a Nafion membrane. Water content varies across a Nafion membrane because of several factors. Perhaps most important is the fact that protons[2] traveling through the pores of Nafion generally drag one or more water molecules along with them. This well-known phenomenon is called *electro-osmotic drag*. The degree to which proton movement causes water movement is quantified by the electro-osmotic drag coefficient n_{drag}, which is defined as the number of water molecules accompanying the movement of each proton ($n_{drag} = n_{H_2O}/H^+$). Obviously, how much water is dragged per proton depends on how much water exists in the Nafion membrane in the first place. It has been measured that $n_{drag} = 2.5 \pm 0.2$ (between 30 and 50°C) in fully hydrated Nafion (when $\lambda = 22$). When $\lambda = 11$, $n_{drag} = \sim 0.9$. Commonly, it is assumed that n_{drag} changes linearly with λ as

$$n_{drag} = n_{drag}^{SAT} \frac{\lambda}{22} \qquad \text{for} \quad 0 \le \lambda \le 22 \tag{4.41}$$

where $n_{drag}^{SAT} \approx 2.5$. Knowledge of the electro-osmotic drag coefficient allows us to estimate the water drag flux from anode to cathode when a net current j flows through the PEMFC:

$$J_{H_2O,drag} = 2n_{drag} \frac{j}{2F} \tag{4.42}$$

where J is the molar flux of water due to electro-osmotic drag (mol/scm^2), j is the operating current density of the fuel cell (A/cm^2), and the quantity $2F$ converts from current density to hydrogen flux. The factor of 2 in front then converts from hydrogen flux to proton flux. As we will see in Chapter 6, the drag coefficient becomes very important in modeling the behavior of Nafion membranes in PEMFCs.

Back Diffusion of Water. In a PEMFC, electro-osmotic water drag moves water from the anode to the cathode. As this water builds up at the cathode, however, *back diffusion* occurs, resulting in the transport of water from the cathode *back* to the anode. This back-diffusion phenomenon occurs because the concentration of water at the cathode is generally

---

[2]Actually, protons travel in the form of hydronium complexes as explained in the text. For simplicity, however, we use the term "proton" in these discussions. Also, it is more straightforward to define the electro-osmotic drag coefficient in terms of the number of water molecules per proton (rather than per hydronium, which contains a water molecule already).

far higher than the concentration of water at the anode (exacerbated by the fact that water is produced at the cathode by the electrochemical reaction). Back diffusion counterbalances the effects of electro-osmotic drag. Driven by the anode/cathode water concentration gradient, the water back-diffusion flux can be determined by

$$J_{H_2O, \text{ back diffusion}} = -\frac{\rho_{\text{dry}}}{M_m} D_\lambda \frac{d\lambda}{dz} \tag{4.43}$$

where ρ_{dry} is the dry density (kg/m^3) of Nafion, M_n is the Nafion equivalent weight (kg/mol), and z is the direction through the membrane thickness.

The key factor in this equation is the diffusivity of water in the Nafion membrane (D_λ). Unfortunately, D_λ is not constant, but a function of water content λ. Since the total water flux in Nafion is simply the addition of electro-osmotic drag and back diffusion, we have

$$J_{H_2O} = 2n_{\text{drag}}^{\text{SAT}} \frac{j}{2F} \frac{\lambda}{22} - \frac{\rho_{\text{dry}}}{M_m} D_\lambda(\lambda) \frac{d\lambda}{dz} \tag{4.44}$$

This combined expression makes it explicitly clear that the water flux in Nafion is a complex function of λ. [We state the water diffusivity as $D_\lambda(\lambda)$ in this equation to emphasize its dependency on water content.]

Summary. Based on the fuel cell operating conditions (humidity and current density), we can estimate the water content in the membrane using Equations 4.34 and 4.44. From the water content estimate, we can then calculate the ion conductivity of the membrane using Equation 4.38. In this fashion, the ohmic losses in a PEMFC may be quantified. This procedure is demonstrated in Example 4.4. In Chapter 6 we will combine these equations with the other fuel cell loss terms to create a complete PEMFC model.

Example 4.4. Consider a hydrogen PEMFC powering an external load at 0.7 A/cm^2. The activities of water vapor on the anode and cathode sides of the membrane are measured to be 0.8 and 1.0, respectively. The temperature of the fuel cell is 80°C. If the Nafion membrane thickness is 0.125 mm, estimate the ohmic overvoltage loss across the membrane.

Solution: We can convert the water activity on the Nafion surfaces to water contents using Equation 4.34:

$$\lambda^A = 0.0043 + 17.81 \times 0.8 - 39.85 \times 0.8^2 + 36.0 \times 0.8^3 = 7.2$$
$$\lambda^C = 0.0043 + 17.81 \times 1.0 - 39.85 \times 1.0^2 + 36.0 \times 1.0^3 = 14.0 \tag{4.45}$$

With these values as boundary conditions, we then solve Equation 4.44. In this equation, we have two unknowns, J_{H_2O} and λ. For convenience, we will set $J_{H_2O} = \alpha N_{H_2} = \alpha(j/2F)$, where α is an unknown that denotes the ratio of water flux to hydrogen flux. After rearrangement, Equation 4.44 becomes

$$\frac{d\lambda}{dz} = \left(2n_{\text{drag}}^{\text{SAT}} \frac{\lambda}{22} - \alpha\right) \frac{jM_m}{2F\rho_{\text{dry}}D_\lambda} \tag{4.46}$$

EQUIVALENT WEIGHT

The equivalent weight of a species is defined by its atomic weight or formula weight divided by its valence:

$$\text{Equivalent weight} = \frac{\text{atomic (formula) weight}}{\text{valence}} \tag{4.47}$$

Valence is defined by the number of electrons that the species can donate or accept. For example, hydrogen has a valence of 1 (H^+). Oxygen has a valence of 2 (O^{2-}). Thus, hydrogen has an equivalent weight of $1.008 \text{ g/mol}/1 = 1.008 \text{ g/mol}$ and oxygen has an equivalent weight of $15.9994 \text{ g/mol}/2 = 7.9997 \text{ g/mol}$. In the case of sulfate radicals (SO_4^{2-}), the formula weight is $(1 \times 32.06) + (4 \times 15.9994) = 96.058 \text{ g/mol}$. Thus the equivalent weight is $96.058 \text{ g/mol}/2 = 48.029 \text{ g/mol}$.

The sulfonic group ($SO_3^- H^+$) in Nafion has a valence of 1 since it can accept only one proton. Thus, the equivalent weight of Nafion is equal to the average weight of the polymer chain structure that can accept one proton. This number is very useful since it facilitates the calculation of sulfonic charge (SO_3^-) concentration in Nafion as

$$C_{SO_3^-} \text{ (mol/m}^3) = \frac{\rho_{\text{dry}} \text{ (kg/m}^3)}{M_m \text{ (kg/mol)}} \tag{4.48}$$

where ρ_{dry} is the dry density of Nafion (kg/m^3) and M_m is the Nafion equivalent weight (kg/mol).

In a similar fashion, water content, $\lambda(H_2O/SO_3^-)$, can be converted to water concentration in Nafion as

$$C_{H_2O} \text{ (mol/m}^3) = \lambda \frac{\rho_{\text{dry}} \text{ (kg/m}^3)}{M_m \text{ (kg/mol)}} \tag{4.49}$$

Typically, Nafion has an equivalent weight of around $\sim$1–1.1 kg/mol and a dry density of $\sim$1970 kg/m^3. Thus, the estimated charge density for Nafion would be

$$C_{SO_3^-} \text{ (mol/m}^3) = \frac{1970 \text{ kg/m}^3}{1 \text{ kg/mol}} = 1970 \text{ mol/m}^3 \tag{4.50}$$

WATER DIFFUSIVITY IN NAFION

As emphasized above, water diffusivity in Nafion (D_λ) is a function of water content λ. Experimentally (using magnetic resonance techniques), this dependence has been measured as

$$D_\lambda = \exp\left[2416\left(\frac{1}{303} - \frac{1}{T}\right)\right]$$

$$\times \left(2.563 - 0.33\lambda + 0.0264\lambda^2 - 0.000671\lambda^3\right) \times 10^{-6}$$

$$\text{for} \quad \lambda > 4 \quad (\text{cm}^2/\text{s}) \tag{4.51}$$

The exponential part describes the temperature dependence, while the polynominal portion describes the λ dependence at the reference temperature of 303 K. This equation is only valid for $\lambda > 4$. For $\lambda < 4$, values extrapolated from Figure 4.14 (dotted line) should be used instead.

Even though this is an ordinary differential equation on λ, we may not solve it analytically since D_λ is a function of λ. However, if we assume λ in the membrane changes from 7.2 to 14.0 according to the boundary conditions, we can see from Figure 4.14 that the water diffusivity is fairly constant over this range. If we assume an average value of $\lambda = 10$, we can estimate D_λ from Equation 4.51 as

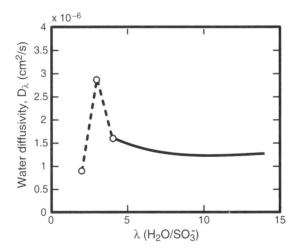

Figure 4.14. Water diffusivity D_λ in Nafion versus water content λ at 303 K (Reprinted with permission from Ref. [8], *Journal of the Electrochemical Society*, 138: 2334, 1991. Copyright 1991 by the Electrochemical Society.).

$$D_\lambda = 10^{-6} \exp\left[2416\left(\frac{1}{303} - \frac{1}{353}\right)\right]$$

$$\times \left(2.563 - 0.33 \times 10 + 0.0264 \times 10^2 - 0.000671 \times 10^3\right)$$

$$= 3.81 \times 10^{-6} \text{ cm}^2/\text{s} \tag{4.52}$$

Now we can evaluate Equation 4.46, yielding the analytical solution

$$\lambda(z) = \frac{11\alpha}{n_{\text{drag}}^{\text{SAT}}} + C \, \exp\left[\frac{j \, M_m n_{\text{drag}}^{\text{SAT}}}{22 F \rho_{\text{dry}} D_\lambda} z\right] = \frac{11\alpha}{2.5}$$

$$+ C \, \exp\left[\frac{(0.7 \text{ A/cm}^2) \times (1.0 \text{ kg/mol}) \times 2.5}{(22 \times 96{,}500 \text{ C/mol}) \times (0.00197 \text{ kg/cm}^3) \times (3.81 \text{ cm}^2/\text{s})} z\right]$$

$$= 4.4\alpha + C \, \exp(109.8z) \tag{4.53}$$

where z is in centimeters, and C is a constant to be determined from the boundary conditions. If we set the anode side as $z = 0$, we have $\lambda(0) = 7.2$ and $\lambda(0.0125) = 14$ from Equation 4.45. Accordingly, Equation 4.53 becomes

$$\lambda(z) = 4.4\alpha + 2.30 \, \exp(109.8z) \qquad \text{where } \alpha = 1.12 \tag{4.54}$$

Now we know that about 1.12 (or 0.56) water molecules are dragged per each hydrogen (or proton). Figure 4.15a shows the result of how λ varies across the membrane in this example. At the start of the problem, we assumed a constant D_λ for λ in the range of 7.2–14. We can confirm that this assumption is reasonable from the results of Figure 4.15.

From Equations 4.38 and 4.54, we can determine the conductivity profile of the membrane:

$$\sigma(z) = \{0.005193[4.4\alpha + 2.30 \, \exp(109.8z)] - 0.00326\}$$

$$\times \exp\left[1268\left(\frac{1}{303} - \frac{1}{353}\right)\right]$$

$$= 0.0404 + 0.0216 \, \exp(109.8z) \tag{4.55}$$

Figure 4.15b shows the result. Finally, we can determine the resistance of the membrane using Equation 4.40:

$$R_m = \int_0^{t_m} \frac{dz}{\sigma[\lambda(z)]} = \int_0^{0.0125} \frac{dz}{0.0404 + 0.0216 \, \exp(109.8z)}$$

$$= 0.150 \, \Omega \cdot \text{cm}^2 \tag{4.56}$$

Thus, the ohmic overvoltage due to the membrane resistance in this PEMFC is approximately:

$$V_{\text{ohm}} = j \times R_m = (0.7 \text{ A/cm}^2) \times (0.15 \, \Omega \cdot \text{cm}^2) = 0.105 \, V \tag{4.57}$$

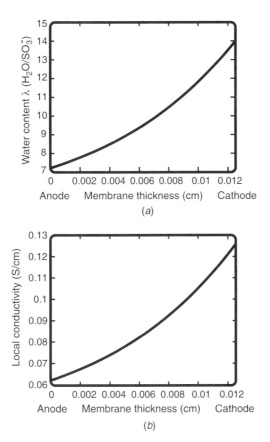

Figure 4.15. Calculated properties of Nafion membrane for Example 4.4. (*a*) Water content profile across Nafion membrane. (*b*) Local conductivity profile across Nafion membrane.

4.5.3 Ionic Conduction in Ceramic Electrolytes

This section explains the underlying physics of ion transport in SOFC electrolytes. As their name implies, SOFC electrolytes are solid, crystalline oxide materials that can conduct ions. The most popular SOFC electrolyte material is yttria stabilized zirconia (YSZ). A typical YSZ electrolyte contains 8% yttria mixed with zirconia. What is the meaning of zirconia and yttria? Zirconia is related to the metal zirconium, and yttria derives its name from another metal, yttrium. Zirconia has the chemical composition ZrO_2; it is the oxide of zirconium. By analogy, yttria, or Y_2O_3, is the oxide of yttrium. A mixture of zirconia and yttria is called yttria-*stabilized* zirconia because the yttria stabilizes the zirconia crystal structure in the cubic phase (where it is most conductive). Even more importantly, however, the yttria introduces high concentrations of oxygen vacancies into the zirconia crystal structure. This high oxygen vacancy concentration allows YSZ to exhibit high ion conductivity.

Adding yttria to zirconia introduces oxygen vacancies due to charge compensation effects. Pure ZrO_2 forms an ionic lattice consisting of Zr^{4+} ions and O^{2-} ions, as shown

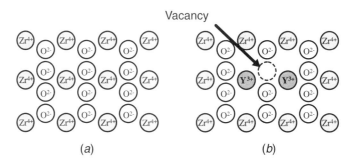

Figure 4.16. View of the (110) plane in (*a*) pure ZrO_2, and (*b*) YSZ. Charge compensation effects in YSZ lead to creation of oxygen vacancies. One oxygen vacancy is created for every two yttria atoms doped into the lattice.

in Figure 4.16*a*. Addition of Y^{3+} ions to this lattice upsets the charge balance. As shown in Figure 4.16*b*, for every two Y^{3+} ions taking the place of Zr^{4+} ions, one oxygen vacancy is created to maintain overall charge neutrality. The addition of 8% (molar) yttria to zirconia causes about 4% of the oxygen sites be to be vacant. At elevated temperatures, these oxygen vacancies facilitate the transport of oxygen ions in the lattice, as shown in Figure 4.8.

As discussed in Section 4.4, a material's conductivity is determined by the combination of *carrier concentration c* and *carrier mobility u*:

$$\sigma = (|z|F)cu \qquad (4.58)$$

In the case of YSZ, carrier concentration is determined by the strength of the yttria doping. Because a vacancy is required for ionic motion to occur within a YSZ lattice, the oxygen vacancies can be considered to be the ionic charge "carriers." Increasing the yttria content will result in increased oxygen vacancy concentration, improving the conductivity. Unfortunately, however, there is an upper limit to doping. Above a certain dopant or vacancy concentration, defects start to interact with each other, reducing their ability to move. Above this concentration, conductivity decreases. Plots of conductivity versus dopant concentration show a maximum at the point where defect interaction or "association" commences. For YSZ, this maximum occurs at about 8% molar yttria concentration. (See Figure 4.17.)

The complete expression for conductivity combines carrier concentration and carrier mobility, as described in Section 4.4.3:

$$\sigma = \frac{c(zF)^2 D}{RT} \qquad (4.59)$$

where carrier mobility is described by D, the *diffusivity* of the carrier in the crystal lattice. Diffusivity describes the ability of a carrier to move, or *diffuse*, from site to site within a crystal lattice. High diffusivities translate into high conductivities because the carriers are able to move quickly through the crystal. The atomic origins and physical explanation

INTRINSIC CARRIERS VERSUS EXTRINSIC CARRIERS

In YSZ and most other SOFC electrolytes, dopants are used to intentionally create high vacancy (or other charge carrier) concentrations. These carriers are known as *extrinsic carriers* because their presence was *extrinsically* created by intentional doping. However, any crystal, even an undoped one, will have at least some natural carrier population. These natural charge carriers are referred to as *intrinsic carriers* because they occur *intrinsically* due to the natural energetics of the crystal. Intrinsic carriers exist because no crystal is perfect (unless it is at absolute zero). All crystals will contain "mistakes" like vacancies that can act as charge carriers for conduction. These mistakes are actually energetically favorable, because they increase the entropy of the crystal. (Recall Section 2.1.4.) For the case of vacancies, an energy balance may be developed which considers the enthalpy cost to create the vacancies versus the entropy benefit they deliver. Solving this balance results in the following expression for intrinsic vacancy concentration as a function of temperature in an ionic crystal:

$$x_V \approx e^{-\Delta h_v/(2kT)} \tag{4.60}$$

where x_V represents the fractional vacancy concentration (expressed as the fraction of lattice sites of the species of interest that are vacant), Δh_v is the formation enthalpy for the vacancy in electron-volts (in other words, the enthalpy cost to "create" a vacancy), k is Boltzmann's constant, and T is the temperature in kelvin. This expression states that the intrinsic concentration of vacancies within a crystal increases exponentially with temperature. However, since Δh_v is typically on the order of 1 eV or larger, intrinsic vacancy concentrations are generally quite low, even at high temperatures. At 800°C, the intrinsic vacancy concentration in pure ZrO_2 is around 0.001, or about one vacancy per 1000 atoms. Compare this to extrinsically doped crystal structures which can attain vacancy concentrations as high as 0.1, or about one vacancy per 10 atoms.

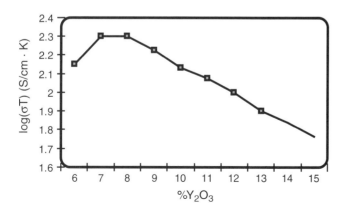

Figure 4.17. YSZ conductivity versus %Y_2O_3 (molar basis) [10]. YSZ conductivity is displayed as $\sigma(\Omega^{-1} \cdot cm^{-1})$ times T (K). In the next section, Figure 4.18 will clarify why it is convenient to multiply σ with T.

behind diffusivity will be detailed in forthcoming sections. For now, however, it is sufficient to know that carrier diffusivity in SOFC electrolytes is exponentially temperature dependent:

$$D = D_0 e^{-\Delta G_{act}/(RT)} \tag{4.61}$$

where D_0 is a constant (cm^2/s), ΔG_{act} is the activation barrier for the diffusion process (J/mol), R is the gas constant, and T is the temperature (K). Combining Equations 4.59 and 4.61 provides a complete expression for conductivity in SOFC electrolytes:

$$\sigma = \frac{c(zF)^2 D_0 e^{-\Delta G_{act}/(RT)}}{RT} \tag{4.62}$$

This equation can be further refined depending on whether the charge carriers are extrinsic or intrinsic:

- For extrinsic carriers, c is determined by the doping chemistry of the electrolyte. In this case, c is a constant and Equation 4.62 can be used as is.
- For intrinsic carriers, c is exponentially dependent on temperature, and Equation 4.62 must be modified as follows:

$$\sigma = \frac{c_{sites}(zF)^2 D_0 e^{-\Delta h_v/(2kT)} e^{-\Delta G_{act}/(RT)}}{RT} \tag{4.63}$$

where c_{sites} stands for the concentration of lattice sites for the species of interest in the material (moles of sites/cm^3).

Experimental observations confirm the relationship described by Equation 4.62. Figure 4.18 shows experimental plots of $\log(\sigma T)$ versus $1/T$ for both YSZ and gadolinea-

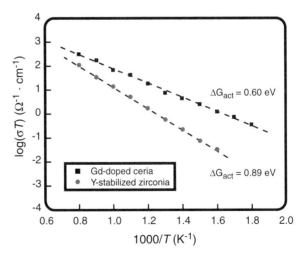

Figure 4.18. Conductivity of YSZ and GDC electrolytes versus temperature.

doped ceria (GDC, another candidate SOFC electrolyte). The multiplication of σ with T assures that the slopes in these plots are indicative of the activation energy for ion migration, ΔG_{act}. The size of ΔG_{act} is often critical for determining the conductivity of SOFC electrolytes. Typically, its value ranges between about 50,000–120,000 J/mol (0.5–1.2 eV).

4.6 MORE ON DIFFUSIVITY AND CONDUCTIVITY (OPTIONAL)

In this optional section, we develop an atomistic picture to explore conductivity and diffusivity in more detail. We find that for conductors where charge transport involves a "hopping"-type mechanism, conductivity and diffusivity are intimately related. Diffusivity measures the intrinsic rate of this hopping process. Conductivity incorporates how this hopping process is modified by the presence of an electric field driving force. Diffusivity is therefore actually the more fundamental parameter.

Diffusivity is a more fundamental parameter of atomic motion because even in the absence of any driving force, hopping of ions from site to site within the lattice still occurs at a rate that is characterized by the diffusivity. Of course, without a driving force, the net movement of ions is zero, but they are still exchanging lattice sites with one another. This is another example of a *dynamic equilibrium*; compare it to the exchange current density phenomenon that we learned about in Chapter 3.

4.6.1 Atomistic Origins of Diffusivity

Using the schematic in Figure 4.19*b*, we can derive an atomistic picture of diffusivity. The atoms in this figure are arranged in a series of parallel atomic planes. We would like to calculate the net flux (net movement) of grey atoms from left to right across the imaginary plane labeled A in Figure 4.19 (which lies between two real atomic planes in the material). Examining atomic plane 1 in the figure, we assume that the flux of grey atoms hopping in the forward direction (and therefore through plane A) is simply determined by the number (concentration) of grey atoms available to hop times the hopping rate:

$$J_{A+} = \tfrac{1}{2}\upsilon c_1 \, \Delta x \qquad (4.64)$$

where J_{A+} is the forward flux through plane A, υ is the hopping rate, c_1 is the volume concentration (mol/cm^3) of grey atoms in plane 1, Δx is the atomic spacing required to convert volume concentration to planar concentration (mol/cm^2), and the $\tfrac{1}{2}$ accounts for the fact that on average only half of the jumps will be "forward" jumps. (On average, half of the jumps will be to the left, half of the jumps will be to the right.)

Similarly, the flux of grey atoms hopping from plane 2 backward through plane A will be given by

$$J_{A-} = \tfrac{1}{2}\upsilon c_2 \, \Delta x \qquad (4.65)$$

where J_{A-} is the backward flux through plane A and c_2 is the volume concentration (mol/cm^3) of grey atoms in plane 2. The net flux of grey atoms across plane A is therefore

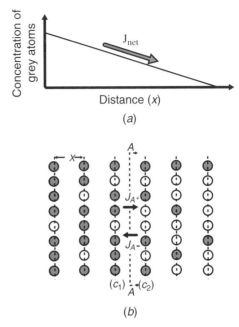

Figure 4.19. (*a*) Macroscopic picture of diffusion. (*b*) Atomistic view of diffusion. The net flux of grey atoms across an imaginary plane A in this crystalline lattice is given by the flux of grey atoms hopping from plane 1 to plane 2 minus the flux of grey atoms hopping from plane 2 to plane 1. Since there are more grey atoms on plane 1 than plane 2, there is a net flux of grey atoms from plane 1 to plane 2. This net flux will be proportional to the *concentration difference* of grey atoms between the two planes.

given by the difference between the forward and backward fluxes through plane A:

$$J_{\text{net}} = \tfrac{1}{2} \upsilon \, \Delta x (c_1 - c_2) \tag{4.66}$$

We would like to make this expression look like our familiar equation for diffusion: $J = -D(dc/dx)$. We can express Equation 4.66 in terms of a concentration gradient as

$$J_{\text{net}} = -\tfrac{1}{2}\upsilon(\Delta x)^2 \frac{(c_2 - c_1)}{\Delta x}$$

$$= -\tfrac{1}{2}\upsilon(\Delta x)^2 \frac{\Delta c}{\Delta x}$$

$$= -\tfrac{1}{2}\upsilon(\Delta x)^2 \frac{dc}{dx} \quad \text{(for small } x) \tag{4.67}$$

Comparison with the traditional diffusion equation $J = -D(dc/dx)$ allows us to identify what we call the diffusivity as

$$D = \tfrac{1}{2}\upsilon(\Delta x)^2 \tag{4.68}$$

We therefore recognize that the diffusivity embodies information about the intrinsic hopping rate for atoms in the material (υ) and information about the atomic length scale (jump distance) associated with the material.

As mentioned previously, the hopping rate embodied by υ is exponentially activated. Consider Figure 4.20b, which shows the free-energy curve encountered by an atom as it hops from one lattice site to a neighboring lattice site. Because the two lattice sites are essentially equivalent, in the absence of a driving force a hopping atom will possess the same free energy in its initial and final positions. However, an activation barrier impedes the motion of the atom as it hops between positions. We might associate this energy barrier with the displacements that the atom causes as it squeezes through the crystal lattice between lattice sites. (See Figure 4.20a, which shows a physical picture of the hopping process.)

In a treatment analogous to the reaction rate theory developed in the previous chapter, we can write the hopping rate as

$$\upsilon = \upsilon_0 e^{-\Delta G_{\mathrm{act}}/(RT)} \tag{4.69}$$

where ΔG_{act} is the activation barrier for the hopping process and υ_0 is the jump attempt frequency.

Based on this activated model for diffusion, we can then write a complete expression for the diffusivity as

$$D = \tfrac{1}{2}(\Delta x)^2 \upsilon_0 e^{-\Delta G_{\mathrm{act}}/(RT)} \tag{4.70}$$

or, lumping all the preexponential constants into a D_0 term,

$$D = D_0 e^{-\Delta G_{\mathrm{act}}/(RT)} \tag{4.71}$$

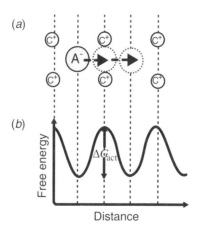

Figure 4.20. Atomistic view of hopping process. (a) Physical picture of the hopping process. As the anion (A^-) hops from its original lattice site to an adjacent, vacant lattice site, it must squeeze through a tight spot in the crystal lattice. (b) Free-energy picture of the hopping process. The tight spot in the crystal lattice represents an energy barrier for the hopping process.

4.6.2   Relationship between Conductivity and Diffusivity (1)

To understand how conductivity relates to diffusivity, we take a look at how an applied electric field will affect the hopping probabilities for diffusion. Consider Figure 4.21, which shows the effect of a linear voltage gradient on the activation barrier for the hopping process. From this picture, it is clear that the activation barrier for a "forward" hop is reduced by $\frac{1}{2}zF\,\Delta x(dV/dx)$ while the activation barrier for the "reverse" hop is increased by $\frac{1}{2}zF\,\Delta x(dV/dx)$. (We are assuming that the activated state occurs exactly halfway between the two lattice positions, or in other words that $\alpha = \frac{1}{2}$.) The forward- and reverse-hopping-rate expressions are therefore

$$
\begin{aligned}
v_+ &= v_0 \; \exp \frac{-[\Delta G_{\text{act}} - \frac{1}{2}zF\,\Delta x(dV/dx)]}{RT} \\[2mm]
v_- &= v_0 \; \exp \frac{-[\Delta G_{\text{act}} + \frac{1}{2}zF\,\Delta x(dV/dx)]}{RT}
\end{aligned}
\tag{4.72}
$$

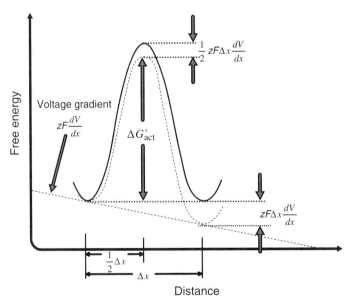

Figure 4.21. Effect of linear voltage gradient on activation barrier for hopping. The linear variation in voltage with distance causes a linear drop in free energy with distance. This reduces the forward activation barrier ($\Delta G'_{\text{act}} < \Delta G_{\text{act}}$). Two adjacent lattice sites are separated by Δx; therefore the total free-energy drop between them is given by $zF\,\Delta x(dV/dx)$. If the activation barrier occurs halfway between the two lattice sites, ΔG_{act} will be decreased by $\frac{1}{2}zF\,\Delta x(dV/dx)$. [In other words, $\Delta G'_{\text{act}} = \Delta G_{\text{act}} - \frac{1}{2}zF\,\Delta x(dV/dx)$.]

This voltage gradient modification to the activation barrier turns out to be small. In fact,

$$\frac{1}{2}\frac{zF}{RT}\Delta x\frac{dV}{dx} \ll 1,$$

so we can use the approximation $e^x \approx 1 + x$ for the second term in the exponentials. This allows us to rewrite the hopping rate expressions as

$$
\begin{aligned}
\upsilon_+ &\approx \upsilon_0 e^{-\Delta G_{\mathrm{act}}/(RT)}\left(1 + \frac{1}{2}\frac{zF}{RT}\Delta x\frac{dV}{dx}\right) \\
\upsilon_- &\approx \upsilon_0 e^{-\Delta G_{\mathrm{act}}/(RT)}\left(1 - \frac{1}{2}\frac{zF}{RT}\Delta x\frac{dV}{dx}\right)
\end{aligned}
\tag{4.73}
$$

Proceeding as before, we can then write the net flux across an imaginary plane A in a material as

$$J_{\mathrm{net}} = J_{A^+} - J_{A^-} = \tfrac{1}{2}\Delta x(c_1\upsilon_+ - c_2\upsilon_-) \tag{4.74}$$

Since we are interested in conductivity this time, we would like to consider a flux that is driven purely by the potential gradient. In other words, we want to get rid of any effects of a concentration gradient by saying that $c_1 = c_2 = c$. Making this modification and inserting the formulae for υ_+ and υ_- give

$$
\begin{aligned}
J_{\mathrm{net}} &= \tfrac{1}{2}\Delta x\upsilon_0 e^{-\Delta G_{\mathrm{act}}/(RT)}\left(\frac{czF}{RT}\Delta x\frac{dV}{dx}\right) \\
&= \tfrac{1}{2}\Delta x\upsilon_0 e^{-\Delta G_{\mathrm{act}}/(RT)}\left(\frac{czF}{RT}\frac{dV}{dx}\right)
\end{aligned}
\tag{4.75}
$$

Recognizing the first group of terms as our diffusion coefficient D, we thus have

$$J_{\mathrm{net}} = \frac{czFD}{RT}\frac{dV}{dx} \tag{4.76}$$

Comparing this to the conduction equation

$$J = \frac{\sigma}{zF}\frac{dV}{dx},$$

we see that σ and D are related by:

$$\sigma = \frac{c(zF)^2 D}{RT} \tag{4.77}$$

For conductors that rely on a diffusive hopping-based charge transport mechanism, this important result relates the observed *conductivity* of the material to the atomistic *diffusivity* of the charge carriers. This equation is our key for understanding the atomistic underpinnings of ionic conductivity in crystalline materials.

4.6.3 Relationship between Diffusivity and Conductivity (2)

Recall from Section 2.4.4 that the introduction of the electrochemical potential gave us an alternate way to understand the Nernst equation. In a similar fashion, looking at charge transport from the perspective of the electrochemical potential gives us an alternate way to understand the relationship between conductivity and diffusivity. Recall the definition of the electrochemical potential (Equation 2.95):

$$\tilde{\mu}_i = \mu_i^0 + RT \ln a_i + z_i F \phi_i$$

If we assume that activity is purely related to concentration ($a_i = c_i/c^0$), then the electrochemical potential can be written as

$$\tilde{\mu}_i = \mu_i^0 + RT \ \ln \frac{c_i}{c^0} + z_i F \phi_i \tag{4.78}$$

The charge transport flux due to a gradient in the electrochemical potential will include both the flux contributions due to the concentration gradient and the flux contributions due to the potential gradient:

$$J_i = -M_{i\mu} \frac{\partial \tilde{\mu}}{\partial x} = M_{i\mu} \left(RT \frac{d[\ln (c_i/c^0)]}{dx} + z_i F \frac{dV}{dx} \right) \tag{4.79}$$

The concentration term in the natural logarithm can be processed by remembering the chain rule of differentiation:

$$\frac{d[\ln (c_i/c^0)]}{dx} = \frac{c^0}{c_i} \frac{d(c_i/c^0)}{dx} = \frac{1}{c_i} \frac{dc_i}{dx} \tag{4.80}$$

Therefore, the total charge transport flux due to an electrochemical potential gradient is really made up of two fluxes, one driven by a concentration gradient and one driven by a voltage gradient:

$$J_i = -\frac{M_{i\mu} RT}{c_i} \frac{dc_i}{dx} - M_{i\mu} z_i F \frac{dV}{dx} \tag{4.81}$$

Comparing the concentration gradient term in this equation to our previous expression for diffusion allows us to identify $M_{i\mu}$ in terms of diffusivity:

$$\frac{M_{i\mu} RT}{c_i} = D \qquad M_{i\mu} = \frac{Dc_i}{RT} \tag{4.82}$$

Comparing the voltage gradient term in this expression to our previous expression for conduction allows us to identify σ in terms of diffusivity:

$$M_{i\mu} z F = \frac{\sigma}{|z|F} \qquad \sigma = \frac{c_i (zF)^2 D}{RT} \tag{4.83}$$

By using the electrochemical potential, we arrive at the same result as before. Interestingly, we did not have to make any assumptions about the mechanism of the transport process this time. Thus, we see that the relationship between diffusivity and conductivity is completely general. (In other words, it does not just apply to hopping mechanisms.) The conductivity and diffusivity of a material are related because the fundamental driving forces for diffusion and conduction are related via the electrochemical potential.

4.7 WHY ELECTRICAL DRIVING FORCES DOMINATE CHARGE TRANSPORT (OPTIONAL)

Our relationship between conductivity and diffusivity allows us to explain why electrical driving forces dominate charge transport.

In metallic electron conductors, the extremely high background concentration of free electrons means that electron concentration is basically invariant across the conductor. This means that there are no gradients in electron chemical potential across the conductor. Additionally, since metal conductors are solid materials, pressure gradients do not exist. Therefore, we find that electron conduction in metals is driven only by voltage gradients.

What about for ion conductors? Like the metallic conductors, most fuel cell ion conductors are also solid state, therefore pressure gradients do not exist. (Even in fuel cells that employ liquid electrolytes, the electrolyte is usually so thin that convection does not contribute significantly). Similarly, the background concentration of ionic charge carriers is also usually large, so that significant concentration gradients do not arise. However, even if large concentration gradients were to arise, we find that the "effective strength" of a voltage gradient driving force is far greater than the effective strength of a concentration gradient driving force. To illustrate this point, lets compare the charge flux generated by a concentration gradient to the charge flux generated by a voltage gradient. The charge flux generated by a concentration gradient (j_c) is given by

$$j_c = zFD \frac{dc}{dx} \tag{4.84}$$

The charge flux generated by a voltage gradient (j_v) is given by

$$j_v = \sigma \frac{dV}{dx} \tag{4.85}$$

Note that the quantity zF is required to convert moles in the diffusion equation into charge in coulombs. As we have learned, σ and D are related by

$$\sigma = \frac{c(zF)^2 D}{RT} \tag{4.86}$$

The maximum possible sustainable charge flux due to a concentration gradient across a material would be

$$j_c = zFD\frac{c_0}{L} \qquad (4.87)$$

where L is the thickness of the material and c_0 is the bulk concentration of charge carriers. The voltage, V, that would be required to produce an equivalent charge flux can be calculated from

$$j_v = j_c$$
$$\frac{c_0(zF)^2D}{RT}\frac{V}{L} = zFD\frac{c_0}{L} \qquad (4.88)$$

Solving for V gives

$$V = \frac{RT}{zF} \qquad (4.89)$$

At room temperature, for $z = 1$, $RT/zF = .0257$ V. Therefore a voltage drop of 25.7 mV across the thickness of the material accomplishes the same thing as the *maximum possible* chemical driving force available from concentration effects. Effectively, the quantity RT/zF sets the strength of the electric driving force relative to the chemical (concentration) driving force. Because RT/zF is small (for the fuel cell temperature range of interest), fuel cell charge transport is dominated by electrical driving forces rather than chemical potential driving forces.

CHAPTER SUMMARY

- Charge transport in fuel cells is predominantly driven by a voltage gradient. This charge transport process is known as conduction.
- The voltage that is expended to drive conductive charge transport represents a loss to fuel cell performance. Known as the ohmic overvoltage, this loss generally obeys Ohm's law of conduction, $V = iR$, where R is the ohmic resistance of the fuel cell.
- Fuel cell ohmic resistance includes the resistance from the electrodes, electrolyte, interconnects, and so on. However, it is usually dominated by the electrolyte resistance.
- Resistance scales with conductor area A, thickness L, and conductivity σ: $R = L/\sigma A$.
- Because resistance scales with area, area-specific fuel cell resistances (ASRs) are computed to make comparisons between different-size fuel cells possible (ASR $= A \times R$).
- Because resistance scales with thickness, fuel cell electrolytes are made as thin as possible.

- Because resistance scales with conductivity, developing high-conductivity electrode and electrolyte materials is critical.

- Conductivity is determined by carrier concentration and carrier mobility: $\sigma_i = (z_i F) c_i u_i$.

- Metals and ion conductors show vastly different structures and conduction mechanisms, leading to vastly different conductivities.

- Ion conductivity even in good electrolytes is generally four to eight orders of magnitude lower than electron conductivity in metals.

- In addition to having high ionic conductivity, electrolytes must be stable in both highly reducing and highly oxidizing environments. This can be a significant challenge.

- The three major electrolyte classes employed in fuel cells are (1) liquid, (2) polymer, and (3) ceramic electrolytes.

- Mobility (and hence conductivity) in aqueous electrolytes is determined by the balance between ion acceleration under an electric field and frictional drag due to fluid viscosity. In general, the smaller the ion and the greater its charge, the higher the mobility.

- Conductivity in Nafion (a polymer electrolyte) is dominated by water content. High water content leads to high conductivity. Nafion conductivity may be determined by modeling the water content in the membrane.

- Conductivity in ceramic electrolytes is controlled by defects ("mistakes") in the crystal lattice. Natural (intrinsic) defect concentrations are generally low, so higher (extrinsic) defect concentrations are usually introduced into the lattice on purpose via doping.

- (Optional section) At the atomistic level, we find that conductivity is determined by a more basic parameter known as diffusivity D. Diffusivity expresses the intrinsic rate of movement of atoms within a material.

- (Optional section) By examining an atomistic picture of diffusion and conduction, we can explicitly relate diffusivity and conductivity: $\sigma = c(zF)^2 D/(RT)$.

- (Optional section) Using the relationship between conductivity and diffusivity, we can understand why voltage driving forces (conduction) dominate charge transport.

CHAPTER EXERCISES

Review Questions

4.1 Why does charge transport result in a voltage loss in fuel cells?

4.2 If a fuel cell's area is increased 10-fold and its resistance is decreased 9-fold, will the ohmic losses in the fuel cell increase or decrease (for a given current density, all else being equal)?

4.3 What are the two main factors that determine a material's conductivity?

4.4   Why are the electron conductivities of metals so much larger than the ion conductivities of electrolytes?

4.5   List at least four important requirements for a candidate fuel cell electrolyte. Which requirement (other than high conductivity) is often the hardest to fulfill?

Calculations

4.6   Redraw Figure 4.4c for a SOFC, where O^{2-} is the mobile charge carrier in the electrolyte.

4.7   Draw a fuel cell voltage profile similar to those shown in Figure 4.4 which simultaneously shows the effects of both activation losses and ohmic losses.

4.8   Given that fuel cell voltages are typically around 1 V or less, what would be the absolute minimum possible functional electrolyte thickness for a SOFC if the dielectric breakdown strength of the electrolyte is 10^8 V/m?

4.9   In Section 4.3.2, we discussed how fuel cell electrolyte resistance scales with thickness (in general as L/σ). Several practical factors were listed that limit the useful range of electrolyte thickness. Fuel crossover was stated to cause an undesirable parasitic loss which can eventually become so large that further thickness decreases *are counterproductive*! In other words, at a given current density, an optimal electrolyte thickness may exist, and reducing the electrolyte thickness below this optimal value will actually *increase* the total fuel cell losses. We would like to model this phenomenon. Assume that the leak current j_{leak} across an electrolyte gives rise to an additional fuel cell loss of the following form: $\eta_{leak} = A \ln j_{leak}$. Furthermore, assume that j_{leak} varies inversely with electrolyte thickness L as $j_{leak} = B/L$. For a given current density j determine the optimal electrolyte thickness that minimizes $\eta_{ohmic} + \eta_{leak}$.

4.10   A 5 cm^2 fuel cell has $R_{elec} = 0.01 \ \Omega$ and $\sigma_{electrolyte} = 0.10 \ \Omega^{-1} \cdot cm^{-1}$. If the electrolyte is 100 μm thick, predict the ohmic voltage losses for this fuel cell at $j = 500$ mA/cm^2.

4.11   Derive Equation 4.32 using Equations 4.22 and 4.24.

4.12   Consider a PEMFC operating at 0.8 A/cm^2 and 70°C. Hydrogen gas at 90°C and 80% relative humidity is provided to the fuel cell at the rate of 8 A. The fuel cell area is 8 cm^2 and the drag ratio of water molecules to hydrogen, α, is 0.8. Find the water activity of the hydrogen exhaust. Assume $p = 1$ atm and that the hydrogen exhaust exits at the fuel cell temperature, 70°C.

4.13   Consider two H_2–O_2 PEMFCs powering an external load at 1 A/cm^2. The fuel cells are running with differently humidified gases: (a) $a_{W,anode} = 1.0$, $a_{W,cathode} = 0.5$; (b) $a_{W,anode} = 0.5$, $a_{W,cathode} = 1.0$. Estimate the ohmic overpotential for both fuel cells if they are both running at 80°C. Assume that they both employ a 125-μm-thick Nafion electrolyte. Based on your results, discuss the relative effects of humidity at the anode versus the cathode.

4.14 **(a)** Calculate the diffusion coefficient for oxygen ions in a pure ZrO_2 electrolyte at $T = 1000°C$ given $\Delta G_{act} = 100$ kJ/mol, $\nu_0 = 10^{13}$ Hz. ZrO_2 has a cubic unit cell with a lattice constant $a = 5$ Å and contains four Zr atoms and eight O atoms. Assume that the oxygen–oxygen "jump" distance $\Delta x = \frac{1}{2}a$.

(b) Calculate the intrinsic carrier concentration in the electrolyte given $\Delta h_v = 1$ eV. (Assume vacancies are the dominant carrier.)

(c) From your answers in (a) and (b), calculate the intrinsic conductivity of this electrolyte at 1000°C.

4.15 You have determined the resistance of a 100-μm-thick, 1.0-cm^2-area YSZ electrolyte sample to be 47.7 Ω at $T = 700$ K and 0.680 Ω at $T = 1000$ K. Calculate D_0 and ΔG_{act} for this electrolyte material given that the material is doped with 8% molar Y_2O_3. Recall from problem 4.14 that pure ZrO_2 has a cubic unit cell with a lattice constant of 5 Å and contains four Zr atoms and eight O atoms. Assume that the lattice constant does not change with doping.

CHAPTER 5

FUEL CELL MASS TRANSPORT

As discussed in the introductory chapter, to produce electricity, a fuel cell must be continually supplied with fuel and oxidant. At the same time, products must be continuously removed so as to avoid "strangling" the cell. The process of supplying reactants and removing products is termed *fuel cell mass transport*. As you will learn, this seemingly simple task can turn out to be quite complicated.

In the previous chapters, we have learned about the electrochemical reaction process (Chapter 3) and the charge transport process (Chapter 4). Mass transport represents the last major fuel cell process to be discussed. After completing this chapter, you will have all the basic tools you need to understand fuel cell operation.

In this chapter, we will concentrate on the movement of reactants and products within a fuel cell. The previous chapter (on charge transport) has already introduced you to many of the fundamental equations that govern the transport of matter from one location to another. Indeed, ionic charge transport is actually just a special subset of mass transport where the mass being transported consists of charged ions. We now deal with the transport of *uncharged species*, thus distinguishing this chapter from the last chapter. Uncharged species are unaffected by voltage gradients, so we must rely instead on convective and diffusive forces for movement. Furthermore, we are concerned mostly with *gas-phase transport* (and occasionally liquid-phase transport). Contrast this to the mostly solid-phase ionic transport discussed in the previous chapter.

Why are we so interested in fuel cell mass transport? The answer is because poor mass transport leads to significant fuel cell performance losses. To understand why poor mass transport can lead to a performance loss, remember that it is the reactant and product concentrations *within the catalyst layer*, not at the fuel cell inlet, that determine fuel cell performance. Thus, reactant depletion (or product accumulation) within the catalyst layer will adversely affect performance. This loss in performance is called a fuel cell "concentration" loss or mass transport loss. Concentration loss is minimized by careful optimization of mass transport in the fuel cell electrodes and fuel cell flow structures.

5.1   TRANSPORT IN ELECTRODE VERSUS FLOW STRUCTURE

This chapter is divided into two major parts: one part on mass transport in fuel cell electrodes and a second part on mass transport in fuel cell flow structures. Why do we make this distinction and what is the difference between them?

The difference between the two domains is one of length scale. More importantly, however, this difference in length scale leads to a difference in transport mechanism. For fuel cell flow structures, dimensions are generally on the millimeter or centimeter scale. Flow patterns typically consist of geometrically well-defined channel arrays that are amenable to the laws of fluid mechanics. Gas transport in these channels is dominated by fluid flow and convection. In contrast, fuel cell electrodes exhibit structure and porosity on the micrometer and nanometer length scale. The tortuous, sheltering geometry of these electrodes insulates gas molecules from the convective forces present in the flow channel. Sheltered from convective flow, gas transport within the electrodes is dominated by diffusion.

CONVECTION VERSUS DIFFUSION

It is important to understand the difference between convection and diffusion:

- *Convection* refers to the transport of a species by bulk motion of a fluid (under the action of a mechanical force).
- *Diffusion* refers to the transport of a species due to a gradient in concentration.

Figure 5.1 illustrates the difference between the two transport modes. Interestingly (and importantly for fuel cells) convection turns out to be far more "effective" at transporting species compared to diffusion. For example, at STP, the maximum likely diffusive O_2 flux across a 500-μm-thick porous electrode is $\approx 4 \times 10^{-5}$ mol/(cm$^2 \cdot$ s). This flux could instead be provided by 0.01 m/s (or less) convective flow of O_2.

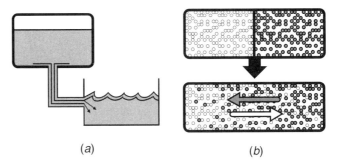

(a) (b)

Figure 5.1. Convection versus diffusion. (a) Convective fluid transport in this system moves material from the upper tank to the lower tank. (b) A concentration gradient between white and grey particles results in net diffusive transport of grey particles to the left and white particles to the right.

Where do the convective forces that dominate transport in the flow channels come from? They are *imposed* by the user (us) who forces fuel or oxidant through the fuel cell at a given rate. The pressure (driving force) required to push a given rate of fuel or oxidant through fuel cell flow channels may be calculated using fluid dynamics. High flow rates can ensure good distribution of reactants (and effective removal of products) across a fuel cell but may require unacceptably high driving pressures or lead to other problems.

Where do the concentration gradients that dominate diffusive transport in the electrode come from? They *develop* due to species consumption (or production) within the catalyst layer. As Figure 5.2 illustrates, a fuel cell anode operating at high current density is consuming H_2 molecules at a voracious rate. This leads to a depletion of H_2 in the vicinity of the catalyst layer, extending out into the electrode. The resulting concentration gradi-

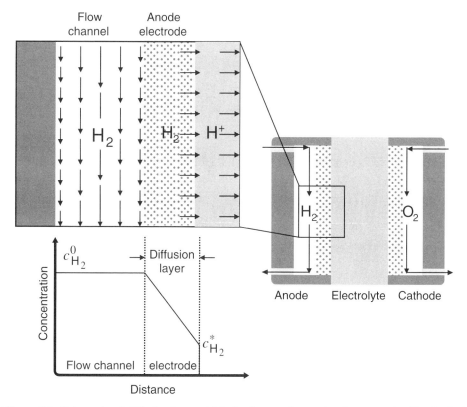

Figure 5.2. Schematic of diffusion layer which develops at the anode of an operating H_2–O_2 fuel cell. Consumption of H_2 gas at the anode–electrolyte interface results in a depletion of H_2 within the electrode. The concentration of H_2 gas falls from its bulk value ($c_{H_2}^0$) at the flow channel to a much lower value ($c_{H_2}^*$) at the catalyst layer. The magnitude of the H_2 gas velocity in the flow channel is schematically illustrated by the size of the flow arrows. Near the channel–electrode interface, the H_2 gas velocity drops toward zero, marking the start of the diffusion layer.

ent provides the driving force for the diffusive transport of H_2 from the electrode to the reaction zones.

The "dividing line," or boundary between convective dominated flow and diffusive dominated flow, often occurs where the fuel cell gas channel and porous electrode meet. Within the flow channel, convection serves to keep the gas stream well mixed, so that concentration gradients do not occur. However, due to frictional effects, the velocity of the moving gas stream tends toward zero at the electrode–channel boundary (as shown in Figure 5.2). In the absence of convective mixing, concentration gradients are then able to form within the stagnant gas of the electrode. We call this stagnant gas region the *diffusion layer*, since it is the region where diffusion dominates mass transport. Because the demarcation line where convective transport ends and diffusive transport begins is necessarily fuzzy, the exact thickness of the diffusion layer is often hard to define. Furthermore, it can change depending on the flow conditions, flow channel geometry, or electrode structure. For example, at very low gas velocities, the diffusion layer may stretch out into the middle of the flow channels. In contrast, at extremely high gas velocities convective mixing may penetrate into the electrode itself, causing the diffusion layer to retreat.

In the following two major sections of this chapter, we will first treat mass transport within the electrode using diffusion. Then, we will treat mass transport within the flow structure using fluid dynamics techniques.

5.2 TRANSPORT IN ELECTRODE: DIFFUSIVE TRANSPORT

In this section, we examine mass transport within the fuel cell electrodes. Technically, we are really treating mass transport within the diffusion layer, but for the purposes of this discussion, we assume that the electrode thickness coincides with the diffusion layer thickness. For most flow situations, this is a reasonable assumption. As mentioned previously, high flow velocities or unusual flow patterns can decrease the diffusion layer; calculating the true diffusion layer thickness in these situations requires sophisticated models. Likewise, low flow velocities can increase the diffusion layer but again require treatment by sophisticated models.

5.2.1 Electrochemical Reaction Drives Diffusion

For most flow scenarios, the mass transport situation within the fuel cell electrode is similar to that shown in Figure 5.3. As illustrated in this figure, an electrochemical reaction on one side of an electrode and convective mixing on the other side of the electrode set up concentration gradients, leading to diffusive transport across the electrode. From this figure, you can see that the electrochemical reaction leads to reactant depletion (and product accumulation) at the catalyst layer. In other words, $c_R^* < c_R^0$ and $c_P^* > c_P^0$, where c_R^*, c_P^* represent the *catalyst layer* reactant and product concentrations respectively and c_R^0, c_P^0 represent the *bulk* (flow channel) reactant and product concentrations, respectively. This reactant depletion (and product accumulation) affects fuel cell performance in two ways, which will now briefly be described:

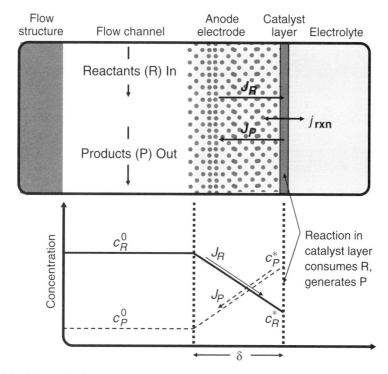

Figure 5.3. Schematic of mass transport situation within typical fuel cell electrode. Convective mixing of reactants and products in the flow channel establishes constant bulk species concentrations outside the diffusion layer (c_R^0 and c_P^0). The consumption/generation of species (at a rate given by j_{rxn}) within the catalyst layer leads to reactant depletion and product accumulation ($c_R^* < c_R^0$ and $c_P^* > c_P^0$). Across the diffusion layer, a reactant concentration gradient is established between c_R^0 and c_R^*, while a product concentration gradient is established between c_P^0 and c_P^*.

1. *Nernstian Losses.* The reversible fuel cell voltage will decrease as predicted by the Nernst equation since the reactant concentration at the catalyst layer is decreased relative to the bulk concentration and the product concentration at the catalyst layer is increased relative to the bulk concentration.

2. *Reaction losses.* The reaction rate (activation) losses will be increased because the reactant concentration at the catalyst layer is decreased relative to the bulk concentration and the product concentration at the catalyst layer is increased relative to the bulk concentration.

The combination of these two loss effects is what we collectively refer to as the fuel cell's concentration (or mass transport) loss. To determine the size of the concentration loss, it is essential to determine exactly *how much* the catalyst layer reactant and product concentrations differ from their bulk values. How do we make this determination? Let's see if we can come up with an answer by taking a closer look at the diffusion process occurring inside a fuel cell electrode.

Consider the fuel cell electrode depicted in Figure 5.4. Imagine that at some time $t = 0$ this fuel cell is "turned on" and it begins producing electricity at a fixed current density j. Initially, the reactant and product concentrations everywhere in this fuel cell are constant (they are given by c_R^0 and c_P^0). As soon as the fuel cell begins producing current, however, the electrochemical reaction leads to depletion of reactants (and accumulation of products) at the catalyst layer. Reactants begin to diffuse toward the catalyst layer from the surrounding area while products begin to diffuse away from the catalyst layer. Over time, the reactant and product concentration profiles will evolve as shown in the figure. Eventually, a steady-state situation will be reached as indicated by the dark lines. At steady state, the reactant and product concentration profiles drop linearly (at least in approximation) with distance across the electrode (diffusion layer). Furthermore, the flux of reactants and products down these concentration gradients will exactly match the consuption/depletion rate of reactants and products at the catalyst layer. (This should make intuitive sense: At steady state the rate of consumption must equal the rate of supply.) Mathematically,

$$j = nF J_{\text{diff}} \tag{5.1}$$

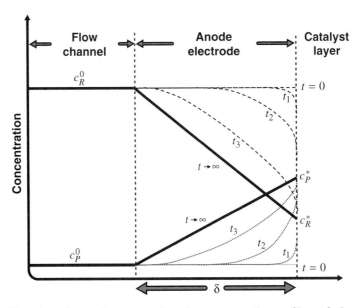

Figure 5.4. Time dependence of reactant and product concentration profiles at fuel cell electrode. The fuel cell begins producing current at time $t = 0$. Starting from constant initial values (c_R^0 and c_P^0), the reactant and product concentration profiles evolve with increasing time as shown for $t_1 < t_2 < t_3$. Eventually the profiles approach a steady-state balance (indicated by the dark solid lines) where concentration varies (approximately) linearly with distance across the diffusion layer. At steady state, the diffusion flux down these linear concentration gradients exactly balances the reaction flux at the catalyst layer.

CALCULATING NOMINAL DIFFUSIVITY

The gas diffusion of a species i depends not only on the properties of i but also on the properties of the species j through which i is diffusing. For this reason, binary gas diffusion coefficients are typically written as D_{ij}, where i is the diffusing species and j is the species through which the diffusion is occurring. For a binary system of two gases, D_{ij} is a strong function of temperature, pressure, and the molecular weights of species i and j. At low pressures, nominal diffusivity can be estimated from the following equation based on the kinetic theory of gases [11]:

$$p \cdot D_{ij} = a \left(\frac{T}{\sqrt{T_{ci} T_{cj}}} \right)^{b} \left(p_{ci} p_{cj} \right)^{1/3} \left(T_{ci} T_{cj} \right)^{5/12} \left(\frac{1}{M_i} + \frac{1}{M_j} \right)^{1/2} \quad (5.2)$$

where p is the total pressure (atm), D_{ij} is the binary diffusion coefficient (cm^2/s), and T is the temperature (K); M_i, M_j are the molecular weights (g/mol) of species i and j and T_{ci}, T_{cj}, p_{ci}, p_{cj} are the critical temperatures and pressures of species i and j. Table 5.1 summarizes T_c and p_c values for some useful gases. The final parameters in Equation 5.2 are a and b. Typically, one can use $a = 2.745 \times 10^{-4}$ and $b = 1.823$ for pairs of nonpolar gases, such as H_2, O_2, and N_2. For pairs involving H_2O (polar) and a nonpolar gas, one can use $a = 3.640 \times 10^{-4}$ and $b = 2.334$. Other equations to estimate diffusivity can be found in the literature.

where j is the fuel cell's operating current density (remember, the current density is a measure of the electrochemical reaction rate) and J_{diff} is the diffusion flux of reactants to the catalyst layer (or the diffusion flux of products away from the catalyst layer). The now-familiar quantity nF is of course required to convert the molar diffusion flux into the units of current density.

TABLE 5.1. Critical Properties of Gases

Substance	Molecular Weight (M)	T_c (K)	p_c (atm)
H_2	2.016	33.3	12.80
Air	28.964	132.4	37.0
N_2	28.013	126.2	33.5
O_2	31.999	154.4	49.7
CO	28.010	132.9	34.5
CO_2	44.010	304.2	72.8
H_2O	18.015	647.3	217.5

Source: From Ref. [11].

CALCULATING EFFECTIVE DIFFUSIVITY

In porous structures, the gas molecules tend to be impeded by the pore walls. The diffusion flux should be corrected to account for the effects of such blockage. Usually this is accomplished by employing a modified or *effective diffusivity*. According to the Bruggemann correction, the effective diffusivity in a porous structure can be expressed as [12]

$$D_{ij}^{\text{eff}} = \varepsilon^{1.5} D_{ij} \tag{5.3}$$

where ε stands for the *porosity* of the porous structure. Porosity represents the ratio of pore volume to total volume. Usually, fuel cell electrodes have porosities of around 0.4, which means 40% of the total electrode volume is occupied by pores. In open space, porosity is 1 and $D_{ij}^{\text{eff}} = D_{ij}$. Often, Equation 5.3 is modified to include tortuosity τ as

$$D_{ij}^{\text{eff}} = \varepsilon^{\tau} D_{ij} \tag{5.4}$$

Tortuosity describes the additional impedance to diffusion caused by a tortuous or convoluted flow path. Highly "mazelike" or meandering pore structures yield high tortuosity values. It is known that tortuosity can vary from 1.5 to 10 depending on pore structure configuration. At high temperatures, however, a different correlation for effective diffusivity proves more accurate [13]:

$$D_{ij}^{\text{eff}} = D_{ij} \frac{\varepsilon}{\tau} \tag{5.5}$$

The diffusion flux J_{diff} can be calculated using the diffusion equation. Recall from the previous chapter (Table 4.1) that diffusive transport may described by

$$J_{\text{diff}} = -D \frac{dc}{dx} \tag{5.6}$$

For the steady-state situation shown in Figure 5.4, this equation becomes (written for a reactant flux)

$$J_{\text{diff}} = -D^{\text{eff}} \frac{c_R^* - c_R^0}{\delta} \tag{5.7}$$

where c_R^* is the catalyst layer reactant concentration, c_R^0 is the bulk (flow channel) reactant concentration, δ is the electrode (diffusion layer) thickness, and D^{eff} is the effective reactant diffusivity within the catalyst layer. (The "effective" diffusivity will be lower than the "nominal" diffusivity due to the complex structure and tortuosity of the electrode. For more on calculating nominal and effective diffusivity, refer to the text box above.) By combining

Equations 5.1 and 5.7, we can then solve for the reactant concentration in the catalyst layer:

$$j = -nFD^{\text{eff}}\frac{c_R^* - c_R^0}{\delta} \tag{5.8}$$

$$c_R^* = c_R^0 - \frac{j\delta}{nFD^{\text{eff}}} \tag{5.9}$$

What this equation says is that the reactant concentration in the catalyst layer (c_R^*) is less than the bulk concentration c_R^0 by an amount that depends on j, δ, and D^{eff}. As j increases, the reactant depletion effect intensifies. Thus, the higher the current density, the worse the concentration losses. However, these concentration losses can be mitigated if the diffusion layer thickness δ is reduced or the effective diffusivity D^{eff} is increased.

5.2.2 Limiting Current Density

It is interesting to consider the situation when the reactant concentration in the catalyst layer drops all the way to zero. This represents the limiting case for mass transport. The fuel cell can never sustain a higher current density than that which causes the reactant concentration to fall to zero. We call this current density the *limiting current density* of the fuel cell. The limiting current density (j_L) can be calculated from Equation 5.8 by setting $c_R^* = 0$:

$$j_L = nFD^{\text{eff}}\frac{c_R^0}{\delta} \tag{5.10}$$

Fuel cell mass transport design strategies focus on increasing the limiting current density. These design strategies include the following:

1. Ensuring a high c_R^0 (by designing good flow structures that evenly distribute reactants)
2. Ensuring that D^{eff} is large and δ is small (by carefully optimizing fuel cell operating conditions, electrode structure, and diffusion layer thickness)

Typical values are about 100–300 μ m for δ and 10^{-2} cm^2/s for D^{eff}. Therefore typical limiting current densities are on the order of 1–10 A/cm^2. This mass transport effect represents the ultimate limit for fuel cells; a fuel cell will never be able to produce a higher current density than that determined by its limiting current density. (Note, however, that other fuel cell losses, e.g., ohmic and activation losses, may reduce the fuel cell voltage to zero well before the limiting current density is ever reached.)

While the limiting current density defines the ultimate fuel cell mass transport limit, concentration losses still occur at lower current densities as well. Recall from Section 5.2.1 that concentration differences in the catalyst layer affect fuel cell performance in two ways: first, by decreasing the Nernst (thermodynamic) voltage and, second, by increasing the activation (reaction rate) loss. We will now examine both of these effects in detail. Surpris-

LIMITING CURRENT DENSITIES AT ANODES AND CATHODES

In general, a limiting current density can be calculated for each reactant species in a fuel cell. For example, in a H_2–O_2 fuel cell, a j_L value can be calculated for both the anode (based on H_2) and the cathode (based on O_2). In both cases, care must be taken to correctly match the reactant species considered with the correct value for n in Equation 5.10. For the case of H_2, 1 mol H_2 will provide $2e^-$, and hence $n = 2$. However, for the case of O_2, 1 mol O_2 will consume $4e^-$, and hence $n = 4$. For most fuel cells, only j_L for oxygen is considered when determining mass transfer losses. Mass transfer limitations due to oxygen transport are typically much more severe than for hydrogen. This is because air (rather than pure oxygen) is typically used and O_2 diffuses more slowly than H_2.

ingly, we will find that both lead to the same result. This result, when generalized, is what we will refer to as the fuel cell's "concentration" overvoltage η_{conc}.

For the sake of clarity and simplicity, we will consider only reactant depletion effects when developing our concentration overvoltage expressions in the following sections. These expressions can be developed in an analogous manner if the product accumulation effects are considered instead.

5.2.3 Concentration Affects Nernst Voltage

The first way that concentration affects fuel cell performance is through the Nernst equation. This is because the real reversible thermodynamic voltage of a fuel cell is determined by the reactant and product concentrations at the reaction sites, not at the fuel inlet. From Chapter 2, recall the form of the Nernst equation (Equation 2.84):

$$E = E^0 - \frac{RT}{nF} \ln \frac{\prod a_{products}^{v_i}}{\prod a_{reactants}^{v_i}} \tag{5.11}$$

For simplicity, we will consider a fuel cell with a single reactant species. As mentioned previously, we will neglect the product accumulation effects in this treatment. We retain our notation from the previous sections: $c_R^* = $ catalyst layer reactant concentration $c_R^0 = $ bulk reactant concentration.

We would like to calculate the incremental voltage loss due to reactant depletion in the catalyst layer (we will call this η_{conc}). In other words, we would like to calculate how much the Nernst potential changes when using c_R^* values instead of c_R^0 values:

$$\eta_{conc} = E_{Nernst}^0 - E_{Nernst}^*$$

$$= \left(E^0 - \frac{RT}{nF} \ln \frac{1}{c_R^0} \right) - \left(E^0 - \frac{RT}{nF} \ln \frac{1}{c_R^*} \right)$$

$$= \frac{RT}{nF} \ln \frac{c_R^0}{c_R^*} \tag{5.12}$$

where E^0_{Nernst} is the Nernst voltage using c^0 values and E^*_{Nernst} is the Nernst voltage using c^* values. Recall that c^0_R can be described in terms of the limiting current density (from Equation 5.10),

$$c^0_R = \frac{j_L \delta}{n F D^{\text{eff}}} \tag{5.13}$$

and that c^*_R can be described in terms of the diffusion equation 5.9:

$$
\begin{aligned}
c^*_R &= c^0_R - \frac{j\delta}{n F D^{\text{eff}}} \\
&= \frac{j_L \delta}{n F D^{\text{eff}}} - \frac{j\delta}{n F D^{\text{eff}}}
\end{aligned}
\tag{5.14}
$$

Thus, the ratio c^0_R / c^*_R can be written as

$$
\begin{aligned}
\frac{c^0_R}{c^*_R} &= \frac{j_L \delta/(n F D^{\text{eff}})}{j_L \delta/(n F D^{\text{eff}}) - j\delta/(n F D^{\text{eff}})} \\
&= \frac{j_L}{j_L - j}
\end{aligned}
\tag{5.15}
$$

Substituting this result into our expression for η_{conc} provides the final result:

$$\eta_{\text{conc}} = \frac{RT}{nF} \ln \frac{j_L}{j_L - j} \tag{5.16}$$

Note that this expression is valid only for $j < j_L$ (j should never be greater than j_L anyway). For $j \ll j_L$, this expression implies that the concentration loss η_{conc} will be minor; however, as $j \to j_L$, η_{conc} increases sharply.

5.2.4 Concentration Affects Reaction Rate

The second way that concentration affects fuel cell performance is through the reaction kinetics. This is because the reaction kinetics also depend on the reactant and product concentrations at the reaction sites. Recall from Chapter 3 that the reaction kinetics may be described by the Butler–Volmer equation 3.33:

$$j = j^0_0 \left(\frac{c^*_R}{c^{0*}_R} e^{\alpha n F \eta_{\text{act}}/(RT)} - \frac{c^*_P}{c^{0*}_P} e^{-(1-\alpha)n F \eta_{\text{act}}/(RT)} \right) \tag{5.17}$$

where c^*_R and c^*_P are arbitrary concentrations and j^0_0 is measured at the *reference* reactant and product concentration values c^{0*}_R and c^{0*}_P. (Note that c^{0*}_R and c^{0*}_P, which are the reference reactant and product concentration values, may be different from c^0_R and c^0_P, the reactant and product bulk concentration values in our fuel cell.)

We are concerned primarily with the high-current-density region, since this is where concentration effects become most pronounced. At high current density, the second term in the Butler–Volmer equation drops out and the expression simplifies to

$$j = j_0^0 \left(\frac{c_R^*}{c_R^{0*}} e^{\alpha n F \eta_{act}/(RT)} \right) \tag{5.18}$$

Written in terms of the activation overvoltage, this becomes

$$\eta_{act} = \frac{RT}{\alpha n F} \ln \frac{j c_R^{0*}}{j_0^0 c_R^*} \tag{5.19}$$

As in the previous section, we would like to calculate the incremental voltage loss due to reactant depletion in the catalyst layer (which we will again call η_{conc}). In other words, we would like to calculate how much the activation overvoltage changes when using c_R^* values instead of c_R^0 values (keeping in mind that c_R^{0*} and c_R^0 are different):

$$\eta_{conc} = \eta_{act}^* - \eta_{act}^0$$

$$= \left(\frac{RT}{\alpha n F} \ln \frac{j c_R^{0*}}{j_0^0 c_R^*} \right) - \left(\frac{RT}{\alpha n F} \ln \frac{j c_R^{0*}}{j_0^0 c_R^0} \right)$$

$$= \frac{RT}{\alpha n F} \ln \frac{c_R^0}{c_R^*} \tag{5.20}$$

where η_{act}^0 is the activation loss using c^0 values and η_{act}^* is the activation loss using c^* values. As before, we can then write the ratio c_R^0/c_R^* as:

$$\frac{c_R^0}{c_R^*} = \frac{j_L}{j_L - j} \tag{5.21}$$

Substituting this result into our expression for η_{conc} provides almost the same final result as before:

$$\eta_{conc} = \frac{RT}{\alpha n F} \ln \frac{j_L}{j_L - j} \tag{5.22}$$

This result differs from our previous expression for the concentration loss (Equation 5.16) only by a factor of α. Because the two effects are virtually identical, we can generalize the total concentration loss as follows:

$$\eta_{conc} = \frac{RT}{n F} \ln \frac{j_L}{j_L - j} + \frac{RT}{\alpha n F} \ln \frac{j_L}{j_L - j}$$

$$= \left(\frac{RT}{n F} \right) \left(1 + \frac{1}{\alpha} \right) \ln \frac{j_L}{j_L - j} \tag{5.23}$$

Written in the most general form, this becomes

$$\eta_{conc} = c \ln \frac{j_L}{j_L - j} \tag{5.24}$$

where c is a constant.

5.2.5   Summary of Fuel Cell Concentration Loss

In the previous sections, we have seen how species depletion/accumulation in the catalyst layer leads to fuel cell performance loss. This performance loss, called the fuel cell concentration loss (or mass transport loss), may be described by the general form

$$\eta_{conc} = c \ln \frac{j_L}{j_L - j} \tag{5.25}$$

where c, a constant, might have the approximate form

$$c = \frac{RT}{nF} \left(1 + \frac{1}{\alpha} \right) \tag{5.26}$$

Interestingly, real fuel cell behavior often exhibits an effective c value which is much larger than that predicted by Equation 5.26 above. Therefore, in many cases, c is obtained empirically.

Figure 5.5 shows the effect of concentration loss on the j–V behavior of a fuel cell. The curves in this figure were generated for various values of j_L (1, 1.5, and 2 A/cm²,

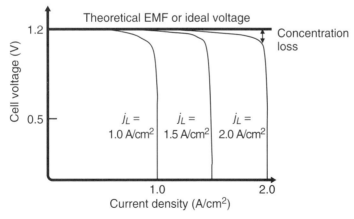

Figure 5.5. Effect of concentration loss on fuel cell performance. Concentration effects in the catalyst layer contribute to a characteristic drop in fuel cell operating voltage as determined by Equation 5.25. The shape of this loss is determined by c and j_L. (Curves calculated for $j_L = 1, 1.5, 2$ A/cm², respectively while c was held constant; c was fixed at 0.0388 V using Equation 5.26 with $T = 300$ K, $n = 2$, $\alpha = 0.5$.)

respectively) while c was held constant ($c = 0.0388$ V using $T = 300$ K, $n = 2$, $\alpha = 0.5$). As the curves clearly indicate, the concentration loss only significantly affects fuel cell performance at high current density (when j approaches j_L). Although the concentration loss appears mainly at high current density, its effect is abrupt and severe. The onset of significant concentration loss marks the practical limit of a fuel cell's operating range. As shown in the figure, increasing j_L can greatly extend a fuel cell's potential operating range; therefore mass transport design is an active area of current fuel cell research. Recall how j_L is defined:

$$j_L = nFD^{\text{eff}}\frac{c_R^0}{\delta} \qquad (5.27)$$

As we have previously discussed, this equation shows that the limiting current density depends on D^{eff}, c_R^0, and δ, where D^{eff} and δ are mostly determined by the electrode. Many constraints exist on electrode design, so it is often difficult to optimize the electrode solely for its mass transport properties. Instead, the flow structure often provides the best opportunities for mass transport optimization. Flow structure design affects the limiting current density because it determines c_R^0, the bulk concentration of reactant (or product) in the flow channel. It is important to realize that c_R^0 is not constant within fuel cell flow channels. (We wish that it were!) Instead, c_R^0 *decreases with distance* along a fuel cell flow channel because the reactants are being consumed. The best flow structure designs minimize this gas depletion effect so that c_R^0 is consistently high across an entire fuel cell device. As we will learn in the next section, maintaining a consistent, high c_R^0 value is often the best way to minimize the concentration losses in a fuel cell.

5.3 TRANSPORT IN FLOW STRUCTURES: CONVECTIVE TRANSPORT

Fuel cell flow structures are designed to distribute reactants across a fuel cell. Perhaps the simplest "flow structure" you could imagine would be a single-chamber structure. To make a single-chamber flow structure, we could encapsulate the entire fuel cell anode in a single compartment, then introduce H_2 gas into one corner. Unfortunately, this single-chamber design would lead to poor fuel cell performance. The H_2 would tend to stagnate inside the chamber, leading to poor reactant distribution and high mass transport losses.

In real fuel cells, mass transport losses are minimized by employing intricate flow structures containing many small flow channels. Compared to a single-chamber design, a design employing many small flow channels keeps the reactants constantly flowing across the fuel cell, encouraging uniform convection, mixing, and homogeneous reactant distribution. Small-flow-channel designs also provide more contact points across the surface of the electrode from which the fuel cell electrical current can be harvested.

To make a fuel cell flow structure, the flow channel design is typically stamped, etched, or machined into a flow field plate. The channels (there can be dozens or even hundreds of them) often snake, spiral, and twist across the flow field plate from a gas inlet at one corner to a gas outlet at another corner. Analyzing convective gas transport in these complex real-world flow structures is only really possible with numerical methods. A common technique is to use a computer simulation tool known as computational fluid dynamics (CFD) modeling, which will be described in Chapter 6. Without using CFD, however, a basic analysis of simple flow scenarios is still possible. This kind of basic analysis, which relies on the principles of fluid mechanics, can still yield great insight into fuel cell mass transport and flow structure design. Therefore, the rest of this chapter focuses on applying fluid mechanical principles to simplified convection in fuel cell flow channels. We begin with a brief review of fluid mechanics.

5.3.1 Fluid Mechanics Review

It is important to realize that when we talk about "fluid" in the context of fuel cell mass transport, we are usually talking about a gas. In the science of fluid mechanics, fluid does not have to mean liquid. A *gas* is a *fluid*. We use fluid mechanics to set up the rules governing how gases *flow* through fuel cell flow channels.

The nature of fluid flow in confined channels is characterized by an important dimensionless number known as the *Reynolds number* Re:

$$\text{Re} = \frac{\rho V L}{\mu} = \frac{V L}{\nu} \tag{5.28}$$

where V is the characteristic velocity of the flow (m/s), L is the characteristic length scale of the flow (m), ρ is the fluid density (kg/m^3), μ is the fluid viscosity ($kg/m \cdot s$ or $N \cdot s/m^2$), and ν is the kinematic viscosity (m^2/s). (The kinematic viscosity is the ratio of μ over ρ.) Physically, the Reynolds number describes the ratio of inertial forces to viscous forces in the fluid flow. Regardless of fluid type, flow velocity, or geometry, flows with the same Reynolds number shows similar viscous behavior.

All fluids have a characteristic *viscosity*. Viscosity measures the resistance to fluid flow. On the microscopic scale, viscosity measures how easily molecules slide past one another when driven by a shear force. It can therefore be thought of as a measure of internal fluid "friction." Mathematically, viscosity relates shear stress τ_{xy} to strain rate $\dot{\varepsilon}_{xy}$. For simple fluids such as water and gases, the relationship between shear stress and strain rate is linear[1]:

$$\tau_{xy} = 2\mu\dot{\varepsilon}_{xy} = 2\mu \cdot \frac{1}{2}\left(\frac{\partial u}{\partial y} + \frac{\partial v}{\partial x}\right) \tag{5.29}$$

where u is the fluid velocity (m/s) in the x direction and v is the fluid velocity (m/s) in y direction.

Considering the microscopic origin of viscosity, it is not surprising that μ is strongly temperature dependent. Viscosity increases with increasing temperature. For dilute gases, the temperature dependence of viscosity can be approximated either by a simple power law,

---

[1]Fluids obeying this equation are called *Newtonian fluids*.

FLOW BETWEEN PLATES

Assume a fluid is present between two parallel plates where the lower is fixed and the upper plate moves to the right at a steady velocity V, as shown in Figure 5.6. Since the plate only moves in the x direction, $u = V$ and $v = 0$. Equation 5.29 for this case becomes

$$\tau_{xy} = 2\mu \cdot \frac{1}{2}\left(\frac{\partial u}{\partial y} + \overset{0}{\cancel{\frac{\partial v}{\partial x}}}\right) = \mu \cdot \frac{du}{dy} = \text{const} \qquad (5.30)$$

Here, τ is constant since the system is in steady state with no acceleration or pressure variation. By solving Equation 5.30, we can obtain the velocity profile in the y direction, $u(y)$, assuming $u(0) = 0$ and $u(H) = V$ (where H is the distance between the plates):

$$u(y) = V\frac{y}{H} \quad \text{and} \quad \tau = \mu \cdot \frac{V}{H} \qquad (5.31)$$

To obtain Equation 5.31, we made the critical assumption that $u(0) = 0$ and $u(H) = V$. In other words, we assumed that the fluid velocity was the same as the plate velocity at both of the fluid/plate boundaries. This is the most widely assumed boundary condition for fluid flow, and it is generally a good assumption. In generalized form, this assumption can be stated as

$$\mathbf{V}_{\text{fluid}} = \mathbf{V}_{\text{solid}} \qquad (5.32)$$

where $\mathbf{V}$ is a vector. This assumption is commonly called the *no-slip condition*. In certain cases, *slip* boundary conditions must instead be used. Situations where slip boundary conditions must be used include gas flow in microchannels or gas flow at extremely low pressures. Such scenarios are generally not relevant to fuel cells.

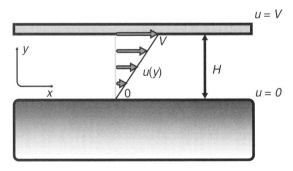

Figure 5.6. Fluid flow between two parallel plates.

$$\frac{\mu}{\mu_0} \approx \left(\frac{T}{T_0}\right)^n \tag{5.33}$$

or by Sutherland's law using the kinetic theory of gases [14],

$$\frac{\mu}{\mu_0} \approx \left(\frac{T}{T_0}\right)^{1.5} \frac{T_0 + S}{T + S} \tag{5.34}$$

In these equations, n, μ_0, T_0, and S can be obtained from experiments or kinetic theory. For most gases of interest, the viscosity values obtained from these equations give less than 3% error over a wide range of temperatures (0–1000°C). Table 5.2 summarizes values for common gases relevant to fuel cells.

Viscosity is also pressure dependent, increasing slowly with increasing pressure. Fuel cells rarely operate at gas pressures higher than 5 atm. At these low pressures, the "low-density limit" for viscosity applies, and the pressure effects on viscosity can be safely ignored. Thus, viscosity pressure effects will not be considered in this text.

Fuel cell gas streams are rarely composed of a single species. Instead, we usually deal with gas *mixtures* (e.g., O_2 and N_2.) The following semiempirical expression provides a good approximation for the viscosity of a gas mixture [16]:

$$\mu_{\text{mix}} = \sum_{i=1}^{N} \frac{x_i \mu_i}{\sum_{j=1}^{N} x_j \Phi_{ij}} \tag{5.35}$$

where Φ_{ij} is a dimensionless number obtained from

$$\Phi_{ij} = \frac{1}{\sqrt{8}} \left(1 + \frac{M_i}{M_j}\right)^{-1/2} \left[1 + \left(\frac{\mu_i}{\mu_j}\right)^{1/2} \left(\frac{M_i}{M_j}\right)^{1/4}\right]^2 \tag{5.36}$$

where N is the total number of species in the mixture, x_i, x_j are the mole fractions of species i and j, and M_i, M_j are the molecular weights (kg/mol) of species i and j.

TABLE 5.2. Parameters for Viscosity Calculation

Gas	$\mu_0 (10^{-6}$ kg/m · s	T_0 (K)	n	S
Air	17.16	273	0.666	111
CO_2	13.7	273	0.79	222
CO	16.57	273	0.71	136
N_2	16.63	273	0.67	107
O_2	19.19	273	0.69	139
H_2	8.411	273	0.68	97
H_2O (vapor)	11.2	350	1.15	1064

Source: From Ref.[15].

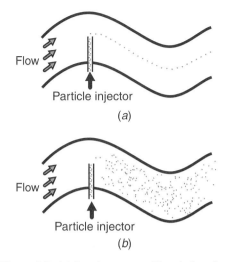

Figure 5.7. (*a*) Laminar versus (*b*) turbulent flow.

Under most conditions, gas flow in fuel cell flow channels is fairly smooth, or *laminar*. At extremely high flow rates, gas flow can become *turbulent* instead. The difference between laminar and turbulent flow is illustrated in Figure 5.7. Turbulent flow is extremely rare in fuel cell flow channels. The boundary between laminar and turbulent flow is determined by the Reynolds number Re. In circular pipes, for example, laminar flow occurs when Re $\leq$ 2000, while turbulent flow occurs for Re $\geq$ 3000.

> ***Example 5.1.*** Consider a fuel cell operating at 80°C. In the cathode, humidified air at 1 atm is supplied with a water vapor mole fraction of 0.2. If the fuel cell employs circular channels with a diameter of 1 mm, find the maximum tolerable air velocity that still ensures laminar flow.
>
> *Solution:* Using Equation 5.33 and Table 5.2, we can determine the viscosity of each gas component in the humidified air stream. For example, the viscosity of N_2 may be calculated as follows:
>
> $$\mu_{N_2}|_{80°C} = \mu_0 \left(\frac{T}{T_0}\right)^n = 16.63 \times 10^{-6} \left(\frac{353.15}{273}\right)^{0.67}$$
> $$= 19.76 \times 10^{-6} \text{ kg/m} \cdot \text{s} \tag{5.37}$$
>
> Similarly, we can obtain $\mu_{O_2}|_{80°C} = 22.92 \times 10^{-6}$ kg/m · s and $\mu_{H_2O}|_{80°C} = 11.32 \times 10^{-6}$ kg/m · s.
> To calculate the total viscosity of the mixture using Equation 5.36, we first assemble the following parameters:

Species	Mole Fraction, x_i	Molecular Weight, M_i	Viscosity, μ_i (10^{-6} kg/m · s)
1. N_2	$0.8 \times 0.79 = 0.632$	28.02	19.76
2. O_2	$0.8 \times 0.21 = 0.168$	32.00	22.92
3. H_2O	0.200	18.02	11.32

Then, we can use Equation 5.36 to produce the following:

Species i	Species j	M_i/M_j	μ_i/μ_j	Φ_{ij}	$x_j \Phi_{ij}$	$\sum\limits_{j=1}^{3} x_j \Phi_{ij}$
1. N_2	1. N_2	1.000	1.000	1.000	0.632	
	2. O_2	0.876	0.862	0.930	0.156	1.059
	3. H_2O	1.555	1.746	1.356	0.271	
2. O_2	1. N_2	1.142	1.160	1.079	0.682	
	2. O_2	1.000	1.000	1.000	0.168	1.146
	3. H_2O	1.776	2.025	1.482	0.296	
3. H_2O	1. N_2	0.643	0.573	0.776	0.491	
	2. O_2	0.563	0.494	0.732	0.123	0.814
	3. H_2O	1.000	1.000	1.000	0.200	

Finally Equation 5.35 gives the mixture viscosity:

$$\mu_{\text{mix}} = \left(\frac{0.632 \times 19.76}{1.059} + \frac{0.168 \times 22.92}{1.146} + \frac{0.200 \times 11.32}{0.814} \right) \times 10^{-6}$$

$$= 17.93 \times 10^{-6} \text{ kg/m} \cdot \text{s}$$

The molecular weight of the mixture is given by

$$M_{\text{mix}} = \sum_{i=1}^{N} x_i M_i = 0.632 \times 28.02 + 0.168 \times 32.00 + 0.200 \times 18.02$$

$$= 26.69 \text{ g/mol}$$

Then, the density of the mixture can be obtained using the ideal gas law:

$$\rho = \frac{p}{RT/M_{\text{mix}}} = \frac{101325 \text{ Pa}}{\dfrac{8.314 \text{ J/mol} \cdot \text{K}}{0.02669 \text{ kg/mol}} (273.15 + 80)} = 0.921 \text{ kg/m}^3 \qquad (5.38)$$

Roughly, laminar flow holds for Re $\leq$ 2000; thus

$$V_{\text{max}} = \frac{\text{Re}^{\text{max}} \mu_{\text{mix}}}{\rho L} = \frac{2000 \times (17.93 \times 10^{-6} \text{ kg/m} \cdot \text{s})}{(0.921 \text{ kg/m}^3) \times (0.001 \text{ m})} = 38.03 \text{ m/s} \qquad (5.39)$$

This is very fast flow considering the channel is only 1 mm in diameter.

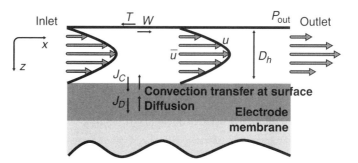

Figure 5.8. Schematic of 2D mass transport in fuel cell flow channel.

5.3.2 Mass Transport in Flow Channels

Pressure Drop in Flow Channels. Figure 5.8 illustrates (in 2D) the typical mass transport situation in a fuel cell flow channel. In this diagram, we have a gas moving from left to right through the flow channel at a mean velocity $\bar{u}$. A pressure difference between the inlet (P_{in}) and the outlet (P_{out}) drives the fluid flow. Increasing the pressure drop between the inlet and the outlet will increase the mean gas velocity in the channel, improving convection.

For circular flow channels, the relationship between pressure drop and mean gas velocity may be calculated from the relation

$$\frac{dp}{dx} = \frac{4}{D}\bar{\tau}_w \tag{5.40}$$

where dp/dx is the pressure gradient, D is the flow channel diameter, and the mean wall shear stress $\bar{\tau}_w$ may be calculated from a nondimensionalized number called the friction factor, f:

$$f = \frac{\bar{\tau}_w}{1/2\rho\bar{u}^2} \tag{5.41}$$

where ρ is the fluid density (kg/m^3) and $\bar{u}$ is the mean flow velocity (m/s). It is found that regardless of channel size or flow velocity, $f \cdot \text{Re} = 16$ for circular channels. Furthermore, for circular channels

$$\text{Re} = \frac{\rho\bar{u}D}{\mu} \tag{5.42}$$

Thus, by combining Equations 5.40, 5.41 and 5.42 and the fact that $f \cdot \text{Re} = 16$, pressure drop and mean gas velocity may be related:

$$\frac{dp}{dx} = \frac{32\bar{u}\mu}{D^2} \tag{5.43}$$

Unfortunately, most fuel cell flow channels are rectangular instead of circular. For rectangular channels, Equation 5.43 cannot be used. For rectangular channels, we must use a "hydraulic diameter" to compute the effective Reynolds number compared to a circular channel:

$$\mathrm{Re}_h = \frac{\rho \bar{u} D_h}{\mu} \tag{5.44}$$

where

$$D_h = \frac{4A}{P} = \frac{4 \times \text{cross-sectional area}}{\text{perimeter}} \tag{5.45}$$

For circular channels, $D_h = D$. Here, D_h can be thought of as the "effective" diameter of a noncircular channel.

For rectangular channels, the relationship between Re_h and f is also more complex than for circular channels. It can be approximated as [17]:

$$f\mathrm{Re}_h = 24(1 - 1.3553\alpha^* + 1.9467\alpha^{*2} - 1.7012\alpha^{*3} + 0.9564\alpha^{*4} - 0.2537\alpha^{*5}) \tag{5.46}$$

where α^* is the aspect ratio of the channel cross section: $\alpha^* = b/a$, where $2a$ and $2b$ are the lengths of the channel sides. Equation 5.46 is plotted as a function of α^* in Figure 5.9.

By determining $\bar{\tau}_w$ for a rectangular channel from Equations 5.41, 5.44, and 5.46, the pressure gradient can then be determined using Equation 5.40 (making sure that D_h is used in place of D).

> **Example 5.2.** Fluid is flowing at a velocity of 1 m/s through a 1-mm-wide, 2-mm-high, 20-cm-long rectangular channel. Find the pressure drop in the channel if the viscosity of the flowing fluid is 17.9×10^{-6} kg/m · s.

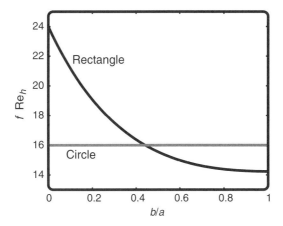

Figure 5.9. Friction factors of circular and rectangular channels.

Solution: We know

$$\frac{dp}{dx} = \frac{4}{D_h}\overline{\tau}_w = \frac{4}{D_h}f\frac{1}{2}\rho\overline{u}^2$$

$$= \frac{4}{D_h}\frac{f\,\mathrm{Re}_h}{\mathrm{Re}_h}\frac{1}{2}\rho\overline{u}^2 = \frac{4}{D_h}\frac{f\,\mathrm{Re}_h\mu}{\rho\overline{u}D_h}\frac{1}{2}\rho\overline{u}^2$$

$$= \frac{2}{D_h^2}f\,\mathrm{Re}_h\mu\overline{u} \tag{5.47}$$

Assume $\alpha^* = b/a = \frac{1}{2}$, and from Equation 5.46

$$f\,\mathrm{Re}_h = 24(1 - 1.3553 \cdot 0.5 + 1.9467 \cdot 0.5^2 - 1.7012 \cdot 0.5^3$$
$$+ 0.9564 \cdot 0.5^4 - 0.2537 \cdot 0.5^5) = 15.56 \tag{5.48}$$

Using

$$D_h = \frac{4 \times (1 \times 2)}{2 \times (1 + 2)} = 1.33 \text{ mm} = 0.00133 \text{ m}$$

Equation 5.47 gives

$$\frac{dp}{dx} = \frac{2}{(0.00133 \text{ m})^2} 15.56 \times 17.9 \times 10^{-6} \text{ kg/m} \cdot \text{s} \times 1 \text{ m/s} = 315 \text{ Pa/m} \tag{5.49}$$

Thus the pressure drop is

$$P_{\text{drop}} = L \times \frac{dp}{dx} = 0.2 \text{ m} \times 315 \text{ Pa/m} = 63 \text{ Pa} \tag{5.50}$$

Convective Mass Transport from Flow Channels to Electrode. As shown in Figure 5.8, although gas is flowing in the *x direction* from left to right along the flow channel, convective mass transport can also occur in the *z direction* from the flow channel into (or out of) the electrode. This type of convective mass transport occurs when the density of a species *i* is different at the electrode surface versus the flow channel bulk. For example, in a fuel cell cathode, water is produced at the electrode. The local density of water at the electrode surface will be greater than the density of water in the flow channel bulk, leading to convective mass transport of water away from the electrode surface. Mathematically, the mass flux due to this form of convective mass transfer may be estimated by

$$J_{C,i} = h_m(\rho_{i,s} - \overline{\rho}_i) \tag{5.51}$$

where $J_{C,i}$ is the convective mass flux (kg/m^2 · s), $\rho_{i,s}$ is the density (kg/m^3) of species *i* at the electrode surface, $\overline{\rho}_i$ is the mean density (kg/m^3) of species *i* in the bulk fluid, and h_m is the mass transfer convection coefficient (m/s). The value of h_m is dependent on the channel geometry, the physical properties of species *i* and *j*, and the wall conditions.

TABLE 5.3. Sherwood Numbers for Laminar Flows in Circular, Rectangular, and Three-Sided Closed Rectangular Ducts

Cross Section		$\alpha = 0.2$	$\alpha = 0.4$	$\alpha = 0.7$	$\alpha = 1.0$	$\alpha = 2.0$	$\alpha = 2.5$	$\alpha = 5.0$	$\alpha = 10.0$
	Sh_D				4.36				
	Sh_F				3.66				
	Sh_D	4.80	3.67	3.08	2.97	3.38	3.67	4.80	5.86
	Sh_F	5.74	4.47	3.75	3.61	4.12	4.47	5.74	6.79
	Sh_D	0.83	1.42	2.02	2.44	3.19	3.39	3.91	4.27
	Sh_F	0.96	1.60	2.26	2.71	3.54	3.78	4.41	4.85

Source: From Ref. [18].
Note: Channel aspect ratio $\alpha = b/a$, where b and a are channel dimensions.

Commonly, h_m can be found from a nondimensional number called the Sherwood (or Nusselt number[2]):

$$h_m = \mathrm{Sh}\frac{D_{ij}}{D_h} \tag{5.52}$$

Sh is the Sherwood number, D_h is the hydraulic diameter, and D_{ij} is the binary diffusion coefficient for species i and j. The Sherwood number depends on channel geometry. Table 5.3 summarizes some values of Sh for geometries commonly encountered in fuel cell flow channels. In most cases, only one wall in a rectangular channel of a fuel cell participates in convective mass transport (the third case represented in the table). The table distinguishes between two different Sherwood numbers. Sh_D values apply when density ρ_i is uniform along a channel; Sh_F values apply when flux $J_{C,i}$ is uniform along a channel. If neither density nor flux is uniform along the channel, Equations 5.51 and 5.52 should not be used.

5.3.3   Gas Is Depleted along Flow Channel

Since either hydrogen or air is consumed continuously along a flow channel, these reactants tend to become depleted, especially near the outlet. Depletion poses an adverse effect on fuel cell performance, since concentration losses increase as the reactant concentrations decrease.

In this section, we will develop a simple 2D mass transport model for a fuel cell cathode. We will use this model to determine how the oxygen density (concentration) decreases along the flow channel using a macroscale mass flux balance.

Consider the simple half PEMFC geometry shown in Figure 5.10. Pure oxygen flows from left to right along the flow channel depicted in this diagram from the fuel cell inlet to the fuel cell outlet. As the gas travels from left to right along the flow channel, it

---

[2]The Nusselt number applies to convective heat transport problems. Due to the similarity between heat and mass transport, both numbers are essentially the same.

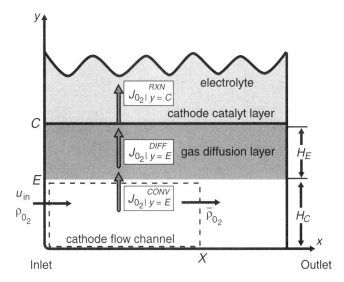

Figure 5.10. Schematic of a 2D fuel cell transport model including diffusion and convection.

is also being consumed. The y-direction flux $J_{O_2}|_{y=E}$ represents the oxygen gas that is removed from the flow channel by convective mass transport into the gas diffusion layer. This oxygen gas then diffuses to the catalyst layer where it reacts to produce the fuel cell current.

For this simple model, we assume the flow channel has a square cross section. We also make a few additional simplifying assumptions:

1. The catalyst layer is infinitely thin.[3]
2. Water exists only in the vapor form.
3. Diffusive mass transport dominates in the diffusion layer. Furthermore, only y-direction diffusion is considered.
4. Convection dominates in the flow channel.

The current density produced by the fuel cell will vary along the x direction because the concentration of oxygen varies along the x direction. We denote the local current density produced by the fuel cell at position X as $j(X)$. From Faraday's law, if the fuel cell is producing a current density $j(X)$ at location X, then the mass flux of oxygen it is consuming is given by

$$\hat{J}_{O_2}|_{x=X, y=C}^{\mathrm{rxn}} = M_{O_2} \frac{j(X)}{4F} \tag{5.53}$$

---

[3]This is a fairly good approximation since real catalyst layers are very thin (~ 10 μm) compared to gas diffusion layers (100–350 μ m) in PEMFCs.

where $\hat{J}_{O_2}$ is the oxygen mass flux (kg/cm^2), $y = C$ denotes the catalyst layer (where the reaction to produce electricity takes place), and M_{O_2} is the molecular weight (kg/mol) of oxygen.

The oxygen flux consumed by the electrochemical reaction must be provided by diffusion in the gas diffusion layer. As we have previously seen, diffusive mass transport is described by Fick's law:

$$\hat{J}_{O_2}|_{x=X,y=E}^{\text{diff}} = -D_{O_2}^{\text{eff}} \frac{\rho_{O_2}|_{x=X,y=C} - \rho_{O_2}|_{x=X,y=E}}{H_E} \tag{5.54}$$

where H_E is the thickness of the diffusion layer. In this equation, we have converted the molar concentrations normally seen in Fick's law into mass concentrations (density ρ is effectively a "mass concentration"). The flux $\hat{J}_{O_2}$ is therefore a mass flux rather than a molar flux.

The oxygen flux due to diffusive transport through the gas diffusion layer is provided by convective mass transport between the flow channel and the gas diffusion layer surface (represented in the diagram by $\hat{J}_{O_2}|_{y=E}^{\text{diff}}$). Recall that this convective mass transport process can be described by Equation 5.51:

$$\hat{J}_{O_2}|_{x=X,y=E}^{\text{conv}} = -h_m \left(\rho_{O_2}|_{x=X,y=E} - \overline{\rho}_{O_2}|_{x=X,y=\text{channel}} \right) \tag{5.55}$$

where h_m is the convection mass transfer coefficient and $\overline{\rho}_{O_2}$ is the average density of oxygen in the flow channel. To maintain flux balance, the oxygen fluxes in Equations 5.53, 5.54, and 5.55 must be the same (steady-state condition). In other words,

$$\hat{J}_{O_2}|_{x=X,y=C}^{\text{rxn}} = \hat{J}_{O_2}|_{x=X,y=E}^{\text{diff}} = \hat{J}_{O_2}|_{x=X,y=E}^{\text{conv}} \tag{5.56}$$

Thus, we can obtain the following relations:

$$\hat{J}_{O_2}|_{x=X,y=E}^{\text{conv}} = M_{O_2} \frac{j(X)}{4F} \tag{5.57}$$

$$\rho_{O_2}|_{x=X,y=C} = \rho_{O_2}|_{x=X,y=E} - M_{O_2} \frac{j(X)}{4F} \frac{H_E}{D_{O_2}^{\text{eff}}} \tag{5.58}$$

$$\rho_{O_2}|_{x=X,y=E} = \overline{\rho}_{O_2}|_{x=X,y=\text{channel}} - M_{O_2} \frac{j(X)}{4F} \frac{1}{h_m} \tag{5.59}$$

Now, we couple the y-direction mass transport of oxygen to the x direction mass transport of oxygen in the flow channel by considering the overall flux balance in the control volume (dotted box) in Figure 5.10. Oxygen is entering into this control volume from the left and leaving to the right. The difference between the amount of oxygen entering on the left and the amount of oxygen leaving on the right yields the amount of oxygen that is leaving out the top into the gas diffusion layer. Mathematically,

$$\underbrace{u_{\text{in}} H_C \overline{\rho}_{O_2}|_{x=0,y=\text{channel}}}_{\substack{\text{amount of gas} \\ \text{entering from left}}} - \underbrace{u_{\text{in}} H_C \overline{\rho}_{O_2}|_{x=X,y=\text{channel}}}_{\substack{\text{amount of gas} \\ \text{leaving from right}}} = \underbrace{\int_0^X \left(\hat{J}_{O_2}|_{y=E}^{\text{conv}} \right) dx}_{\substack{\text{amount of gas} \\ \text{leaving out the top}}} \tag{5.60}$$

Equation 5.57 then allows us to relate the gas leaving out the top of the control volume to the current density produced by the fuel cell:

$$\int_0^X \left(\hat{J}_{O_2}\big|_{y=E}^{\text{conv}} \right) dx = \int_0^X \frac{M_{O_2} j(x)}{4F} dx \tag{5.61}$$

Remember, we are seeking an expression for the x direction oxygen profile at the catalyst layer. (In other words, we want to find $\rho_{O_2}|_{x=X,y=C}$.) Starting with Equation 5.58, $\rho_{O_2}|_{x=X,y=C}$ may be determined by plugging in Equations 5.59, 5.60, and 5.61. This yields

$$\rho_{O_2}|_{x=X,y=C} = \overline{\rho}_{O_2}|_{x=0,y=\text{channel}} - \frac{M_{O_2}}{4F} \left(\frac{j(X)}{h_m} + \frac{H_E j(X)}{D_{O_2}^{\text{eff}}} + \int_0^X \frac{j(x)}{u_{\text{in}} H_C} dx \right) \tag{5.62}$$

For an exact solution, Equation 5.62 can then be solved in combination with the Tafel equation. However, to avoid mathematical complication, we assume that the current density j is constant along the x direction. (This assumption is not quite true. The oxygen concentration changes along the x direction, and thus the local current density will also change. For small oxygen concentration changes, however, the current density effect will be minor.) Using the constant-current-density assumption, Equation 5.62 becomes

$$\rho_{O_2}|_{x=X,y=C} = \overline{\rho}_{O_2}|_{x=0,y=\text{channel}} - M_{O_2} \frac{j}{4F} \left(\frac{1}{h_m} + \frac{H_E}{D_{O_2}^{\text{eff}}} + \frac{X}{u_{\text{in}} H_C} \right) \tag{5.63}$$

Using Equation 5.52, we can determine h_m based on the constant-flux Sherwood number Sh_F for the flow channel:

$$h_m = \frac{\text{Sh}_F D_{O_2}}{H_C} \tag{5.64}$$

Plugging this result into Equation 5.63 yields a final expression for the oxygen profile:

$$\rho_{O_2}|_{x=X,y=C} = \overline{\rho}_{O_2}|_{x=0,y=\text{channel}} - M_{O_2} \frac{j}{4F} \left(\frac{H_C}{\text{Sh}_F D_{O_2}} + \frac{H_E}{D_{O_2}^{\text{eff}}} + \frac{X}{u_{\text{in}} H_C} \right) \tag{5.65}$$

Equation 5.65 tells us that oxygen density decreases linearly as X increases.[4] In other words, the oxygen concentration is depleted linearly as the gas moves along the channel. The three terms in the parentheses represent the effects of channel size H_C, diffusion layer thickness H_E, and inlet flow velocity u_{in} in that order. Supplying more oxygen (increasing u_{in}) improves mass transport, thus increasing the oxygen density at the catalyst layer. Similarly, decreasing the diffusion layer thickness H_E also increases the oxygen density at the catalyst layer. The effect of channel size H_C is a little tricky to calculate, since H_C appears

---

[4]See problem 5.8.

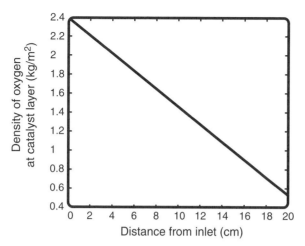

Figure 5.11. Oxygen density profile predicted from Equation 5.65 for the following case: electrode porosity $\varepsilon = 0.4$; inlet gas pressure $p = 2$ atm; model temperature $T = 80°C$; current density $j = 1$ A/cm^2; inlet gas velocity $u_{in} = 10$ cm/s; channel height $H_C = 0.1$ cm; electrode thickness $H_E = 0.035$ cm; Sherwood number $Sh_F = 2.71$.

in both the first and third terms in the parentheses. However, if we assume that the total gas supply to the fuel cell (either by volume or mass) is constant, we have

$$N_{total} = u_{in} H_C = \text{const} \qquad (5.66)$$

Thus, if the oxygen supply rate is constant, $u_{in} H_C$ in the last term is fixed. In this case, decreasing the channel size H_C will increase the oxygen density. An example oxygen profile prediction given by Equation 5.65 is displayed in Figure 5.11.

5.3.4 Flow Structure Design

Flow Structure Materials. In the most general terms, the flow structure serves two main purposes: (1) It supplies the reactant gases and removes the reaction products and (2) it harvests the electrical current generated by the fuel cell. In spite of these seemingly simple tasks, flow structures are subject to a challenging set of materials selection criteria [19]:

- High electrical conductivity
- High corrosion resistance
- High chemical compatibility
- High thermal conductivity
- High gas tightness
- High mechanical strength

- Low weight and volume
- Ease of manufacturability
- Cost effectiveness

The most commonly used material for low-temperature fuel cell flow plates is graphite. Graphite satisfies most of the criteria discussed above except: (1) ease of manufacturability, (2) cost, and (3) high mechanical strength. These criteria are not fulfilled because of costly machining requirements and the intrinsic brittleness of the material. Surprisingly, the machining of graphite is so expensive that graphite plates can take up to half the cost of a fuel cell system [20]. Alternatives to graphite include corrosion-resistant metals such as stainless steel [21, 22]. In general, metal plates offer less expensive fabrication and higher mechanical strength compared to graphite plates. Thin metal flow plates can significantly reduce the volume and weight of a fuel cell system. One critical issue concerning metal plates is the formation of surface metal oxides. Even a thin metal oxide layer will increase the contact resistance between the flow plate and the electrode, resulting in degraded fuel cell performance [21]–[24]. This problem has been partially overcome by the use of corrosion-resistant surface coatings [23, 24], although the long-term stability of such coatings needs improvement.

Flow plates in high-temperature fuel cells are made from ceramics such as lanthanum chromite (for high temperatures) or ferritic stainless steel (for moderate temperatures). In SOFCs and MCFCs, flow plate stability and durability are critical, since the high operating temperature facilitates degradation. Also, any thermal mismatch between the plate material and the electrode material will be a source of serious mechanical stress during thermal cycles. Thus, the thermal properties of the flow plate should be carefully matched to the rest of the fuel cell system. Certain SOFC designs, such as tubular SOFCs, do not require flow plates and avoid the issue of high-temperature sealing. These designs are discussed in Chapter 9.

Flow Structure Patterns. As was previously mentioned, flow plates contain dozens or even hundreds of fine channels (or "grooves") to distribute the gas flow over the surface of the fuel cell. The shape, size, and pattern of flow channels can significantly affect the performance of a fuel cell. Choosing the right flow pattern is especially critical for PEMFCs. In PEMFCs, flow field design efforts often focus on the water removal capability of the cathode side. Poorly designed flow field plates leave certain regions flooded with liquid water, thus blocking gas access and reducing the output current of the cell. Such blocked regions not only reduce performance but can actually cause irreversible damage to the fuel cell. This is because cell polarity can be locally reversed in gas-starved regions, leading to corrosion and material degradation [25].

Although a wide variety of flow patterns are employed by research groups and developers, most fall under three basic flow pattern archetypes (Figure 5.12):

1. Parallel flow
2. Serpentine flow
3. Interdigitated flow

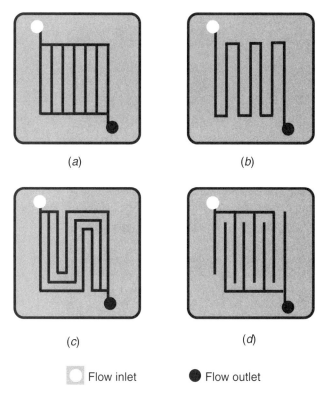

$$(a) \qquad (b)$$

$$(c) \qquad (d)$$

○ Flow inlet ● Flow outlet

Figure 5.12. Major flow channel geometries: (*a*) parallel; (*b*) serpentine; (*c*) parallel serpentine; (*d*) interdigitated. Flow channel geometries seek to provide homogeneous distribution of reactants across an electrode surface while minimizing pressure drop losses and maximizing water removal capability.

Parallel Flow. In a parallel configuration, flow evenly enters each straight channel and exits through the outlet. (See Figure 5.13*a*.) A significant advantage of the parallel pattern is the low overall pressure drop between gas inlet and outlet. However, when the width of the flow field is relatively large, flow distribution in each channel may not be uniform. This causes water buildup in certain channel areas, leading to increased mass transfer losses (and a corresponding current density decrease). Several fuel cell developers (IFC, Energy Partners) employ this channel type in their portable fuel cell systems.

Serpentine Flow. This is the most common geometry found in fuel cell prototypes. The advantage of the serpentine pattern lies in the water removal capability. Only one flow path exists in the pattern, so liquid water is forced to exit the channel. (See Figure 5.13*b*.) Unfortunately, in large-area cells, a serpentine design leads to a large pressure drop. Several variations of the serpentine design have been investigated, such as the parallel–serpentine configuration. This hybrid design, combining the advantages of serpentine and parallel patterns, is famously used in Ballard PEMFC stacks.

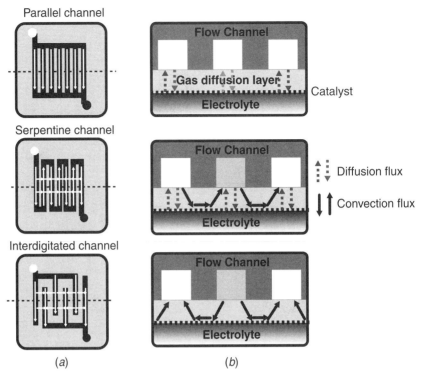

Figure 5.13. Gas transport modes in various flow channel geometries. Each channel type induces a different convective transport scheme in the electrode.

Interdigitated Flow. The interdigitated design promotes forced convection of the reactant gases through the gas diffusion layer. (See Figure 5.13c). Subject to much recent attention, research shows that this design provides far better water management, leading to improved mass transport [26]. The forced convection through the gas diffusion layer leads to significant pressure drop losses. However, there is evidence that this major disadvantage might be partially overcome by employing extremely small rib spacing [27].

In addition to the channel pattern, channel shape and size can also significantly affect performance [23], [27]–[30]. These parameters are best explored using computer numerical simulations. One such simulation technique, known as CFD modeling, will be discussed in the forthcoming chapter.

CHAPTER SUMMARY

- *Mass transport* governs the supply and removal of reactants and products in a fuel cell.
- Poor mass transport leads to a loss in fuel cell performance due to reactant depletion (or product clogging) effects.

- Mass transport in fuel cell electrodes is typically dominated by diffusion. Mass transport in fuel cell flow structures is typically dominated by convection.

- *Convection* refers to the transport of a species by the bulk motion of a fluid. Diffusion refers to the transport of a species due to a gradient in concentration.

- Diffusive transport limitations in the electrode lead to a limiting current density j_L. The limiting current density corresponds to the point where the reactant concentration falls to zero in the fuel cell catalyst layer. A fuel cell can never sustain a current density higher than j_L.

- Reactant depletion affects both the Nernstian cell voltage and the kinetic reaction rate. Depletion leads to a similar loss in both cases. This "concentration loss" can be generalized as $\eta_{conc} = c[j_L/(j_L - j)]$, where c is a constant that depends on the geometry and mass transport properties of the fuel cell.

- Concentration losses are most effectively minimized by careful consideration of the convective transport situation in the fuel cell flow channels.

- Convection in fuel cell flow channels is characterized by the Reynolds number Re, a nondimensional parameter that characterizes the viscous behavior of the flow. Usually, gas flow in fuel cells is laminar.

- Viscosity μ characterizes the resistance of a fluid to flow. Viscosity can be thought of as a measure of the "internal friction" in the fluid.

- The viscosity of a gas mixture is dependent on the temperature and composition of the mixture.

- A pressure difference is required to drive gas flow through a channel.

- The pressure drop in a flow channel is mainly caused by friction between fluid and the channel walls. This friction is quantified by wall shear stress $\overline{\tau}_w$. Pressure drops can be determined using the friction factor f, which is dependent on the Reynolds number and channel geometry.

- Although gases in fuel cell flow channels move *along* the flow channel, they can also be transported *between* the flow channel and the electrode. This is known as convective mass transport. Convective mass transport is characterized by a convective mass transfer coefficient h_m, which may be calculated from the Sherwood number, Sh.

- A simple 2D fuel cell mass transport model can be constructed to show how reactant gases are depleted in a flow channel from the inlet to the outlet. In general, increasing the gas flow velocity, decreasing the channel size, or decreasing the diffusion layer thickness will improve the mass transport situation along the length of the flow channel.

- Choice of the flow field pattern significantly affects the size of the mass transport losses. Due to the liquid water formation in the cathode, PEMFCs require flow fields with high water removal capability.

- Serpentine or parallel–serpentine designs are the most commonly used flow field types. They provide a decent compromise between pressure drop and water removal capability.

CHAPTER EXERCISES

Review Questions

5.1   Everything else being equal, would the concentration losses in a fuel cell using "synthetic air" (21% oxygen, 79% helium) be higher or lower than the concentration losses in a fuel cell using real air ($\approx$21% oxygen, $\approx$79% nitrogen)? Defend your answer.

5.2   Discuss why cathode flow channel design is less important for SOFCs than for PEMFCs. *Hint:* Consider the typical operating temperature of a SOFC and its effect on j_L.

5.3   Discuss the factors that determine j_L. List at least three ways to increase j_L.

Calculations

5.4   Using Equation 5.10, calculate the limiting current density for a fuel cell cathode running on air at STP. Assume only O_2 and N_2 and ignore the presence of water vapor. Assume the diffusion layer is 500 μ m thick and has a porosity of 40%.

5.5   Generate a series of plots similar to the ones shown in Figure 5.5 but for different values of c while holding j_L constant at 2.0 A/cm^2. Generate plots for c values of 0.1, 0.05, and 0.01, respectively.

5.6   Consider a fuel cell operating at 800°C, 1 atm. In the cathode, humidified air is supplied with the mole fraction of water vapor equal to 0.1. If the fuel cell employs circular flow channels with a diameter of 1 mm, find the maximum velocity of air that can be used while still maintaining laminar flow. Compare your result to Example 5.1.

5.7   Estimate the maximum fuel cell area that can be operated at 1 A/cm^2, under the condition from Example 5.1. Assume a stoichiometric number of 2. Assume the fuel cell is made of a single straight flow channel with ribs on both sides. The thickness of each rib is half the size of the channel. Discuss why channel flow in fuel cells is almost always considered to be laminar.

5.8   Plot the oxygen distribution along the channel (the x direction at the catalyst layer) for the fuel cell flow model developed in Section 5.3.3, assuming $u_{\text{in}} = 1$ m/s, $H_C = 1$ mm, and an operating temperature of 80°C. Estimate D_{O_2,H_2O} and $D^{\text{eff}}_{O_2,H_2O}$ using Equations 5.2 and 5.3 assuming $\varepsilon = 0.4$ and $p = 1$ atm.

5.9   Following a procedure similar to that illustrated by the model developed in Section 5.3.3, derive an equation for the water vapor density distribution along a fuel cell flow channel (at the catalyst layer).

5.10   Find the oxygen density distribution along the channel (at the catalyst layer) for the fuel cell model developed in Section 5.3.3 assuming constant voltage but not constant current. *Hint:* Use the Tafel equation to set up an ordinary differential equation for $j(X)$.

CHAPTER 6

FUEL CELL MODELING

In the last four chapters you have acquired the necessary tools to describe the basic operation of a fuel cell. Now it is time to complete the picture. In this chapter you will put all your tools together to build a complete fuel cell model. Your model will include thermodynamics (Chapter 2), reaction kinetics (Chapter 3), charge transport (Chapter 4), and mass transport (Chapter 5). Do not worry if putting all these things together sounds intimidating. In fact, it is surprisingly simple! You will be amazed at the predictive power provided by even a modest fuel cell model. Furthermore, modeling offers a great opportunity to see how the material you have learned in the last four chapters fits together into a cohesive unit.

After discussing the big picture in the context of a simple fuel cell model, we will delve into the details of several more sophisticated modeling approaches. One example is a flux balance-based approach which we use to model both a PEMFC and an SOFC. Still more complex is the CFD approach to fuel cell modeling. Computational fluid dynamics modeling allows the detailed interactions between flow structure geometry, fluid dynamics, multiphase flow, and electrochemical reaction to be simulated numerically. These more sophisticated modeling techniques can provide predictive capability and may one day allow fuel cell designers to better optimize fuel cells electronically before ever testing them in the laboratory.

6.1 PUTTING IT ALL TOGETHER: A BASIC FUEL CELL MODEL

If you recall from the first chapter of this book, we noted that the real voltage output of a fuel cell could be written by starting with the thermodynamically predicted voltage and then subtracting the various overvoltage losses:

$$V = E_{\text{thermo}} - \eta_{\text{act}} - \eta_{\text{ohmic}} - \eta_{\text{conc}} \qquad (6.1)$$

where V = operating voltage of fuel cell

E_{thermo} = thermodynamically predicted voltage of fuel cell

η_{act} = activation losses due to reaction kinetics

η_{ohmic} = ohmic losses from ionic and electronic resistance

η_{conc} = concentration losses due to mass transport

In the last four chapters, we determined basic expressions for each of the quantities in Equation 6.1. For example, in Chapter 3 we learned how the activation loss η_{act} could be described by the Butler–Volmer equation (or the simpler Tafel equation). We were even able to draw a graph which showed the effect of the activation loss on fuel cell performance. In Chapters 4 and 5 we were able to draw graphs describing the effects of charge transport and mass transport on fuel cell performance. As Equation 6.1 illustrates, overall fuel cell performance is simply given by the combined effect of all these various losses. Pictorially, the concept is illustrated in Figure 6.1. By starting with the thermodynamically predicted fuel cell voltage and then graphically subtracting out the losses due to activation, ohmic resistance, and concentration effects, we are left with the net fuel cell performance. Mathematically (using the simplest expressions developed in Chapters 3–5 for η_{act}, η_{ohmic} and η_{conc}), the net fuel cell j–V behavior can be written as

$$V = E_{\text{thermo}} - (a_A + b_A \ln j) - (a_C + b_C \ln j) - (j\,\text{ASR}_{\text{ohmic}}) - \left(c \ln \frac{j_L}{j_L - j} \right) \quad (6.2)$$

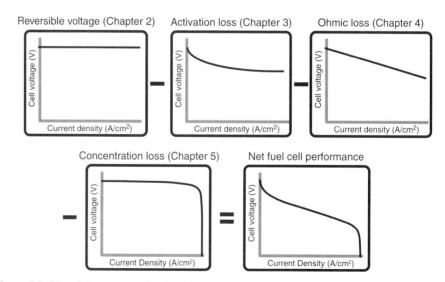

Figure 6.1. Pictorial summary of major factors that contribute to fuel cell performance. The overall fuel cell j–V performance can be determined by starting from the ideal thermodynamic fuel cell voltage and subtracting out the losses due to activation, conduction, and concentration.

where:

$\eta_{\text{act}} = (a_A + b_A \ln j) + (a_C + b_C \ln j)$: activation losses from both anode (A) and the cathode (C) based on natural logarithm form of Tafel Equation 3.41

$\eta_{\text{ohmic}} = j\text{ASR}_{\text{ohmic}}$: ohmic resistance loss based on current density and ASR (Equation 4.11)

$\eta_{\text{conc}} = c \ln[j_L/(j_L - j)]$: combined fuel cell concentration loss based on Equation 5.25, where c is an empirical constant

Because we use the Tafel approximation for the fuel cell kinetics, this model is only valid when $j \gg j_0$. For detailed modeling of the low-current-density region, the full form of the Butler–Volmer equation is required. In its most general form, this simple model has seven "fitting constants": a_A, a_C, b_A, b_C, c, $\text{ASR}_{\text{ohmic}}$, and j_L. However, for H_2–O_2 fuel cells, the anode kinetic losses can often be neglected compared to the cathode kinetic losses (eliminating a_A and b_A). Also, if the "first-principles" values of a, b, and c are used, we know that they are really related to the two more fundamental constants α and j_0. In the extremely streamlined case, then, as few as four parameters (α_A, j_{0A}, $\text{ASR}_{\text{ohmic}}$, and j_L) are required.

In reality, we find that one additional term is usually needed to reflect the true behavior of most fuel cell systems. This additional term, j_{leak}, is associated with the parasitic loss due to current leakage, gas crossover, and unwanted side reaction. In almost all fuel cell systems, some current is lost due to these parasitic processes. You might recall that we have already talked a little bit about gas crossover in previous chapters. The net effect of this parasitic current loss is to offset the fuel cell's operating current by an amount given by j_{leak}. In other words, the fuel cell has to produce extra current to compensate for the current that is lost due to parasitic effects. Pictorially, this loss effect is illustrated in Figure 6.2.

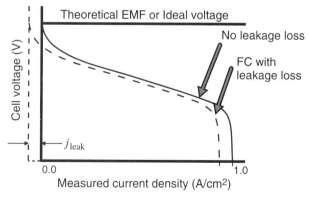

Figure 6.2. Pictorial illustration of the effect of a leakage current loss on overall fuel cell performance. A leakage current effectively "offsets" a fuel cell's i–V curve, as shown by the dotted curve in the figure. This has a significant effect on the open circuit voltage of the fuel cell (y-axis intercept) which is reduced below its thermodynamically predicted value.

TABLE 6.1. Summary of Typical Parameters for Low-Temperature PEMFC versus High-Temperature SOFC

Parameter	Typical Value for PEMFC	Typical Value for SOFC
Temperature	350 K	1000 K
E_{thermo}	1.22 V	1.06 V
$j_0(H_2)$	0.10 A/cm^2	10 A/cm^2
$j_0(O_2)$	10^{-4} A/cm^2	0.10 A/cm^2
$\alpha(H_2)$	0.50	0.50
$\alpha(O_2)$	0.30	0.30
ASR_{ohmic}	0.01$\Omega \cdot$ cm^2	0.04$\Omega \cdot$ cm^2
j_{leak}	10^{-2} A/cm^2	10^{-2} A/cm^2
j_L	2 A/cm^2	2 A/cm^2
c	0.10 V	0.10 V

Mathematically,

$$j_{gross} = j + j_{leak} \tag{6.3}$$

where j_{gross} is the gross current produced at the fuel cell electrodes, j_{leak} is the parasitic current that is wasted, and j is the actual fuel cell operating current that we can measure and use. In our fuel cell model, η_{act} and η_{conc} should be based on j_{gross} since the reaction

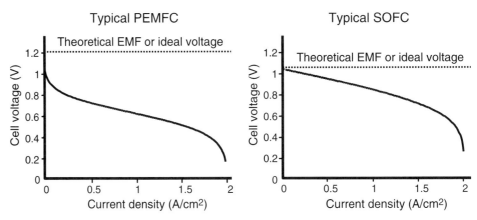

Figure 6.3. Comparison of our simple model results for typical PEMFC versus typical SOFC. As shown by the shape of the curves, the PEMFC benefits from a higher thermodynamic voltage but suffers from larger kinetic losses. SOFC performance is dominated by ohmic and concentration losses. The input parameters used to generate these model results are summarized in Table 6.1.

kinetics and species concentrations are affected by the leakage current. However, η_{ohmic} should be based on j, since only the operating current of the fuel cell is actually conducted through the cell. (The leakage current is wasted by side-reactions or nonelectrochemical reactions at the electrodes and does not give rise to real current flow across the cell.) Thus, we can rewrite our fuel cell model in the following final form:

$$V = E_{\text{thermo}} - [a_A + b_A \ln(j + j_{\text{leak}})] - [a_C + b_C \ln(j + j_{\text{leak}})]$$

$$- (j\text{ASR}_{\text{ohmic}}) - \left(c \ln \frac{j_L}{j_L - (j + j_{\text{leak}})} \right) \qquad (6.4)$$

The most noticeable effect of leakage current is to reduce a fuel cell's open-circuit voltage below its thermodynamically predicted value. At high current density, the limiting current density will also be reduced by the leakage current. However, at midrange current densities, the leakage current effects tend to be minor or insignificant. Careful inspection of the two curves in Figure 6.2 illustrates this effect.

The simple fuel cell model described by Equation 6.4 can be used for virtually unlimited numbers of "what-if" scenarios. For example, the model can be used to contrast the j–V behavior of a typical low-temperature (e.g., polymer electrolyte membrane) fuel cell versus a typical high-temperature (e.g., solid-oxide) fuel cell. In a typical H_2–O_2 PEMFC, activation losses are significant due to the low reaction temperature, but ohmic losses are relatively small due to the high conductivity of the polymer electrolyte. In contrast, ohmic losses tend to dominate H_2–O_2 SOFC performance while the activation losses are minor due to the high reaction temperature.

Typical parameters for H_2–O_2 PEMFCs and SOFCs are summarized in Table 6.1. Using these parameters as inputs into our simple model (Equation 6.4) produces the contrasting j–V behaviors shown in Figure 6.3. The large j_0 values in the SOFC model require the use of the full Butler–Volmer Equation for η_{act}. Alternatively, since j_0 is so large in the SOFC, the small η_{act} approximation of the Butler–Volmer equation can be successfully used. (Recall from Equation 3.38 that this approximation gives $\eta_{\text{act}} \approx [RT/(nFj_0)j]$.)

6.2 A 1D FUEL CELL MODEL

Having discussed a simple fuel cell model in the previous section, we now introduce a more sophisticated 1D model for SOFCs and PEMFCs. This model is based on the *flux balance* concept. Flux balance allows us to keep track of all the species that flow in, out, and through a fuel cell. Flux-balance-based models are popular in the fuel cell literature. The model that we will develop in this section is really just a simplified version of the popular literature models developed in the last decade [8], [31]–[36].

Flux-balance-based models are suited to both PEMFCs and SOFCs. Generally, PEMFCs are more difficult to model because water can be transported through the membrane, complicating the flux balance. Also, in PEMFCs, water is present as a *liquid*. Liquid water is far more difficult to model than water vapor. Remember that in SOFCs all the reactants and products exist as gases (including water); this makes the modeling easier. However, SOFC modeling can be complicated by other issues such as nonisothermal behavior and thermal-expansion-induced mechanical stress. While these issues can be integrated into a structural SOFC model, the complexity swiftly becomes daunting. In the present models, therefore, we will focus only on fuel cell species transport. By keeping track of species concentration profiles inside a model fuel cell, we can extract electrochemical losses and the j–V curve.

6.2.1   Flux Balance in Fuel Cells

A 1D flux balance fuel cell model starts as a very careful bookkeeping exercise. To generate an accurate model, the fluxes of all chemical species going into, out of, and through the fuel cell must be detailed. Figure 6.4 illustrates the high-level flux detail needed in our 1D fuel cell model. In this diagram, individual fluxes are numbered consecutively. While the exact meaning of each flux term is unimportant for now, this diagram essentially allows us to keep track of the H_2O and H_2 flowing into/out of the anode, the H_2O, N_2 and O_2 flowing into/out of the cathode, and the H_2O and H^+ (for PEMFC) or O^{2-} (for SOFC) flowing across the electrolyte membrane.

The fluxes in Figure 6.4 can be related to one another using the principle of *flux balance*. Flux balance expresses the idea that *what comes in must go out*. In fuel cells, all fluxes can be related to a single characteristic flux—the *current density*, or charge flux of the fuel cell. Here is an example of how the current density (flux 14 in Figure 6.4*a*) can be related to the other fluxes in a PEMFC. Based on an examination of the fluxes in Figure 6.4*a*, we can write

$$\text{Flux } 14 = \text{flux } 5 = \text{flux } 1 - \text{flux } 4 = \text{flux } 8 - \text{flux } 13 \tag{6.5}$$

In other words, the current density produced by the fuel cell must equal the proton flux across the electrolyte, which must equal the hydrogen flux into the anode catalyst layer, which must equal the oxygen flux into the cathode catalyst layer. Mathematically,

$$\frac{j}{2F} = \frac{J_{H^+}}{2} = J_{H_2}^A = 2J_{O_2}^C = S_{H_2O}^C \tag{6.6}$$

where j, F, and J stand for current density (A/cm^2), Faraday's constant (96,484 C/mol), and molar flux (mol/s $\cdot$ cm^2), respectively; $J_{H_2}^A$ stands for the *net* flux of H_2 in the anode (in other words, the flux of hydrogen coming in minus the flux of hydrogen going out). Since the net hydrogen flux is the difference between what goes in and what goes out, it represents hydrogen that is consumed inside the fuel cell by the reaction. Likewise, $J_{O_2}^C$ stands for the net flux of oxygen at the cathode. Also, note that the water generation rate $S_{H_2O}^C$ (mol/s $\cdot$ cm^2) at the cathode is equal to the net hydrogen flux. (For each mole of hydrogen that is consumed, 1 mol of water will be produced.)

In an analogous manner, the following water flux balance must also be satisfied:

$$\underset{\text{anode}}{\text{flux } 2 - \text{flux } 3} = \underset{\text{membrane}}{\text{flux } 6 - \text{flux } 7} = \underset{\text{cathode}}{\text{flux } 12 - \text{flux } 9 - \text{flux } 5} \tag{6.7}$$

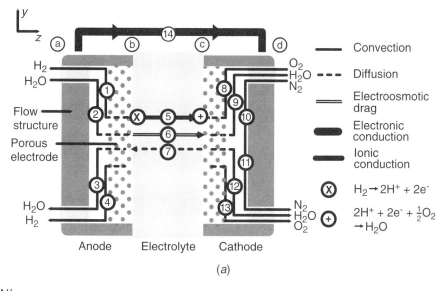

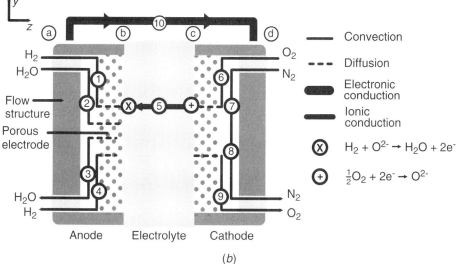

Figure 6.4. Flux details for (a) 1D PEMFC model and (b) 1D SOFC model. (a) In a PEMFC, water (H_2O) and protons (H^+) transport through the electrolyte. (b) In a SOFC, oxygen ions (O^{2-}) transport through the electrolyte.

In other words, the net water flux into the anode catalyst layer must be equal to the net water flux across the electrolyte (given by the balance between the electro-osmotic drag and back-diffusion water fluxes), which must be equal to the net water flux out of the cathode catalyst layer. Note that the water generation at the cathode (flux 5) also must be included for correct flux balance. Mathematically,

$$J^A_{H_2O} = J^M_{H_2O} = J^C_{H_2O} - \frac{j}{2F} \tag{6.8}$$

where $J^A_{H_2O}$, $J^M_{H_2O}$, and $J^C_{H_2O}$ represent the net flux into the anode catalyst layer, across the electrolyte, and out of the cathode catalyst layer, respectively, and $j/2F$ represents the water generation rate at the cathode due to electrochemical reaction.

For convenience (see Example 4.4), we introduce an unknown, α, which represents the ratio between the water flux across the membrane and the charge flux across the membrane:

$$\alpha = \frac{J^M_{H_2O}}{j/2F} \tag{6.9}$$

Using Equations 6.9 and 6.6, we can write Equation 6.8 in terms of J and α:

$$J^C_{H_2O} = \frac{j}{2F}(1 + \alpha) \tag{6.10}$$

Now, by combining Equations 6.6, 6.8, 6.9, and 6.10, all the fluxes in the fuel cell may be connected together through j and α:

$$\frac{j}{2F} = \frac{J^M_{H^+}}{2} = J^A_{H_2} = 2J^C_{O_2} = \frac{J^A_{H_2O}}{\alpha} = \frac{J^M_{H_2O}}{\alpha} = \frac{J^C_{H_2O}}{1 + \alpha} \tag{6.11}$$

This is the master flux balance equation for our PEMFC model. The flux balance principle captured by this equation relates to what are known as the *conservation laws*. To arrive at Equation 6.11, we have used the laws of *mass conservation*, *species conservation*, and *charge conservation*.

In an analogous manner, we can set up a flux balance equation for a SOFC as shown in Figure 6.4*b*:

$$\frac{j}{2F} = J^M_{O^{2-}} = J^A_{H_2} = 2J^C_{O_2} = -J^A_{H_2O} \tag{6.12}$$

The overall flux balance for a SOFC is simpler than that for a PEMFC since only oxygen ions (O^{2-}) transport through the electrolyte. Since a SOFC generates water at the anode, the water flux at the anode is equal to the current density. Also, the water flux at the cathode will be zero.

When we set up the governing equations for the anode, membrane, and cathode of our fuel cell models, they will all be connected by Equation 6.11 (for a PEMFC) or 6.12 (for a SOFC). Current density j is usually the known quantity in the flux balance. Solving our model equations as a function of j will provide detailed information on the oxygen concentration in the cathode catalyst layer and the water (or O^{2-}) concentration profile in the electrolyte membrane. From this information, we can calculate the activation and ohmic overvoltages for the fuel cell, allowing us to determine the operating voltage.

6.2.2   Simplifying Assumptions

Possessing a flux balance for the species in the fuel cell, it is almost time to write equations describing how the species move and interact inside the fuel cell. These equations are called governing equations. If we wanted to include all the possible processes occurring inside our fuel cell, we would have to write governing equations for all the items listed in Table 6.2. Modeling all of these different phenomena for all these different species in all these different domains would be daunting. Fortunately, by making the following simplifying assumptions, most of the items in Table 6.2 can be ignored in our current model:

1. Convective transport is ignored. Except for special cases, it is extremely difficult to obtain an analytical solution for convection. Convection is typically the dominant mass transport phenomena in fuel cells. However, since our model is a 1D model, we can safely ignore convection. As Figures 6.4 indicates, convective transport is mostly along the Y axis, but in our 1D model we consider transport only along the Z axis.

2. Diffusive transport in the flow channels is ignored. In the flow channels, diffusion is far less dominant than convection. Since we are already ignoring convection, diffusion in flow channels can be ignored too. (We will not ignore diffusion in the electrodes, however.)

3. We assume that all the ohmic losses come from the electrolyte membrane. For most fuel cells, this is a reasonable assumption, because the ohmic losses from ionic conduction in the electrolyte tend to dominate the other ohmic losses. (See Chapter 4.) This assumption means that we can ignore any conduction phenomena occurring in the electrode, catalyst layer, and flow channels.

4. We ignore the anode reaction kinetics. In H_2–O_2 fuel cells, the anode activation losses are usually much smaller than the cathode activation losses since oxygen reduction is the most sluggish process. (See Chapter 3.) We assume that the kinetic losses in our fuel cell model are determined by the oxygen concentration at the cathode catalyst layer (see the following text box).

5. We assume that the catalyst layers are extremely thin or act as "interfaces" (no thickness). With this assumption, we can ignore all convection, diffusion, and conduction processes in the catalyst layer, focusing instead only on the reaction kinetics. This is a reasonable assumption for most PEMFCs since the catalyst layer is extremely thin ($\sim 10\ \mu$m) compared to the electrode (100–350 μm). In most SOFCs, however, the catalyst layer and electrode form a single unified body. Ionic conduction and electrochemical reactions may happen throughout the entire thickness of the electrode. Usually, however, reactions are localized to a very thin region of the catalyst/electrode bordering the electrolyte. In this case, our assumption is still justified.

6. The last and fairly bold assumption we make is that water exists only as water vapor. For SOFCs, this assumption is justified; only water vapor will exist at typical SOFC operating temperatures. In PEMFCs, however, we would expect both water vapor and liquid water to be present. Unfortunately, however, it is difficult to model

TABLE 6.2. Description of Full PEMFC (or SOFC, in italics) Model

Domains	Convection	Diffusion	Conduction	Electrochemical Reaction
Anode				
Flow channels	(1) H_2, $H_2O_{(g)}$, $H_2O_{(l)}$	(2) H_2, $H_2O_{(g)}$, $H_2O_{(l)}$	(3) e^-	—
	(1) H_2, $H_2O_{(g)}$	*(2) H_2, $H_2O_{(g)}$*	*(3) e^-*	—
Electrode	(1) H_2, $H_2O_{(g)}$, $H_2O_{(l)}$	(6) H_2, $H_2O_{(g)}$, $H_2O_{(l)}$	(3) e^-	—
	(1) H_2, $H_2O_{(g)}$	*H_2, $H_2O_{(g)}$*	*(3,5) e^-, O^{2-}*	*(5) $H_2 + O^{2-} \rightarrow H_2O + 2e^-$*
Catalyst	(1) H_2, $H_2O_{(g)}$, $H_2O_{(l)}$	(5) H_2, $H_2O_{(g)}$, $H_2O_{(l)}$	(3,5) e^-, H^+	(4) $H_2 \rightarrow 2H^+ + 2e^-$
	(1) H_2, $H_2O_{(g)}$	*(5) H_2, $H_2O_{(g)}$*	*(3,5) e^-, O^{2-}*	*$H_2 + O^{2-} \rightarrow H_2O + 2e^-$*
Electrolyte	—	(6) $H_2O_{(l)}$	(6) H^+, $H_2O_{(l)}$[a]	—
	—	—	*O^{2-}*	—
Cathode				
Catalyst	(1) N_2, O_2, $H_2O_{(g)}$, $H_2O_{(l)}$	(5) N_2, O_2, $H_2O_{(g)}$, $H_2O_{(l)}$	(3,5) e^-, H^+	(6) $2H^+ + \frac{1}{2}O_2 + 2e^- \rightarrow H_2O_{(l)}$
	(1) N_2, O_2	*(5) N_2, O_2*	*(3,5) e^-, O^{2-}*	*$\frac{1}{2}O_2 + 2e^- \rightarrow O^{2-}$*
Electrode	(1) N_2, O_2, $H_2O_{(g)}$, $H_2O_{(l)}$	(6) N_2, O_2, $H_2O_{(g)}$, $H_2O_{(l)}$	(3) e^-	—
	(1) N_2, O_2	*N_2, O_2*	*(3,5) e^-, O^{2-}*	*(5) $\frac{1}{2}O_2 + 2e^- \rightarrow O^{2-}$*
Flow channels	(1) N_2, O_2, $H_2O_{(g)}$	(2) N_2, O_2, $H_2O_{(g)}$, $H_2O_{(l)}$	(3) e^-	—
	(1) N_2, O_2	*(2) N_2, O_2*	*(3) e^-*	—

Note: Six key assumptions, numbered 1–6 in parentheses, lead to the simplified model shown in Table 6.3.

[a]To be precise, this water transport phenomenon is due to electro-osmotic drag (see Chapter 4). For convenience, it has been categorized as conduction due to its close relationship with proton conduction.

SOFC STRUCTURE AFFECTS MODELING ASSUMPTIONS

In anode-supported SOFC structures, several of the modeling assumptions listed above prove problematic. Because the components in a SOFC are quite brittle, the anode electrode, the cathode electrode, or the electrolyte must be made thick enough to act as a support. Thus, three structures exist—anode supported, cathode-supported, and electrolyte-supported SOFCs. When modeling anode-supported SOFC structures, the modeling assumptions listed above cannot be used. For example, we may not ignore anodic reaction losses for anode supported SOFCs. This is because hydrogen diffusion limitations in thick anode structures can lead to severe mass transport constraints and therefore high anodic reaction losses despite fast anode reaction kinetics. The assumptions described above in the text should be used only for cathode- and electrolyte-supported SOFCs.

the combined transport of a liquid and gas mixture. (Combined liquid–gas transport models are known as *two-phase flow* models. Developing a two-phase flow model for a PEMFC is currently an area of active research.) By ignoring two-phase flow, we will introduce significant error into our PEMFC cathode water distribution results. This will affect the cathode overvoltage results, making our model less realistic. The departure from reality is most pronounced at high current density, when significant amounts of liquid water are produced at the cathode. In real fuel cells, this leads to flooding, a phenomenon that our model cannot capture.

The simplifying assumptions listed above significantly reduce our modeling requirements, as shown in Table 6.3.

6.2.3 Governing Equations

We must now assign reasonable governing equations for each domain in Table 6.3. Actually, we have already learned all the required governing equations in previous chapters. By solving these governing equations, we can determine how the concentrations of H_2, O_2, H_2O, and N_2 vary across our fuel cell (in the z direction). From these concentration profiles, we can then calculate the mass transport overvoltage η_{conc}, activation overvoltage η_{act}, and ohmic overvoltage η_{ohmic} at different current density levels j. With this information, we are then able to construct a j–V curve.

Electrode Layer. We start by writing the governing equations for the electrodes. In the electrodes, we must model diffusion processes for H_2, O_2, H_2O, and N_2. We start with a modified form of the basic diffusion model that was described by Equation 5.7:

$$J_i = \frac{-pD_{ij}^{\text{eff}}}{RT}\frac{dx_i}{dz} \tag{6.13}$$

where x_i stands for the mole fraction of species i and p is the total gas pressure (Pa) at the electrode, which satisfies $p_i = px_i$. This equation is more convenient than Equation 5.7

TABLE 6.3. Description of Simplified PEMFC (or SOFC, in italics) Model

Domains	Convection	Diffusion	Conduction	Electrochemical Reaction
Anode				
Flow channels	—	—	—	—
Electrode	—	$H_2, H_2O_{(g)}$	—	—
	—	$H_2, H_2O_{(g)}$	—	—
Catalyst	—	—	—	—
	—	—	—	$H_2 + O^{2-} \rightarrow H_2O_{(g)} + 2e^-$
Electrolyte	—	$H_2O_{(g)}$	$H^+, H_2O_{(g)}$	—
			O^{2-}	
Cathode				
Catalyst	—	—	—	$2H^+ + \frac{1}{2}O_2 + 2e^- \rightarrow H_2O_{(g)}$
	—	—	—	$\frac{1}{2}O_2 + 2e^- \rightarrow O^{2-}$
Electrode	—	$N_2, O_2, H_2O_{(g)}$	—	—
	—	N_2, O_2	—	—
Flow channels	—	—	—	—

Note: The items to be modeled in this table are described by governing equations which are developed in the next section.

because it is based on gas pressures instead of concentrations. It can be derived directly from Equation 5.7 by using the ideal gas law ($p_i = c_i RT$). Recall how the effective diffusivity D_{ij}^{eff} is obtained using Equations 5.2–5.5 based on the measured/assumed porosity of the electrode structure.

Equation 6.13 is sufficient to describe diffusion processes involving two gas species. At PEMFC cathodes, however, three gas species are typically present (N_2, O_2, and H_2O). In such cases, we need to apply a multicomponent diffusion model such as the Maxwell–Stefan equation. However, since there is no N_2 diffusion flux in fuel cells (no generation or consumption of N_2), we will simply ignore the nitrogen flux. This sacrifices model accuracy but allows us to use a simple binary diffusion model based on the oxygen and water fluxes only. Students interested in employing the more accurate multicomponent models are directed to the explanatory text box below.

DIFFUSION MODELS FOR FUEL CELLS

Binary Diffusion Model

In simple cases, the rate of diffusion is directly proportional to a gradient in concentration (as explained in Chapter 5):

$$J_i = -D_{ij} \frac{dc_i}{dx} \tag{6.14}$$

This equation is called Fick's law of binary diffusion. It works well for binary systems where only two species (i and j) are involved in diffusion. A good example of a binary system is a stream of humidified hydrogen. In a mixture of hydrogen and water vapor, the only possible diffusion processes are hydrogen diffusion (species i) in water vapor (species j) or vice versa. The *binary diffusivity D_{ij}* can be calculated using Equation 5.2. Fick's law of binary diffusion also works when species j diffuses in species i; in this case

$$J_j = -D_{ji} \frac{dc_j}{dz} \tag{6.15}$$

From the definition of the diffusion flux, the relationship $J_i + J_j = 0$ always holds, which results in $D_{ij} = D_{ji}$. (See problem 6.6.)

Maxwell–Stefan Model

Multicomponent diffusion applies when three or more species are involved in a diffusion process. At low density, multicomponent gas diffusion can be approximated by the Maxwell–Stefan equation [37]:

$$\frac{dx_i}{dz} = RT \sum_{j \neq i} \frac{x_i J_j - x_j J_i}{p D_{ij}^{\text{eff}}} \tag{6.16}$$

(continued)

This equation allows us to calculate the z-profile of a species i by summing up the effects due to the interactions with the j other species making up the mixture. In this equation, x_i and x_j stand for the mole fractions of species i and j, J_i and J_j stand for the molar fluxes of species i and j (mol/m² · s), R is the gas constant (J/mol · K), T is the temperature (K), p is the total gas pressure (atm), and D_{ij}^{eff} is the effective binary diffusivity (m²/s). Even though we do not use the Maxwell–Stefan model in our text due to mathematical complication, you may find it useful in more sophisticated models [8].

Electrolyte. Having used the diffusion equations to describe gas transport in the electrodes, we now write the governing equations for species transport in the electrolyte. The governing equation we use depends on whether we are modeling a SOFC or a PEMFC.

For SOFCs, we only need to worry about the O^{2-} flux across the electrolyte. From our flux balance Equation 6.12 we can relate the O^{2-} flux to the current density:

$$J_{O^{2-}}^M = \frac{j}{2F} \tag{6.17}$$

Then, we can determine the ohmic voltage loss from Equation 4.11:

$$\eta_{\text{ohmic}} = j(\text{ASR}_{\text{ohmic}}) = j\left(\frac{t^M}{\sigma}\right) \tag{6.18}$$

where t^M is the thickness of the electrolyte. To calculate the electrolyte conductivity σ, we use a simplified form of Equation 4.62:

$$\sigma = \frac{Ae^{-\Delta G_{\text{act}}/(RT)}}{T} \tag{6.19}$$

where $A(\text{K}/\Omega \cdot \text{cm})$ and ΔG_{act} (J/mol) are usually obtained from experiment.

For PEMFCs, we know the proton flux from Equation 6.11. In addition to the proton flux, however, we also need to consider the water flux in the electrolyte. Water causes the electrolyte conductivity to vary spatially. Therefore, we need to be able to calculate the water profile in the electrolyte. In a Nafion membrane, two water fluxes exist: back diffusion and electro-osmotic drag. Revisiting Equation 4.44, we can account for both of these fluxes, resulting in the following combined water flux balance within the membrane:

$$J_{H_2O}^M = 2n_{\text{drag}} \frac{j}{2F} \frac{\lambda}{22} - \frac{\rho_{\text{dry}}}{M_m} D_\lambda \frac{d\lambda}{dz} \tag{6.20}$$

Keep in mind that the water content λ in this equation is not constant, but a function of z [$\lambda = \lambda(z)$]. By obtaining the water profile $\lambda(z)$, we can estimate the resistance of the electrolyte. A detailed explanation and an example of this process have been provided in Section 4.5.2.

Catalyst. The governing equations for the catalyst are quite straightforward. As discussed previously, we consider only the cathode reaction kinetics. Since the oxygen partial pressure at the cathode is the dominant factor in determining the cathodic overvoltage, we can use the simplified form of the Butler–Volmer equation from Section 5.2.4:

$$\eta_{\text{cathode}} = \frac{RT}{4\alpha F} \ln \frac{j c_{O_2}^0}{j_0 c_{O_2}} \qquad (6.21)$$

Here, the 4 in the denominator represents the electron valence number for an oxygen molecule. For an ideal gas ($p = cRT$), the above equation becomes

$$\eta_{\text{cathode}} = \frac{RT}{4\alpha F} \ln \frac{j}{j_0} p^C x_{O_2} \qquad (6.22)$$

where p^C is the total pressure at the cathode and x_{O_2} is the oxygen mole fraction at the cathode catalyst layer. Note that we use atm as the unit of pressure p and the reference pressure p_0, which is 1 atm, disappears.

6.2.4 Examples

Having developed simplified governing equations for our 1D fuel cell model in the previous sections, we are now ready to introduce a few examples, showing how we can obtain $j–V$ curve predictions from our model for both a SOFC and a PEMFC.

1D SOFC Model Example. For the 1D SOFC example, we will use Figure 6.4b for our model. From Equation 6.13, we can describe H_2 and H_2O transport in the anode as

$$J_{H_2}^A = \frac{-p^A D_{H_2,H_2O}^{\text{eff}}}{RT} \frac{dx_{H_2}}{dz}$$

$$J_{H_2O}^A = \frac{-p^A D_{H_2,H_2O}^{\text{eff}}}{RT} \frac{dx_{H_2O}}{dz} \qquad (6.23)$$

Using Equation 6.12, we can relate $J_{H_2}^A$ and $J_{H_2O}^A$ to the fuel cell current density j. When we integrate Equation 6.23, however, we need to provide boundary conditions. Fortunately, we know (or can impose) the values of x_{H_2} and x_{H_2O} at the fuel cell inlet (interface "a" in Figure 6.4b.) These inlet values serve as our boundary conditions. Solving Equation 6.23 gives linear profiles for the hydrogen and water concentrations in the anode:

$$x_{H_2}(z) = x_{H_2}|_a - z \frac{jRT}{2Fp^A D_{H_2,H_2O}^{\text{eff}}}$$

$$x_{H_2O}(z) = x_{H_2O}|_a + z \frac{jRT}{2Fp^A D_{H_2,H_2O}^{\text{eff}}} \qquad (6.24)$$

Solving for the hydrogen and water concentrations at the anode–membrane interface (interface "b" in Figure 6.4b) yields

$$x_{H_2}|_b = x_{H_2}|_a - t^A \frac{jRT}{2Fp^A D^{\text{eff}}_{H_2,H_2O}}$$

$$x_{H_2O}|_b = x_{H_2O}|_a + t^A \frac{jRT}{2Fp^A D^{\text{eff}}_{H_2,H_2O}} \tag{6.25}$$

where t^A represents anode thickness. Following a similar procedure, we can also obtain the oxygen profile at the cathode:

$$x_{O_2}|_c = x_{O_2}|_d - t^C \frac{jRT}{4Fp^C D^{\text{eff}}_{O_2,N_2}} \tag{6.26}$$

Note that we ignore the nitrogen profile since the nitrogen flux is zero (nitrogen is neither produced nor consumed in the fuel cell). Having determined the oxygen concentration at the cathode catalyst layer, we can combine Equations 6.26 and 6.22 to calculate the cathode overpotential:

$$\eta_{\text{cathode}} = \frac{RT}{4\alpha F} \ln \left[\frac{j}{j_0 p^C \left\{ x_{O_2}|_d - t^C jRT / \left(4Fp^C D^{\text{eff}}_{O_2,N_2} \right) \right\}} \right] \tag{6.27}$$

Because we account for the oxygen concentration in this equation, we are effectively accounting for both the *activation losses* and the *concentration losses* at the same time. All that remains, then, is to calculate the ohmic losses. From Equations 6.18 and 6.19, we can calculate the ohmic loss as

$$\eta_{\text{ohmic}} = j(\text{ASR}_{\text{ohmic}}) = j\frac{t^M}{\sigma} = j\frac{t^M T}{A e^{-\Delta G_{\text{act}}/(RT)}} \tag{6.28}$$

Finally, we obtain the fuel cell voltage as

$$V = E_{\text{thermo}} - \eta_{\text{ohmic}} - \eta_{\text{cathode}}$$

$$= E_{\text{thermo}} - j\frac{t^M T}{A e^{-\Delta G_{\text{act}}/(RT)}} - \frac{RT}{4\alpha F} \ln \left[\frac{j}{j_0 p^C \left\{ x_{O_2}|_d - t^C [jRT/(4Fp^C D^{\text{eff}}_{O_2,N_2})] \right\}} \right] \tag{6.29}$$

where E_{thermo} is the thermodynamically predicted fuel cell voltage.

We now apply Equation 6.29 to predict the performance of a realistic SOFC. For example, consider the parameter values and conditions shown in Table 6.4. We compute the output voltage for this SOFC at a current density of 500 mA/cm^2:

$$\eta_{\text{ohmic}} = 0.5 \text{ A/cm}^2 \frac{(0.00002 \text{ m})(1073 \text{ K})}{(9 \times 10^7 \text{ S} \cdot \text{K/m}) e^{-(100 \text{ kJ/mol})/(8.314 \text{ J/mol·K} \times 1073 \text{ K})}}$$

$$= (0.5 \text{ A/cm}^2)(0.176 \text{ } \Omega \text{ cm}^2) = 0.088 \text{ V} \tag{6.30}$$

TABLE 6.4. Physical Properties of SOFC Used in Example

Physical properties	Values
Thermodynamic voltage, E_{thermo} (V)	1.0
Temperature, T (K)	1073
Hydrogen inlet mole fraction, $x_{H_2}|_a$	0.95
Oxygen inlet mole fraction, $x_{O_2}|_d$	0.21
Cathode pressure, p^C (atm)	1
Anode pressure, p^A (atm)	1
Effective hydrogen (or water) diffusivity, D_{H_2,H_2O}^{eff} (m^2/s)	1×10^{-4}
Effective oxygen diffusivity, D_{O_2,N_2}^{eff} (m^2/s)	2×10^{-5}
Transfer coefficient, α	0.5
Exchange current density, j_0 (A/cm^2)	0.1
Electrolyte constant, A(K/$\Omega \cdot$ m)	9×10^7
Electrolyte activation energy, ΔG_{act} (kJ/mol)	100
Electrolyte thickness, t^M (μm)	20
Anode thickness, t^A (μm)	50
Cathode thickness t^C (μm)	800
Gas constant, R (J/mol $\cdot$ K)	8.314
Faraday constant, F (C/mol)	96,485

$$\eta_{cathode} = \frac{(8.314 \text{ J/mol} \cdot \text{K})(1073 \text{ K})}{4 \times 0.5 \times 96485 \text{ C/mol}} \ln \left[\frac{0.5 \text{ A/cm}^2}{0.1 \text{ A/cm}^2 \cdot 1 \text{ atm}} \right. \tag{6.31}$$

$$\left. \times \frac{1}{0.210 - 0.0008 \text{ m} \frac{5000 \text{ A/m}^2 \times 8.314 \text{ J/(mol·K)} \times 1073 \text{ K}}{(4 \times 96,485 \text{ C/mol}) \times 101,325 \text{ Pa} \times 0.00002 \text{ m}^2/\text{s}}} \right]$$

$$= 0.158 \text{ V}$$

$$V = 1.0 \text{ V} - 0.088 \text{ V} - 0.158 \text{ V} = 0.754 \text{ V} \tag{6.32}$$

By iteratively following this procedure of a range of current densities, we can easily construct a complete $j-V$ curve.

1D PEMFC Model Example. Now we will explore the PEMFC model shown in Figure 6.4a. Just like in a SOFC anode, we must account for hydrogen and water in the PEMFC anode. From Equation 6.13, we obtain the model equations

$$J_{H_2}^A = \frac{-p^A D_{H_2,H_2O}^{\text{eff}}}{RT} \frac{dx_{H_2}}{dz}$$

$$J_{H_2O}^A = \frac{-p^A D_{H_2,H_2O}^{\text{eff}}}{RT} \frac{dx_{H_2O}}{dz} \tag{6.33}$$

These equations look exactly like the SOFC anode Equations 6.23. One significant and important difference, however, is that $J_{H_2O}^A$ is unknown in our PEMFC model since we do not know α in the flux balance Equation 6.11. Using this flux balance information, where α is an unknown, the above equations have the following solutions:

$$x_{H_2}(z) = x_{H_2}|_a - z \frac{jRT}{2Fp^A D_{H_2,H_2O}^{\text{eff}}} \tag{6.34}$$

$$x_{H_2O}(z) = x_{H_2O}|_a - z \frac{\alpha^* jRT}{2Fp^A D_{H_2,H_2O}^{\text{eff}}} \tag{6.35}$$

Note that we add asterisk to the unknown α to avoid confusion with the transfer coefficient (which is also represented by α). From the above equations, we can calculate the hydrogen and water concentrations at the anode–membrane interface (interface "b" in Figure 6.4a):

$$x_{H_2}|_b = x_{H_2}|_a - t^A \frac{jRT}{2Fp^A D_{H_2,H_2O}^{\text{eff}}} \tag{6.36}$$

$$x_{H_2O}|_b = x_{H_2O}|_a - t^A \frac{\alpha^* jRT}{2Fp^A D_{H_2,H_2O}^{\text{eff}}} \tag{6.37}$$

In a similar manner we can obtain the oxygen and water concentrations at the cathode–membrane interface "c":

$$x_{O_2}|_c = x_{O_2}|_d - t^C \frac{jRT}{4Fp^C D_{O_2,H_2O}^{\text{eff}}} \tag{6.38}$$

$$x_{H_2O}|_c = x_{H_2O}|_d + t^C \frac{(1+\alpha^*)jRT}{2Fp^C D_{O_2,H_2O}^{\text{eff}}} \tag{6.39}$$

As before, we have ignored the nitrogen flux to simplify the model. Similar to the anode solution, the cathode solution also contains the unknown α^*. Just like the SOFC model, once we obtain the oxygen concentration at interface "c," we can calculate the cathodic overpotential via Equation 6.27.

The biggest challenge of our PEMFC model is to find the ohmic overpotential. The critical issue is to obtain the water profile in the membrane, since the water profile lets us calculate the membrane resistance. We can obtain the water profile in the membrane along with the unknown α^* by solving the membrane water flux equation 6.20. Equations 6.37 and 6.39 serve as our boundary conditions.

The solution to Equation 6.20 has been previously worked out in Chapter 4 (Equation 4.53 in Example 4.5.2):

$$\lambda(z) = \frac{11\alpha^*}{n_{\text{drag}}^{\text{SAT}}} + C \exp\left(\frac{j M_m n_{\text{drag}}^{\text{SAT}}}{22 F \rho_{\text{dry}} D_\lambda} z \right) = \frac{11\alpha^*}{2.5}$$

$$+ C \exp\left(\frac{j \ (\text{A/cm}^2) \times 1.0 \ \text{kg/mol} \times 2.5}{22 \times 96{,}500 \ \text{C/mol} \times 0.00197 \ \text{kg/cm}^3 \times D_\lambda \ (\text{cm}^2/\text{s})} z \ (\text{cm}) \right)$$

$$= 4.4\alpha^* + C \exp\left(\frac{0.000598 \cdot j \ (\text{A/cm}^2) \cdot z \ (\text{cm})}{D_\lambda \ (\text{cm}^2/\text{s})} \right) \qquad (6.40)$$

Using this equation, we can obtain the water content λ at the anode–membrane interface "b" and the cathode–membrane interface "c" as

$$\lambda|_b = \lambda(0) = 4.4\alpha^* + C \qquad (6.41)$$

$$\lambda|_c = \lambda(t^M) = 4.4\alpha^* + C \exp\left(\frac{0.000598 \cdot j \ (\text{A/cm}^2) \cdot t^M \ (\text{cm})}{D_\lambda \ (\text{cm}^2/\text{s})} \right) \qquad (6.42)$$

where t^M represents the membrane thickness. So far we have two unknowns—C in the above equation and α^* from Equations 6.37 and 6.39. To make further progress, we need to relate the water fluxes in Equations 6.37 and 6.39 to the water contents in Equations 6.41 and 6.42.

As explained in Chapter 4.5.2, the Nafion water content is a nonlinear function of the surrounding water vapor pressure. As it is quite complicated to solve these nonlinear equations, we introduce two more simplifying assumptions.

1. Water content in the Nafion membrane increases *linearly* with water activity. Thus, we use the following linearized form of Equation 4.34:

$$\lambda = 14 a_W \qquad \text{for} \quad 0 < a_W \leq 1 \qquad (6.43)$$

$$\lambda = 12.6 + 1.4 a_W \quad \text{for} \quad 1 < a_W \leq 3 \qquad (6.44)$$

   This piecewise equation linearly approximates the real water content versus water activity behavior shown in Figure 4.11.

2. Water diffusivity in Nafion is constant. This is a fairly reasonable assumption since the water diffusivity does not change much over most water content ranges.

Since $a_W|_b = p^C x_{\text{H}_2\text{O}}|_b / p_{\text{SAT}}$, combining Equations 6.43 and 6.37 gives

$$\lambda|_b = 14 a_W|_b = 14 \frac{p^C}{p_{\text{SAT}}} \left(x_{\text{H}_2\text{O}}|_a - t^A \frac{\alpha^* j RT}{2 F p^A D_{\text{H}_2,\text{H}_2\text{O}}^{\text{eff}}} \right) \qquad (6.45)$$

Similarly, combining Equations 6.39 and 6.44 for the cathode side yields

$$\lambda|_c = 12.6 + 1.4 a_W|_c = 12.6 + 1.4 \frac{p^C}{p_{\text{SAT}}} \left(x_{\text{H}_2\text{O}}|_d + t^C \frac{(1 + \alpha^*) j R T}{2 F p^C D^{\text{eff}}_{\text{O}_2,\text{H}_2\text{O}}} \right) \qquad (6.46)$$

In the above two equations, we have assumed that $a_W < 1$ for "b" and $a_W > 1$ for "c." At "b," water is consumed to provide water flux to Nafion, and at "c," water is generated. Since water is depleted at "b" and produced at "c," the water activity assumptions are reasonable.

Using the system of equations that we have set up, we will now work a practical example. Consider the specific fuel cell properties listed in Table 6.5. Incorporating these properties into Equations 6.45 and 6.46 gives

TABLE 6.5. Physical Properties of PEMFC Used in Example

Physical properties	Values
Thermodynamic voltage, E_{thermo} (V)	1.0
Operating current density, j (A/cm^2)	0.5
Temperature, T (K)	343
Vapor saturation pressure, p_{SAT} (atm)	0.307
Hydrogen mole fraction, x_{H_2}	0.9
Oxygen mole fraction, x_{O_2}	0.19
Cathode water mole fraction, $x_{\text{H}_2\text{O}}$	0.1
Cathode pressure, p^C (atm)	3
Anode pressure, p^A (atm)	3
Effective hydrogen (or water) diffusivity, $D^{\text{eff}}_{\text{H}_2,\text{H}_2\text{O}}$ (cm^2/s)	0.149
Effective oxygen (or water) diffusivity, $D^{\text{eff}}_{\text{O}_2,\text{H}_2\text{O}}$ (cm^2/s)	0.0295
Water diffusivity in Nafion, D_λ (cm^2/s)	3.81×10^{-6}
Transfer coefficient, α	0.5
Exchange current density, j_0 (A/cm^2)	0.0001
Electrolyte thickness, t^M (μm)	125
Anode thickness, t^A (μm)	350
Cathode thickness t^C (μm)	350
Gas constant, R (J/mol K)	8.314
Faraday constant, F (C/mol)	96,485

$$\lambda|_b = 14 \frac{3 \text{ atm}}{0.307 \text{ atm}}$$

$$\times \left(0.1 - 0.00035 \text{ m} \frac{\alpha^* \times 0.5 \text{A}/0.0001 \text{ m}^2 \cdot 8.314 \text{ J/mol K} \times 343 \text{ K}}{(2 \times 96{,}485 \text{ C/mol})(3 \times 101{,}325 \text{ Pa})(0.149 \times 0.0001 \text{ m}^2/\text{s})} \right)$$

$$= 13.68 - 0.781\alpha^* \tag{6.47}$$

$$\lambda|_c = 12.6 + 1.4 \frac{3 \text{ atm}}{0.307 \text{ atm}}$$

$$\times \left(0.1 + 0.00035 \text{ m} \frac{(1 + \alpha^*) \times 0.5 \text{A}/0.0001 \text{ m}^2 \times 8.314 \text{ J/mol} \cdot \text{K} \times 343 \text{ K}}{(2 \cdot 96{,}485 \text{ C/mol})(3 \times 101{,}325 \text{ Pa})(0.0295 \times 0.0001 \text{ m}^2/\text{s})} \right)$$

$$= 14.36 + 0.394\alpha^* \tag{6.48}$$

and Equations 6.41 and 6.42 become

$$\lambda|_b = \lambda(0) = 4.4\alpha^* + C \tag{6.49}$$

$$\lambda|_c = 4.4\alpha^* + C \exp \left(\frac{0.000598 \times 0.5 \text{ A/cm}^2 \times 0.0125 \text{ cm}}{3.81 \times 10^{-6}} \right)$$

$$= 4.4\alpha^* + 2.667C \tag{6.50}$$

Now, we can equate Equation 6.47 with Equation 6.49 and Equation 6.48 with Equation 6.50 to find $\alpha = 2.25$ and $C = 2.0$.

From Equations 4.38 and and 6.40, we can then determine the conductivity profile of the membrane:

$$\sigma(z) = \left\{ 0.005193 \left[4.4\alpha + C \exp \left(\frac{0.000598 \times 0.5}{3.81 \times 10^{-6}} z \right) \right] - 0.00326 \right\}$$

$$\times \exp \left[1268 \left(\frac{1}{303} - \frac{1}{343} \right) \right]$$

$$= 0.0784 + 0.0169 \exp(78.48z) \tag{6.51}$$

Finally, we can determine the resistance of the membrane using Equation 4.40:

$$R_m = \int_0^{t_m} \frac{dz}{\sigma(z)} = \int_0^{0.0125} \frac{dz}{0.0784 + 0.0169 \exp(78.48z)}$$

$$= 0.117 \, \Omega \cdot \text{cm}^2 \tag{6.52}$$

Thus, the ohmic overvoltage due to the membrane resistance in this PEMFC is approximately

$$\eta_{\text{ohmic}} = j \times \text{ASR}_m = 0.5 \text{ A/cm}^2 \times 0.117 \, \Omega \, \text{cm}^2 = 0.0585 \text{ V} \tag{6.53}$$

and we can compute the cathodic overvoltage using Equation 6.27 as

$$\eta_{cathode} = \frac{(8.314 \text{ J/mol} \cdot \text{K})(343 \text{ K})}{4 \times 0.5 \times 96485 \text{ C/mol}} \ln \left[\frac{0.5 \text{ A/cm}^2}{0.0001 \text{ A/cm}^2 \times 3 \text{ atm}} \right.$$

$$\left. \times \frac{1}{\left(0.19 - 0.00035m \frac{5000 \text{ A/m}^2 \times 8.314 \text{ J/mol·K} \times 343 \text{ K}}{(4 \times 96,485 \text{ C/mol})(3 \times 101,325 \text{ Pa})(0.0295 \times 10^{-4} \text{ m}^3/\text{s})} \right)} \right]$$

$$= 0.135 \text{ V} \tag{6.54}$$

Finally, we find the fuel cell voltage as

$$V = 1.0 \text{ V} - 0.0585 \text{ V} - 0.135 \text{ V} = 0.806 \text{ V} \tag{6.55}$$

Gas Depletion Effects: Modifying the 1D SOFC Model. So far in our example models, we have assumed an infinite supply of hydrogen and oxygen at the fuel cell inlets. Physically, this is represented by assigning constant mole fractions for the species at boundaries "a" and "d" in Figure 6.4b. Now, however, we will consider a more realistic case where oxygen can be *depleted* at these boundaries depending on the relative rates of oxygen supply and consumption. For simplicity, we illustrate this modification with our SOFC model, although a similar modification could also be applied to the PEMFC model. Also, we consider only oxygen depletion effects. Hydrogen depletion is not considered since our model ignores the anodic overvoltage losses in the first place. At the cathode outlet (boundary "d") we may derive the expression

$$x_{O_2}|_d = \frac{J^C_{O_2,\text{outlet}}}{J^C_{O_2,\text{outlet}} + J^C_{N_2,\text{outlet}}} \tag{6.56}$$

where the denominator represents the total species flux at the fuel cell cathode outlet. This equation simply tells us that the oxygen mole fraction at the boundary is given by the ratio of the outlet oxygen flux to the total outlet gas flux. As oxygen is consumed in the fuel cell, the mole fraction of oxygen will decrease at "d." Although we fix the inlet flux values in our model, the outlet flux will change according to usage of oxygen (which in turn corresponds to the operating current density).

We will now replace $J^C_{O_2,\text{outlet}}$ and $J^C_{N_2,\text{outlet}}$ with known values. From the SOFC flux balance equation 6.12, we know that

$$J^C_{O_2,\text{outlet}} = J^C_{O_2,\text{inlet}} - J^C_{O_2} = J^C_{O_2,\text{inlet}} - \frac{j}{4F} \tag{6.57}$$

Commonly, in fuel cell operation, the oxygen inlet flux $J^C_{O_2,\text{inlet}}$ (and the hydrogen inlet flux) are regulated according to the *stoichiometric number*. The concept of a stoichiometric number is briefly introduced in the text box below. From the definition of the stoichiometric number,

$$J_{O_2,inlet}^C = \lambda_{O_2} J_{O_2}^C \tag{6.58}$$

Plugging the above equation into Equation 6.57 allows us to solve for $J_{O_2,outlet}^C$ in terms of the stoichiometric number:

$$J_{O_2,outlet}^C = (\lambda_{O_2} - 1)J_{O_2}^C = (\lambda_{O_2} - 1)\frac{j}{4F} \tag{6.59}$$

Finding $J_{N_2,outlet}^C$ is easier. Since there is no nitrogen consumption,

$$J_{N_2,outlet}^C = J_{N_2,inlet}^C = \omega J_{O_2,inlet}^C = \omega\lambda_{O_2} J_{O_2}^C = \omega\lambda_{O_2}\frac{j}{4F} \tag{6.60}$$

where ω represents the molar ratio of nitrogen to oxygen in air (typically $\omega = 0.79/0.21 = 3.76$).

Now, we plug Equations 6.59 and 6.60 into Equation 6.56 and solve for $x_{O_2}|_d$:

$$\begin{aligned}
x_{O_2}|_d &= \frac{(\lambda_{O_2} - 1)[j/(4F)]}{(\lambda_{O_2} - 1)[j/(4F)] + \omega\lambda_{O_2}[j/(4F)]} \\
&= \frac{\lambda_{O_2} - 1}{(1 + \omega)\lambda_{O_2} - 1} \tag{6.61}
\end{aligned}$$

When $\lambda_{O_2} = 1$, Equation 6.61 tells us that $x_{O_2}|_d = 0$ since all the oxygen is consumed in the fuel cell.

STOICHIOMETRIC NUMBER

As described in Section 2.5.2, it is common to operate a fuel cell at a certain *stoichiometric number* to maximize fuel cell efficiency. The stoichiometric number λ reflects the rate at which a reactant is provided to a fuel cell relative to the rate at which it is consumed. For example, $\lambda = 2$ means that twice as much reactant as needed is being provided to a fuel cell. Choosing an optimal λ is a delicate task. A large λ is wasteful, resulting in parasitic power consumption and/or lost fuel. As λ decreases toward 1, however, reactant depletion effects become more severe. Obviously, two stoichiometric numbers must be specified in fuel cells—one for hydrogen and one for oxygen. For our SOFC model, we can define the hydrogen and oxygen stoichiometric numbers based on the ratios of the inlet to consumption fluxes:

$$\lambda_{H_2} = \frac{J_{H_2,inlet}}{J_{H_2}^A} \qquad \lambda_{O_2} = \frac{J_{O_2,inlet}}{J_{O_2}^C} \tag{6.62}$$

We can incorporate gas depletion effects into our SOFC model by simply plugging Equation 6.61 into 6.29, giving us the following final model equation:

$$V = E_{\text{thermo}} - \eta_{\text{ohmic}} - \eta_{\text{cathode}}$$

$$= E_{\text{thermo}} - j\frac{t^M T}{A e^{-\Delta G_{\text{act}}/(RT)}}$$

$$-\frac{RT}{4\alpha F}\ln\left[\frac{j}{j_0 p^C\left(\dfrac{\lambda_{O_2}-1}{(1+\omega)\lambda_{O_2}-1}-t^C\dfrac{jRT}{4Fp^C D_{O_2,N_2}^{\text{eff}}}\right)}\right] \quad (6.63)$$

Using the same table of fuel cell parameters as in the previous SOFC example with $\lambda_{O_2} = 1.5$ and $j = 500$ mA/cm^2, this modified model gives:

$$\eta_{\text{cathode}} = \frac{(8.314 \text{ J/mol} \cdot \text{K})(1073 \text{ K})}{4 \cdot 0.5 \cdot 96485 \text{ C/mol}}\ln\left[\frac{0.5 \text{ A/cm}^2}{0.1 \text{ A/cm}^2 \cdot 1 \text{ atm}}\right.$$

$$\left.\times\frac{1}{\left(\dfrac{1.5-1}{(1+3.76)1.5-1}-0.0008 \text{ m}\dfrac{0.5 \text{ A/cm}^2\times8.314 \text{ J/mol·K}\times1073 \text{ K}}{(4\times96,485 \text{ C/mol})(101,325 \text{ Pa})(0.00002 \text{ m}^2/\text{s})}\right)}\right]$$

$$= 0.228 \text{ V} \quad (6.64)$$

$$V = 1.0 \text{ V} - 0.088 \text{ V} - 0.228 \text{ V} = 0.684 \text{ V} \quad (6.65)$$

Note how we obtain a much higher cathodic overvoltage compared to the first example. This is because the low λ_{O_2} value ($\lambda_{O_2} = 1.5$) causes significant gas depletion effects ($x_{O_2}|_d = 0.21$ in the first example versus $x_{O_2}|_d = 0.0814$ in the current example).

6.2.5 Additional Considerations

As additional levels of detail are introduced, fuel cell models quickly become more difficult. For the case of the 1D model, recall how we made a series of simplifying assumptions in Section 6.2.2 to keep the system manageable. By relaxing some of these assumptions, a more accurate fuel cell model can be generated. However, this accuracy comes at the cost of greatly increased complexity.

Ambitious fuel cell models may incorporate thermal or mechanical effects. Thermal fuel cell modeling is extremely difficult. Numerous heat flows must be considered, including convective heat transfer via the fuel and air streams, conductive heat transfer through the fuel cell structures, heat absorption/release from phase changes of water, entropy losses from the electrochemical reaction, and heating due to the various overvoltages. Mechanical modeling is likewise challenging.

In most cases, these issues are implemented using sophisticated computer software programs based on numerical methods. In the next section, we introduce a fuel model based on CFD which includes most of the issues we ignored earlier in this chapter.

6.3   FUEL CELL MODELS BASED ON COMPUTATIONAL FLUID DYNAMICS (OPTIONAL)

Computational fluid dynamics modeling is a broad field of research. The intricacies of the field are beyond the scope of this book. Our purpose here is to only briefly introduce the subject. In this section, we will use CFD to simulate a PEMFC with serpentine flow channels. Rather than discuss the detailed governing equations and theory behind CFD, we instead present this serpentine flow channel example to illustrate the utility, advantages, and limitations of the CFD technique. For those students interested in the details of CFD modeling, further discussion may be found in Appendix E.

Figure 6.5 shows a CFD model of our example serpentine channel fuel cell. The complex flow geometry embodied by this fuel cell would be difficult, if not impossible, to model analytically. Fortunately, it is quite amenable to computer-based numerical modeling. Referring to Figure 6.5, note that this fuel cell employs a single serpentine channel pattern for both the anode and cathode flow structures. The cathode structure (air side) is located on top and the anode structure (hydrogen side) is on the bottom. Inlet and outlet gas locations are marked on the figure. Tables 6.6 summarizes the major physical properties used in this fuel cell model.

Figure 6.6 shows the $j–V$ curve obtained from the CFD model. This $j–V$ curve does not look much different from the curves obtained by simple analytical fuel cell models. In addition to this $j–V$ curve, however, our CFD model permits us to investigate and visualize the effects of geometry. This is where the true power of CFD becomes apparent. For example, we can use our CFD model to examine the oxygen distribution across the serpentine channel pattern as shown in Figures 6.7 and 6.8. Figure 6.7 shows a cross-sectional cut across the center of the serpentine pattern. The cathode side is on the top. As the air is introduced from the inlet on the left and travels to the outlet on the right, note how the oxygen concentration gradually drops. As a result, fuel cell performance is inhomo-

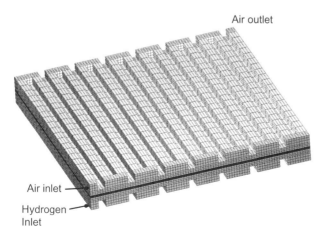

Figure 6.5. Isometric view of serpentine flow channel fuel cell model (500 μm channel feature size). Since no repetitive unit exists, the entire physical domain is modeled.

TABLE 6.6. Physical Properties Used in CFD Fuel Cell Model

Properties	Values
Fuel cell area	14×14 mm
Electrode thickness, t_g	0.25 mm
Catalyst thickness, t_c	0.05 mm
Membrane thickness, t_m	0.125 mm
Flow channel width, w_f	0.5 mm
Flow channel height, t_f	0.5 mm
Rib width, w_r	0.5 mm
Relative humidity of inlet gases	100%
Temperature, T	50°C
Hydrogen inlet flow rate	1.8 A/cm^2 equivalent
Air inlet flow rate	1.9 A/cm^2 equivalent
Outlet pressure	1 atm

Note: The gas flow rates are expressed in terms of equivalent current density.

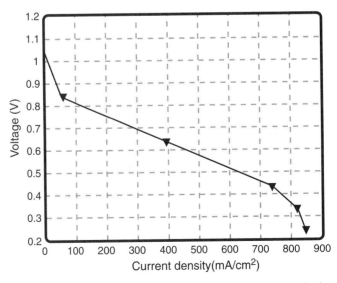

Figure 6.6. Cell j–V curves for serpentine flow channel model. Activation, ohmic, and concentration losses are clearly observed.

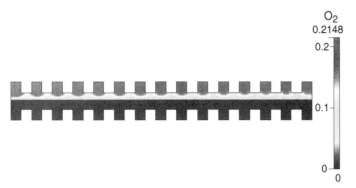

Figure 6.7. Oxygen concentration in cathode at 0.8 V overvoltage. This cross-sectional cut across the center of the serpentine pattern illustrates how the oxygen concentration in the flow channel slowly decreases from inlet to outlet.

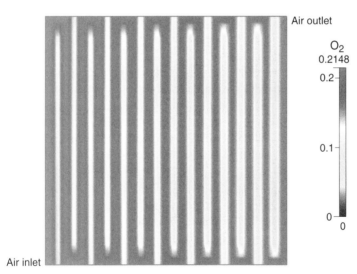

Figure 6.8. Oxygen concentration in cathode at 0.8 V overvoltage. The plan view shows the oxygen concentration profile across the cathode surface. Low oxygen concentration is observed under the channel ribs due to the blockage of oxygen flux.

geneous. Less current is produced near the outlet as the oxygen stream becomes depleted. Figure 6.8 illustrates how the channel rib structures also cause oxygen depletion. The channel ribs block the diffusion flux, leading to local "dead zones." Our CFD model provides performance enhancement hints. For example, a multichannel design and/or narrower ribs might alleviate the oxygen depletion problems.

In a 1D or 2D fuel cell model, these geometric effects are difficult to observe. The visualization tools provided by CFD modeling provide a highly intuitive way to understand and explore geometric effects in fuel cells. CFD models are especially useful when experimental investigation is difficult or impractical. When used in combination with experiment, CFD models can add significant speed and power to the fuel cell design process.

CHAPTER SUMMARY

Fuel cell models are used to understand and predict fuel cell behavior. Simple models can be used to understand basic trends (e.g., what happens when temperature increases or pressure decreases). Sophisticated models can be used as design guides (e.g., what happens when the diffusion layer thickness is reduced from 500 to 100 μm). All fuel cell models incorporate *assumptions*. When interpreting model results, major assumptions and limitations must be taken into account.

- There are three major fuel cell losses: activation losses (η_{act}), ohmic losses (η_{ohmic}), and concentration losses (η_{act}).
- A simple fuel cell model can be developed by starting with the thermodynamic fuel cell voltage and then deducting the three major loss terms: $V = E_{thermo} - \eta_{act} - \eta_{ohmic} - \eta_{conc}$.
- To accurately reflect the behavior of most fuel cells, an additional loss term, known as the leakage loss j_{leak}, must be introduced.
- The leakage loss j_{leak} is associated with the parasitic loss due to current leakage, gas crossover, unwanted side reaction, and so on. The net effect of this parasitic current loss is to offset a fuel cell's operating current to the left by an amount given by j_{leak}. This has the effect of reducing a fuel cell's open-circuit voltage below the thermodynamically predicted value.
- The basic fuel cell model requires four parameters. Two parameters (α and j_0) describe the kinetic losses, one parameter (ASR_{ohmic}) describes the ohmic losses, and one parameter (j_L) describes the concentration losses.
- A wide variety of different fuel cell behaviors can be explored by varying only a few basic parameters.
- All models include assumptions. The number and type of assumptions determine the complexity and accuracy of the model.
- More sophisticated fuel cell models use conservation laws and governing equations to relate fuel cell behavior to basic physical principles.
- The governing equations of a fuel cell model are related to one another by flux balance and conservation laws. Proper boundary conditions are required to generate solutions.
- In a SOFC, proper model assumptions can be significantly impacted by the electrode and electrolyte geometry.
- In PEMFCs, proper modeling of water distribution is critical.
- The CFD fuel cell models use numerical methods to simulate fuel cell behavior. Computational fluid dynamics modeling permits detailed investigation and visualization of electrochemical and transport phenomena. It is especially useful when experimental

investigation is difficult or impractical and illustrates tremendous promise and power as a fuel cell design tool.

CHAPTER EXERCISES

Review Questions

6.1   Match the following five scenarios to the five corresponding hypothetical j–V curves in Figure 6.9:

   (a)   A SOFC limited by an extremely high electrolyte resistance

   (b)   A PEMFC suffering from a large leakage current loss

   (c)   A PEMFC severely limited by poor reaction kinetics

   (d)   A PEMFC with an extremely low ohmic resistance

   (e)   A SOFC suffering from reactant starvation

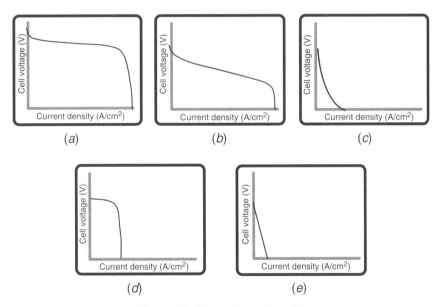

Figure 6.9.   Curves for problem 6.1.

6.2   From an efficiency standpoint, which fuel cell in Figure 6.3 would be more desirable, the PEMFC or the SOFC?

6.3   **(a)**   The thickness of the catalytically active area in a SOFC electrode is determined by a subtle balance between ionic resistance, electronic resistance, gas transport resistance, and charge transfer resistance. Answer the following questions:

1. When the gas transport resistance of the electrode increases (due to, e.g., smaller electrode pores and a thicker electrode), does the catalytically active area (a) become thicker, (b) become thinner, or (c) is it unaffected? Why?

2. When the ionic resistance of the electrode increases, does the catalytically active area (a) become thicker, (b) become thinner, or (c) is it unaffected? Why?

3. When the electronic resistance of the electrode increases, does the catalytically active area (a) become thicker, (b) become thinner, or (c) is it unaffected? Why?

4. When the charge transfer resistance of the electrode increases, does the catalytically active area (a) become thicker, (b) become thinner, or (c) is it unaffected? Why?

(b) Based on the observation of typical SOFC j–V curves, discuss which one of the resistances mentioned above usually dominates and therefore determines the thickness of the catalytically active area.

(c) Most PEMFCs are designed with a thicker catalyst layer on the cathode side. Why?

Calculations

6.4 This problem estimates the effect of j_{leak} on the open-circuit voltage of a fuel cell. Assume a simple fuel cell model which depends only on the activation losses at the cathode (i.e., do not include the effects of ohmic or concentration losses). For a typical pure H_2–O_2 PEMFC cathode, $j_0 \approx 10^{-3}$ and $\alpha \approx 0.5$. Using these values, determine the approximate drop in open-circuit voltage caused by a leakage current $j_{leak} = 10$ mA/cm^2 (assume STP). *Hint:* To solve this question properly, carefully consider which approximation of the Butler–Volmer equation you should use. Cross-check your final answer with the approximation assumptions.

6.5 This problem has several parts. By following each part, you will develop a simple fuel cell model similar to the one discussed in the text.

(a) Calculate E_{thermo} for a PEMFC running on atmospheric pressure H_2 and atmospheric air at 330 K.

(b) Calculate a_c and b_c (the constants for the natural log form of the Tafel equation for the cathode of this PEMFC) if $j_0 = 10^{-3}$ A/cm^2, $n = 2$, and $\alpha = 0.5$.

(c) Calculate ASR$_{ohmic}$ if the membrane has a conductivity of 0.1 $\Omega^{-1} \cdot$ cm^{-1} and a thickness of 100 μm. Assume there are no other contributions to cell resistance.

(d) Calculate the effective binary diffusion coefficient for O_2 in air in the cathode electrode. Neglect the effect of water vapor (consider only O_2 and N_2) and assume the cathode electrode has a porosity of 20%.

(e) Calculate the limiting current density in the cathode given $\delta = 500$ μm.

(f) To complete your model, assume c (the geometric constant in the concentration loss equation) has a value of 0.10 V. Assume $j_{leak} = 5$ mA/cm^2. Neglect all anode effects. Using some type of software package, plot the j–V and power density curves for your model.

(g) What is the maximum power density for your simulated fuel cell? At what current density does the power density maximum occur?

(h) Assuming 90% fuel utilization, what is the total efficiency of your simulated fuel cell at the maximum power density point?

6.6 Show that $D_{ij} = D_{ji}$ using the fact that $J_i + J_j = 0$ and $x_i + x_j = 1$. *Hint:* Use the equation $J_i = \rho D_{ij}(dx_i/dz)$.

6.7 Show that the Maxwell–Stefan equation 6.16 satisfies $x_1 + x_2 + \cdots + x_N = 1$.

6.8 **(a)** Plot the complete j–V curve for the 1D SOFC model example (without the gas depletion modification) in the text (Section 6.2.4.1).

(b) Plot the ohmic overvoltage and cathodic overvoltage versus current density. Find the limiting current density from the j–V curve.

6.9 **(a)** Plot the complete j–V curve for the 1D SOFC model example in the text assuming that all the properties are unchanged as shown in Table 6.4 except that the operating temperature is now 873 K.

(b) Plot the ohmic overvoltage and the cathodic overvoltage versus current density. Compare your results with problem 6.8. Which overvoltage (ohmic or cathodic) shows a larger change?

6.10 **(a)** Using the 1D SOFC model, plot the j–V curve of an *electrolyte-supported* SOFC that has a 200-μm-thick electrolyte, a 50-μm-thick cathode, and a 50-μm-thick anode. Ignore the anodic overpotential and use the properties provided in Table 6.4.

(b) Repeat the process in (a) assuming that the fuel cell operating temperature is 873 K. Explain why an electrolyte-supported SOFC may not be suitable for lower temperature operation.

6.11 In the text, our 1D SOFC model did not incorporate anodic overvoltage. In this problem, we consider it.

(a) Using a linear approximation of Butler–Volmer equations for the anode as

$$j = j_0 \frac{p}{p_0} \frac{2\alpha F}{RT} \eta_{\text{act}} \tag{6.66}$$

show that the anodic overvoltage can be modeled as

$$\eta_{\text{anode}} = \frac{RT}{2\alpha F} \frac{j}{j_0 p^A \left(x_{\text{H}_2} \Big|_a - T^A \frac{jRT}{2Fp^A D_{\text{H}_2,\text{H}_2\text{O}}^{\text{eff}}} \right)} \tag{6.67}$$

(b) Based on the information from Tables 6.1 and 6.4, plot the anodic and cathodic overvoltages for this model SOFC.

6.12 **(a)** Plot the j–V curve for an *anode supported* SOFC that has a 1000-μm-thick anode and a 50-μm-thick cathode. Consider both the anodic and the cathodic overvoltages by using Equations 6.66 and 6.67. Use the properties provided in Table 6.4.

  (b) Plot the anodic overvoltage and cathodic overvoltage for this fuel cell.

  (c) Find the limiting current for each overvoltage curve. Which electrode shows more loss? Explain the consequence of ignoring the anodic overpotential in an anode-supported SOFC.

6.13 **(a)** Plot the complete j–V curve for the 1D PEMFC model example from the text (Section 6.2.4.2).

  (b) Plot the ohmic overvoltage versus current density. Is the curve linear? If not, explain why.

6.14 **(a)** Plot the complete j–V curve for the final 1D SOFC example in the text, where oxygen gas depletion effects are considered. Assume the oxygen stoichiometric number is 1.2.

  (b) Assume this fuel cell employs an air pump which consumes 10% of the fuel cell power to deliver an oxygen stoichiometric number of 1.2. When the stoichiometric number is set to 2.0, the pump consumes 20% of fuel cell power. Ignore all other sources of parasitic load. Which operation mode provides more power? Discuss your answer by carefully calculating the power density curves for each of the two operating modes.

CHAPTER 7

FUEL CELL CHARACTERIZATION

Characterization techniques permit the quantitative comparison of fuel cell systems, distinguishing good fuel cell designs from poor ones. The most effective characterization techniques also indicate *why* a fuel cell performs well or poorly. Answering these "why" questions requires sophisticated testing techniques that can pinpoint performance bottlenecks. In other words, the best characterization techniques discriminate between the various sources of loss within a fuel cell: fuel crossover, activation, ohmic, and concentration losses.

As mentioned in previous chapters, in situ testing is critically necessary. Usually, the performance of a fuel cell system cannot be determined simply by summing the performance of its individual components. Besides the losses due to the components themselves, the interfaces between components often contribute significantly to the total losses in a fuel cell system. Therefore it is important to characterize all aspects of a fuel cell while it is assembled and running under realistic operating conditions.

In this chapter, the most popular and effective fuel cell characterization techniques are introduced and discussed. We focus on in situ electrical characterization techniques because these techniques provide a wealth of information about operational fuel cell behavior. In spite of our emphasis on in situ testing, there are many useful ex situ characterization techniques that can supplement or accentuate the information provided by in situ testing. Therefore, some of these techniques are also discussed.

7.1 WHAT DO WE WANT TO CHARACTERIZE?

We start this chapter with a list of the various fuel cell properties we might want to characterize:

- Overall performance (i–V curve, power density)
- Kinetic properties (η_{act}, j_0, α, electrochemically active surface area)
- Ohmic properties (R_{ohmic}, electrolyte conductivity, contact resistances, electrode resistances, interconnect resistances)
- Mass transport properties (j_L, D_{eff}, pressure losses, reactant/product homogeneity)
- Parasitic losses (j_{leak}, side reactions, fuel crossover)
- Electrode structure (porosity, tortuosity, conductivity)
- Catalyst structure (thickness, porosity, catalyst loading, particle size, electrochemically active surface area, catalyst utilization, triple-phase boundaries, ionic conductivity, electrical conductivity)
- Flow structure (pressure drop, gas distribution, conductivity)
- Heat generation/heat balance
- Lifetime issues (lifetime testing, degradation, cycling, startup/shutdown, failure, corrosion, fatigue)

This list is certainly not comprehensive. Nevertheless, it gives a sense of the literally dozens, if not hundreds, of properties, effects, and issues that contribute to the overall performance and behavior of a fuel cell. Some of these play just a minor role, while others can have a huge effect. How do we know on which properties to focus? Which ones are most important to characterize? Essentially, the answers to these questions depend on your interests, your goals, and your desired level of detail.

In this chapter, we will focus our efforts on just a few of the most widely used characterization techniques. We organize our goals with a reminder of the two main reasons to characterize fuel cells:

1. To separate good fuel cells from bad fuel cells
2. To understand why a given fuel cell preforms the way it does

Separating the good from the bad is fairly straightforward. This separation is usually obtained by measuring j–V performance; the fuel cell that delivers the highest voltage at the current density of interest wins. Of course, fuel cell j–V performance can change dramatically depending on factors like the operating conditions and testing procedures. To ensure that j–V performance comparisons are fair, identical operating conditions, testing procedures, and device histories must be applied. In addition, j–V performance is the ultimate "acid test" for new fuel cell innovations. For example, say you develop a marvelous new ultra high-conductivity electrolyte or an incredible new fuel cell catalyst. This is great— but until you put your material into a working fuel cell and show that it delivers high performance, the scientific community will reserve their applause.

It is considerably more difficult to understand *why* a given fuel cell performs the way it does. Generally, the best way to tackle this problem is to think about a fuel cell's performance in terms of the various major loss categories: activation loss, ohmic loss, concentration loss, and leakage loss. If we can somehow determine the relative sizes of each of these losses, then we are close to understanding our fuel cell's problems. For example, if we find that concentration losses are killing performance, then a redesigned flow structure might solve the problem. In another instance, testing may reveal that our fuel cell has an abnormally large ohmic resistance. In this case, we probably want to check the electrolyte, the electrical contacts, the conductive coatings, or the interconnects.

As these examples illustrate, diagnostic fuel cell testing needs to be able to separate the various fuel cell losses, η_{act}, η_{ohmic}, and η_{conc}. In the ideal case, characterization techniques should even determine the underlying fundamental properties of the fuel cell, such as j_0, α, $\sigma_{electrolyte}$, and D_{eff}.

In the next several sections, we work toward this characterization goal. Starting with basic fuel cell tests that give overall quantitative information about fuel cell performance, we then move to more sophisticated characterization techniques that distinguish between various fuel cell losses. With refinement and care, some of these tests can even be used to determine such fundamental properties as j_0 or D_{eff}.

7.2 OVERVIEW OF CHARACTERIZATION TECHNIQUES

We divide fuel cell characterization techniques into two types:

1. **Electrochemical Characterization Techniques (In Situ).** These techniques use the electrochemical variables of voltage, current, and time to characterize the performance of fuel cell devices under operating conditions.

2. **Ex Situ Characterization Techniques.** These techniques characterize the detailed structure or properties of the individual components comprising the fuel cell, but generally only components removed from the fuel cell environment in a nonassembled, nonfunctional form.

Within the area of in situ electrochemical characterization, we discuss four major methods:

1. **Current–Voltage (j–V) Measurement.** The most ubiquitous fuel cell characterization technique, a j–V measurement provides an overall quantitative evaluation of fuel cell performance and fuel cell power density.

2. **Current Interrupt Measurement.** This method separates the contributions to fuel cell performance into ohmic and nonohmic processes. Versatile, straightforward, and fast, current interrupt can be used even for high-power fuel cell systems and is easily implemented in parallel with j–V curve measurements.

3. **Electrochemical Impedance Spectroscopy (EIS).** This more sophisticated technique can distinguish between ohmic, activation, and concentration losses. However, the results may be difficult to interpret. In addition, EIS is relatively time consuming, and it is difficult to implement for high-power fuel cell systems.

4. **Cyclic Voltammetry (CV).** This is another sophisticated technique that provides insight into fuel cell reaction kinetics. Like EIS, CV can be time consuming and results may be difficult to interpret. It may require specialized modification of the fuel cell under test and/or use of additional test gases such as argon or nitrogen.

In the area of ex situ characterization, we discuss the following methods:

1. **Porosity Determination.** Effective fuel cell electrode and catalyst structures must have high porosity. Several characterization techniques determine the porosity of sample structures, although many of them are destructive tests. More sophisticated techniques even produce approximate pore size distributions.

2. **Brunauer–Emmett–Teller (BET) Surface Area Measurement.** Fuel cell performance critically depends on the use of extremely high surface area catalysts. Some electrochemical techniques yield approximate surface area values; however, the BET method allows highly accurate ex situ surface area determinations for virtually any type of sample.

3. **Gas Permeability.** Even highly porous fuel cell electrodes may not be very gas-permeable if the pores do not lead anywhere. Understanding mass transport in fuel cell electrodes, therefore, requires permeability measurements in addition to porosity determination. While fuel cell electrodes and catalyst layers should be highly permeable, electrolytes should be gas tight. Gas permeability testing of electrolytes is critical to the development of ultrathin membranes, where gas leaks can prove catastrophic.

4. **Structure Determinations.** A wide variety of microscopy and diffraction techniques are used to investigate the structure of fuel cell materials. By structure, we mean grain size, crystal structure, orientation, morphology, and so on. This determination is especially critical when new catalysts, electrodes, or electrolytes are being developed or when new processing methods are used.

5. **Chemical Determinations.** In addition to characterizing physical structure, characterizing the chemical composition of fuel cell materials is also critical. Fortunately, many techniques are available for chemical composition and analysis. Often, the hardest part is deciding which is best for a given situation.

7.3 IN SITU ELECTROCHEMICAL CHARACTERIZATION TECHNIQUES

In the following section, we detail the most commonly used in situ electrochemical characterization techniques. All in situ electrochemical fuel cell characterization techniques rely on the measurement of current and voltage. Of course, these tests often involve the variation of other variables besides current and voltage. For example, we may want to vary temperature, gas pressure, gas flow rate, or humidity. In all these cases we are trying to answer the following question: What effect does a given variable have on fuel cell current and voltage? Current and voltage are the "end indicators" of fuel cell performance.

7.3.1 Fundamental Electrochemical Variables: Voltage, Current, and Time

In an electrochemical experiment, the three fundamental variables are voltage (V), current (i), and time (t). We can measure or control the voltage of our system, we can measure or control the current of our system, and we can do either as a function of time.

That's it. From an electrical characterization standpoint, there is nothing else we can do. Furthermore, since current and voltage are intimately related in a fuel cell, *we cannot independently vary both of them at the same time*. If we choose to control voltage, then the electrochemistry of our system sets the current. If we instead choose to control current, then the electrochemistry of our system sets the voltage. Because of this interdependence between current and voltage, there are really only two fundamental types of electrochemical characterization techniques: *potentiostatic* techniques and *galvanostatic* techniques:

1. **Potentiostatic Techniques.** The *voltage* of a system is controlled by the user and the *resulting current* response is measured. "*Static*" is an unfortunate historical misnomer. Potentiostatic techniques can either be steady state (where the control voltage is constant in time) or dynamic (where the control voltage varies with time).
2. **Galvanostatic Techniques.** The *current* of a system is controlled by the user and the *resulting voltage* response is measured. Galvanostatic techniques can also be steady state (where the control current is constant in time) or dynamic (where the control current varies with time).

Both potentiostatic and galvanostatic techniques can be applied to fuel cells. For example, fuel cell $j–V$ curves are generally acquired using steady-state potentiostatic or galvanostatic measurements. In fact, at steady state, it does not matter whether a potentiostatic or galvanostatic measurement is used to record a fuel cell's $j–V$ curve—the measurements represent two sides of the same coin. In the steady state condition, a potentiostatic and a galvanostatic measurement of a system made at the same point will yield the identical result. In other words, if a steady-state galvanostatic measurement of a fuel cell yields 0.5 V at an imposed current of 1.0 A, then the steady-state potentiostatic measurement of the same fuel cell should yield a current of 1.0 A at an imposed voltage of 0.5 V.

For short time periods or under non-steady-state conditions, potentiostatic and galvanostatic measurements may deviate from one another. Often, this deviation is because a system has not had enough time to *relax* to its steady-state condition. Actually, deviations from the steady state due to slow relaxation processes can be exploited to help understand fuel cell behavior. This is where the more sophisticated dynamic techniques enter in. One technique which exploits the dynamic behavior of a fuel cell is known as the current interrupt measurement. We will briefly contrast the difference between a true steady-state $j–V$ measurement and a current interrupt measurement:

- **Steady-State $j–V$ Measurement.** The current of the fuel cell is held fixed in time and the *steady-state value* of the fuel cell voltage is recorded after a long equilibration time. Or, the voltage of the fuel cell is held fixed in time and the *steady-state value* of the fuel cell current is recorded after a long equilibration time.
- **Current Interrupt Measurement.** A current is abruptly imposed (or withdrawn) at time $t = 0$ and the system voltage resulting *time-dependent approach* to steady state is measured.

While time-invariant techniques can give useful information about the steady-state properties of fuel cells, it is the dynamic (time-variant) techniques that give truly powerful insight into the various loss components that contribute to performance. In addition to current interrupt, two other powerful dynamic techniques, cyclic voltammetry and electrochemical impedance spectroscopy, are also detailed in this chapter. We briefly compare these two dynamic techniques:

- **Cyclic Voltammetry.** In this dynamic technique, the voltage applied to a system is swept linearly with time back and forth across a voltage window of interest. The resulting cyclic current response is measured as a function of time but is plotted as a function of the cyclic voltage sweep.
- **Electrochemical Impedance Spectroscopy.** In this dynamic technique a sinusoidal perturbation (usually a voltage perturbation) is applied to a system and the amplitude and phase shift of the resulting current response are measured. Measurements can be conducted over a wide range of frequencies, resulting in the construction of an *impedance spectrum.*

All of these techniques require a basic fuel cell testing platform and some standard electrochemical measurement equipment. Therefore, before going into further detail on the techniques themselves, we will take a brief look at the basic fuel cell test station requirements.

7.3.2 Basic Fuel Cell Test Station Requirements

Figure 7.1 presents a diagram of a basic test station used for in situ fuel cell characterization measurements. This diagram is specifically for a PEMFC, but a similar setup could be implemented for any type of fuel cell. Since fuel cell performance strongly depends on the

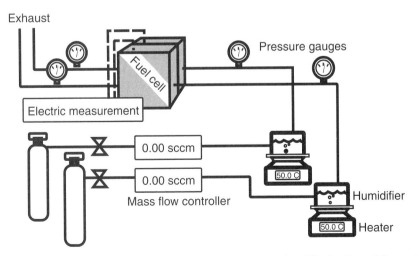

Figure 7.1. Typical fuel cell test station. Pressures, temperatures, humidity levels, and flow rates of gases are controlled.

operating conditions, a good test setup allows flexible control over the operating pressures, temperatures, humidity levels, and flow rates of the reactant gases.

Mass flow controllers, pressure gauges, and temperature sensors allow the operating conditions of the fuel cell to be continually monitored during testing. Electrochemical measurement equipment, usually including a potentiostat/galvanostat and an impedance analyzer, is attached to the fuel cell. These measurement devices have at least two leads; one connects to the fuel cell cathode while the other connects to the fuel cell anode. Often a third lead is provided for a reference electrode. Most commercially available potentiostats can perform a wide range of potentiostatic/galvanostatic experiments, including j–V curve measurements, current interrupt, and cyclic voltammetry. Electrochemical impedance spectroscopy often requires a dedicated impedance analyzer or an add-on unit in addition to the potentiostat.

With a complete fuel cell test station like the one shown in Figure 7.1, there are literally dozens of possible characterization experiments that can be conducted. One of the first measurements you will probably want to take is a j–V curve.

7.3.3 Current–Voltage Measurement

As previously introduced, the performance of a fuel cell is best summarized by its current–voltage response, or j–V curve (recall Figure 1.9). The j–V curve shows the voltage output of the fuel cell for a given current density loading. High-performance fuel cells will exhibit less loss and therefore a higher voltage for a given current load. Fuel cell j–V curves are usually measured with a potentiostat/galvanostat system. This system draws a fixed current from the fuel cell and measures the corresponding output voltage. By slowly stepping the current demand, the entire j–V response of the fuel cell can be determined.

In taking j–V curve measurements of fuel cells, the following important points must be considered:

- Steady state must be ensured.
- The test conditions should be noted.

These points will now be addressed.

Steady State. Reliable j–V curve measurements require a steady-state system. Steady state means that the voltage and current readings do not change with time. When current is demanded from a fuel cell, the voltage of the cell drops to reflect the higher losses associated with producing current. However, this voltage drop is not instantaneous. Instead, it can take seconds, minutes, or even hours for the voltage to relax all the way to a steady-state value. This delay is due to subtle changes, such as temperature changes and reactant concentration changes that take time to propagate through the fuel cell. Usually, the larger the fuel cell, the slower the approach to steady state. It is not unusual for a large automotive or residential fuel cell stack to require 30 min to reach steady state after an abrupt current or voltage change. Current or voltage measurements recorded before a fuel cell reaches steady state will be artificially high or artificially low.

For large fuel cell systems, $j-V$ curve testing can be a tedious, time-consuming process. Often, measurements are made galvanostatically; the fuel cell is subjected to a given current load and the voltage response is monitored until it no longer changes significantly in time. This voltage is recorded. Then, the current load is increased to a new predetermined value and the procedure is repeated. Frequently, time constraints only permit 10–20 points along the fuel cell's $j-V$ curve to be acquired. While the data are coarse, it is generally sufficient to outline the fuel cell's performance.

For small fuel cell systems, slow-scan $j-V$ curve measurements can be acquired. In a slow-scan galvanostatic measurement, the current demanded from the fuel cell is *gradually* scanned in time from zero to some predetermined limit. The voltage of the fuel cell will continuously drop as the current is ramped. The resulting graph of current versus voltage represents a pseudo-steady-state version of the fuel cell's $j-V$ curve *if the current scan is slow enough*. The question is, how does one know if the current scan is sufficiently slow? The answer is found by conducting a series of $j-V$ measurements at several different scan speeds. If the scan speed is too fast, the $j-V$ curve will be artificially high. If decreasing the scan speed no longer affects the $j-V$ curve, the speed is sufficiently slow.

Test Conditions. Test conditions will dramatically affect fuel cell performance. Therefore, care must be taken to fully document measurement operating conditions, testing procedures, device histories, and so on. A "bad" PEMFC operating at 80°C on humidified oxygen and hydrogen gases under 5 atm pressure may show better $j-V$ curve performance than a "good" PEMFC operating at 30°C on dry air and hydrogen at atmospheric pressure. However, if the two fuel cells are tested under identical conditions, the truly good fuel cell will become apparent.

The most important testing conditions to document are now briefly discussed:

- **Warm-up.** To ensure that a fuel cell system is well equilibrated, it is customary to conduct a standardized warm-up procedure prior to cell characterization. A typical warm-up procedure might involve operating the cell at a fixed current load for 30–60 min prior to testing. Failure to properly warm up a fuel cell system can result in highly nonstationary (non-steady-state) behavior.

- **Temperature.** It is important to document and maintain a constant fuel cell temperature during measurement. Both the gas inlet and exit temperatures should be measured as well as the temperature of the fuel cell itself. Sophisticated techniques even allow temperature distributions across a fuel cell device to be monitored in real time. In general, increased temperature will improve performance due to improved kinetics and conduction processes. (For PEMFCs, this is only true up to about 80°C, above which membrane drying becomes an issue.)

- **Pressure.** Gas pressures are generally monitored at both the fuel cell inlets and outlets. This allows the internal pressure of the fuel cell to be determined as well as the pressure drop within the cell. Increased cell pressure will improve performance. (However, increasing the pressure requires additional energy "input" from compressors, fans, etc.)

- **Flow Rate.** Flow rates are generally set using mass flow controllers. During a $j-V$ test, there are two main ways to handle reactant flow rates. In the first method, flow rates are held constant during the entire test at a flow rate that is sufficiently high so that even at the largest current densities there is sufficient supply. This method is known as the *fixed-flow-rate condition*. In the second method, flow rates are adjusted stoichiometrically with the current so that the ratio between reactant supply and current consumption is always fixed. This method is known as the *fixed stoichiometry condition*. Fair *j–V* curve comparisons should be done using the same flow-rate method. Increased flow usually improves performance. (For PEMFCs, increasing the flow rate of extremely humid or extremely dry gases can upset the water balance in the fuel cell and actually decrease performance.)

- **Compression Force.** For most fuel cell assemblies there is an optimal cell compression force which leads to best performance; thus cell compression force should be noted and monitored. Cells with lower compression forces can suffer increased ohmic loss, while cells with higher compression forces can suffer increased pressure or concentration losses.

Interpreting j–V Curve Measurements. Generally, *j–V* curve measurements are used to quantitatively describe the overall performance of a fuel cell system. At first glance, it appears impossible to individually separate the various loss contributions (e.g., activation, ohmic, concentration losses) from the *j–V* curve. Nevertheless, careful data analysis permits approximate activation losses to be isolated using the Tafel equation.

At low current densities, the ohmic loss is usually small compared to the activation loss. Thus, the ohmic loss can be ignored and the approximate activation loss can be calculated directly from the data. If plotted on a log scale, the low-current-density *j–V* curve regimen shows *linear* behavior, as expected from the Tafel equation (3.41). The transfer coefficient and the exchange current density can be obtained by fitting a line through the data. The line can be extended throughout the *j–V* curve, allowing the approximate activation loss contribution to be identified at each current density. Figure 7.2 briefly illustrates the process.

7.3.4 Electrochemical Impedance Spectroscopy

While the *j–V* curve provides general quantification of fuel cell performance, a more sophisticated test is required to accurately differentiate between all the major sources of loss in a fuel cell. Electrochemical impedance spectroscopy is the most widely used technique for distinguishing the different losses.

EIS Basics. Like resistance, impedance is a measure of the ability of a system to *impede* the flow of electrical current. Unlike resistance, impedance can deal with time- or frequency-dependent phenomena. Recall how we define resistance R from Ohm's law as the ratio between voltage and current:

$$R = \frac{V}{i} \tag{7.1}$$

In an analogous manner, impedance Z is given by the ratio between a time-dependent voltage and and a time-dependent current:

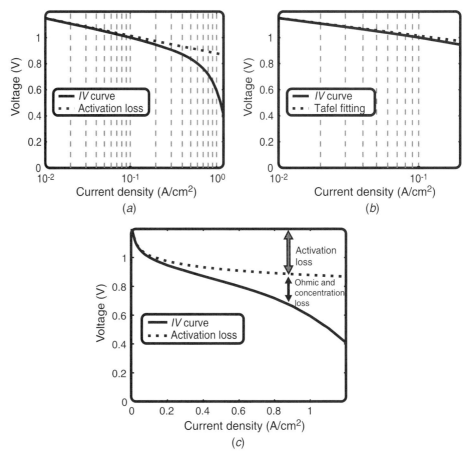

Figure 7.2. (*a*) Typical log-scaled j–V curve. The activation loss contribution is plotted by the dashed line. (*b*) The low-current-density regimen of the j–V curve shows linear behavior on a log scale. Fitting this line to the Tafel equation gives the transfer coefficient and the exchange current density. (*c*) Activation loss is plotted throughout the j–V curve. The difference between the activation loss and the j–V curve represents the sum of ohmic and concentration losses

$$Z = \frac{V(t)}{i(t)} \tag{7.2}$$

Impedance measurements are usually made by applying a small sinusoidal voltage perturbation, $V(t) = V_0 \cos(wt)$, and monitoring the system's resultant current response, $i(t) = i_0 \cos(wt - \phi)$. In these expressions, $V(t)$ and $i(t)$ are the potential and current at time t, V_0 and i_0 are the amplitudes of the voltage and current signals, and w is the radial frequency. The relationship between radial frequency w (expressed in radians per second) and frequency f (expressed in hertz) is

$$w = 2\pi f \tag{7.3}$$

In general, the current response of a system may be shifted in phase compared to the voltage perturbation. This phase shift effect is described by ϕ. A graphical representation of the relationship between a sinusoidal voltage perturbation and a phase-shifted current response is shown in Figure 7.3 (for a linear system).

Following Equation 7.2 we can write the sinusoidal impedance response of a system as

$$Z = \frac{V_0 \cos(wt)}{i_0 \cos(wt - \phi)} = Z_0 \frac{\cos(wt)}{\cos(wt - \phi)} \tag{7.4}$$

Alternatively, we can use complex notation to write the impedance response of a system in terms of a real and an imaginary component:

$$Z = \frac{V_0 e^{jwt}}{i_0 e^{(jwt - j\phi)}} = Z_0 e^{j\phi} = Z_0(\cos\phi + j\,\sin\phi) \tag{7.5}$$

The impedance of a system can therefore be expressed in terms of an impedance magnitude Z_0 and a phase shift ϕ, or in terms of a real component ($Z_{\text{real}} = Z_0 \cos\phi$) and an imaginary component ($Z_{\text{imag}} = Z_0 \sin\phi j$). Note that j in these expressions represents the imaginary number ($j = \sqrt{-1}$), not the current density! Typically, impedance data are plotted in terms of the real and imaginary components of impedance (Z_{real} on the x axis and $-Z_{\text{imag}}$ on the y axis). Such graphical representations of impedance data are known as *Nyquist plots*. Because impedance measurements are made at dozens or even hundreds of different frequencies, Nyquist plots generally summarize the impedance behavior of a system over many orders of magnitude in frequency.

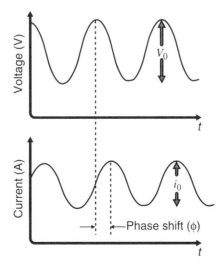

Figure 7.3. A sinusoidal voltage perturbation and resulting sinusoidal current response. The current response will possess the same period (frequency) as the voltage perturbation but will generally be phase shifted by an amount ϕ.

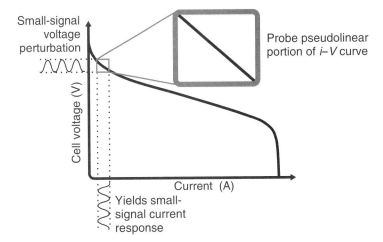

Figure 7.4. Application of a small signal voltage perturbation confines the impedance measurement to a pseudolinear portion of a fuel cell's $i-V$ curve.

System linearity is required for facile impedance analysis. In a linear system, doubling the current will double the voltage. Obviously, electrochemical systems are not linear. (Consider Butler–Volmer kinetics, which predicts an exponential relationship between voltage and current.) We circumvent this problem by using small-signal voltage perturbations in our impedance measurements. As Figure 7.4 illustrates, if we sample a small enough portion of a cell's $i-V$ curve, it will *appear linear*. In normal EIS practice, a 1-20-mV AC signal is applied to the cell. This signal is generally small enough to confine us to a pseudolinear segment of the cell's $i-V$ curve.

EIS and Fuel Cells. Before we get into the details of impedance theory, we will present a brief example illustrating the power of EIS for fuel cell characterization. Consider a hypothetical fuel cell which suffers from three loss effects:

1. Anode activation loss
2. Ohmic electrolyte loss
3. Cathode activation loss

Figure 7.5 shows what the EIS Nyquist plot for this fuel cell might look like. Don't worry about understanding this spectrum yet. The key thing to note is that two semicircular peaks are visible in the plot. For the hypothetical fuel cell in this example, the size of these two semicircles can be attributed to the magnitude of the two (anode and cathode) activation losses. Looking more closely at the diagram, you will see that the three x-axis intercepts defined by the semicircles mark off three impedance regions which are denoted by Z_Ω, Z_{fA}, and Z_{fC}. The size of these three impedances correspond to the relative size of η_{ohmic}, $\eta_{\text{act,anode}}$, and $\eta_{\text{act,cathode}}$ in our fuel cell. Thus, in this hypothetical EIS example, it is clear

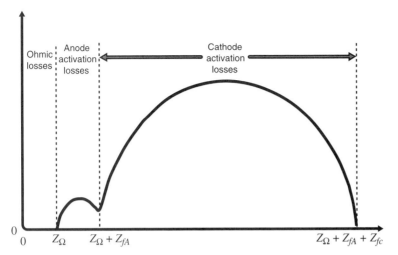

Figure 7.5. Example Nyquist plot from a hypothetical fuel cell. The three regions marked on the impedance plot are attributed to the ohmic, anode activation, and cathode activation losses. The relative size of the three regions provides information about the relative magnitude of the three losses in this fuel cell.

that the cathode activation loss dominates the fuel cell's performance while the ohmic and anode activation losses are small.

How were we able to generate this spectrum using EIS and how could we assign the various intercepts in the spectrum to the various loss processes in the fuel cell? This requires a discussion on impedance theory and equivalent circuit modeling.

EIS and Equivalent Circuit Modeling. The processes that occur inside a fuel cell can be modeled using circuit elements. For example, we can assign groups of resistors and capacitors to describe the behavior of electrochemical reaction kinetics, ohmic conduction processes, and even mass transport. Such circuit-based representations of fuel cell behavior are known as *equivalent circuit models*. If we measure a fuel cell's impedance spectrum and compare it to a good equivalent circuit model, it is then possible to extract information about the reaction kinetics, ohmic conduction processes, mass transport, and other properties.

We now introduce the common circuit elements used to describe fuel cell behavior. We will then build a sample equivalent circuit model of a fuel cell using these circuit elements for illustration. We begin with the ohmic conduction processes.

Ohmic Resistance. The equivalent circuit representation of an ohmic conduction process is rather straightforward; it is a simple resistor!

$$Z_\Omega = R_\Omega \tag{7.6}$$

As was mentioned previously, impedance data are generally plotted on a Nyquist diagram. Recall from the complex definition of impedance that the impedance of a system can be represented in terms of its real component ($Z_0 \cos \phi$) and its imaginary component ($Z_0 \sin \phi$):

$$Z = Z_0 \cos \phi + j Z_0 \sin \phi \tag{7.7}$$

A Nyquist diagram plots the real component of impedance versus the imaginary component of impedance (actually, the negative of the imaginary component of impedance) over a range of frequencies. For the case of a simple resistor, the imaginary component of resistance is zero, ϕ is zero, and the impedance does not change with frequency. The Nyquist plot for a resistor is therefore a single point on the real axis (x-axis) with value R. The equivalent circuit and corresponding Nyquist diagram of a simple resistor are given in Figure 7.6.

Electrochemical Reaction. The equivalent circuit representation of an electrochemical reaction is more complicated. Figure 7.7 depicts the typical electrochemical reaction interface. As illustrated in this figure, the impedance behavior of the reaction interface can be modeled as a parallel combination of a resistor and a capacitor (R_f and C_{dl}). Here, R_f, the *Faradaic resistance*, models the kinetics of the electrochemical reaction while C_{dl}, the *double-layer capacitance*, reflects the capacitive nature of the interface. We will briefly discuss both C_{dl} and R_f.

The easiest to visualize is C_{dl}. As Figure 7.7 illustrates, during an electrochemical reaction, a significant separation of charge occurs across the reaction interface, with electron accumulation in the electrode matched by ion accumulation in the electrolyte. The charge separation causes the interface to behave like a capacitor. The strength of this capacitive behavior is reflected in the size of C_{dl}. For a perfectly smooth electrode–electrolyte interface, C_{dl} is typically on the order of 30 $\mu F/cm^2$ interfacial area. However, with high-surface-area fuel cell electrodes, C_{dl} can be orders of magnitude larger.

The impedance response of a capacitor is purely imaginary. The equation relating voltage and current for a capacitor is

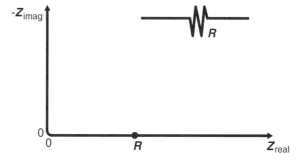

Figure 7.6. Circuit diagram and Nyquist plot for a simple resistor. The impedance of a resistor is a single point of value R on the real impedance axis (x axis). The impedance of a resistor is independent of frequency.

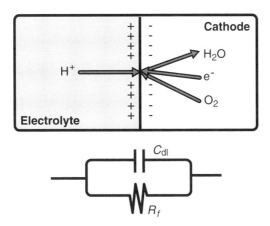

Figure 7.7. Physical representation and proposed equivalent circuit model of an electrochemical reaction interface. The impedance behavior of an electrochemical reaction interface can be modeled as a parallel combination of a capacitor and a resistor. The capacitor (C_{dl}) describes the charge separation between ions and electrons across the interface. The resistor (R_f) describes the kinetic resistance of the electrochemical reaction process.

$$i = C\frac{dV}{dt} \tag{7.8}$$

For a sinusoidal voltage perturbation ($V = V_0 e^{jwt}$) this gives

$$i(t) = C\frac{d(V_0 e^{jwt})}{dt} = C(jw)V_0 e^{jwt} \tag{7.9}$$

which yields an impedance of

$$Z = \frac{V(t)}{i(t)} = \frac{V_0 e^{jwt}}{C(jw)V_0 e^{jwt}} = \frac{1}{jwC} \tag{7.10}$$

If this capacitor is placed in series with a resistor, the net impedance will be given by the sum of the impedances of the two elements. In other words, series impedances, like series resistances, are additive:

$$Z_{series} = Z_1 + Z_2 \tag{7.11}$$

For a capacitor and resistor in series, the net impedance would be

$$Z = R + \frac{1}{jwC} \tag{7.12}$$

The equivalent circuit diagram and corresponding Nyquist impedance plot of the resistor–capacitor series combination is shown in Figure 7.8. One drawback of the Nyquist

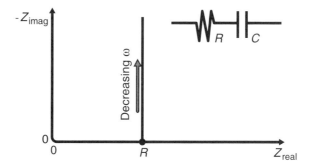

Figure 7.8. Circuit diagram and Nyquist plot for a series RC. The impedance is a vertical line that increases with decreasing w. The real component of the impedance is given by the value of the resistor. As frequency decreases, the imaginary component of the impedance (as given by the capacitor) dominates the response of the circuit.

plot is that you cannot tell what frequency was used to record each point. In Figure 7.8, we mitigate this disadvantage by noting the general frequency trend for reference.

For the case of the reaction interface shown in Figure 7.7, the capacitor and resistor are in parallel rather than in series. Before we talk about parallel impedances, however, we will discuss the Faradaic resistance R_f in more detail.

To understand how the reaction process can be modeled by R_f, recall the Tafel simplification of reaction kinetics (Equation 3.40):

$$\eta_{\text{act}} = -\frac{RT}{\alpha n F} \ln i_0 + \frac{RT}{\alpha n F} \ln i \tag{7.13}$$

Note that we have replaced current density j by raw current i to facilitate the impedance calculation. For a small-signal sinusoidal perturbation, the impedance response $Z = V(t)/i(t)$ can be approximated as $Z = dV/di$. (In other words, the impedance is the instantaneous slope of the i–V response at the point of interest.) Thus the impedance of a Tafel-like kinetic process may be calculated as

$$Z_f = \frac{d\eta}{di} = \frac{RT}{\alpha n F}\frac{1}{i} \tag{7.14}$$

Substituting $i = i_0 e^{\alpha n F \eta_{\text{act}}/(RT)}$ into this expression yields

$$Z_f = R_f = \left(\frac{RT}{\alpha n F}\right)\frac{1}{i_0 e^{\alpha n F \eta_{\text{act}}/(RT)}} \tag{7.15}$$

Notice that Z_f has no imaginary component and therefore can be represented as a pure resistor ($Z_f = R_f$). The size of R_f depends on the kinetics of the electrochemical reaction. A high R_f indicates a *highly resistive* electrochemical reaction. A large i_0 or a large activation overvoltage (η_{act}) will decrease R_f, decreasing the kinetic resistance of the reaction.

As was previously mentioned, the total impedance of our electrochemical interface model is given by the parallel combination of the capacitive double layer impedance and the resistive Faradaic impedance. Just like combining parallel resistances, the parallel combination of two impedance elements is given by

$$\frac{1}{Z_{\text{parallel}}} = \frac{1}{Z_1} + \frac{1}{Z_2} \tag{7.16}$$

For our case, this becomes

$$\frac{1}{Z} = \frac{1}{R_f} + jwC_{\text{dl}} \tag{7.17}$$

Thus

$$Z = \frac{1}{1/R_f + jwC_{\text{dl}}} \tag{7.18}$$

The equivalent circuit and corresponding Nyquist diagram of this reaction interface model is given in Figure 7.9. Note that the impedance shows a characteristic semicircular response. The leftmost point on the diagram corresponds to the highest frequency; frequency then steadily decreases as the points progress from left to right across the diagram. In most electrochemical systems, the real component of impedance will almost always increase (or remain constant) with decreasing frequency.

The high-frequency intercept of the semicircle in Figure 7.9 is zero, while the low-frequency intercept is R_f. Thus, the diameter of the semicircle provides information about the size of the activation resistance of a fuel cell. A fuel cell with highly facile reaction kinetics will show a small impedance loop. In contrast, a *blocking electrode* (one where $R_f \rightarrow \infty$ because the electrode "blocks" the electrochemical reaction) shows an

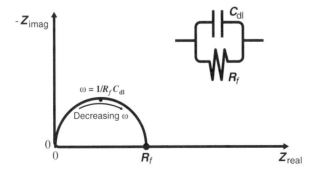

Figure 7.9. Circuit diagram and Nyquist plot for a parallel RC. This semicircular impedance response is typical of an electrochemical reaction interface. The high-frequency intercept of the semicircle is zero, while the low-frequency intercept of the impedance semicircle is R_f. The diameter of the semicircle (R_f) gives information about the reaction kinetics of the electrochemical interface. A small loop indicates facile reaction kinetics while a large loop indicates sluggish reaction kinetics.

impedance response similar to the pure capacitor in Figure 7.8. Examination of the limits in Equation 7.18 for $w \to \infty$ and $w \to 0$ confirms these observations. At intermediate frequencies, the impedance response contains both real and imaginary components. The frequency at the apex of the semicircle is given by the RC time constant of the interface: $w = 1/(R_f C_{dl})$. From this value, C_{dl} may be determined.

The impedance behavior illustrated in Figure 7.9 can be understood intuitively by examining the RC circuit model. At extremely high frequencies, capacitors act as short circuits; at extremely low frequencies, capacitors act as open circuits. Thus, at high frequency, the current can be completely shunted through the capacitor and the effective impedance of the model is zero. In contrast, at extremely low frequencies, all of the current is forced to flow through the resistor and the effective impedance of the model is given by the impedance of the resistor. For intermediate frequencies, the situation is somewhere in-between, and the impedance response of the model will have both resistive and capacitive elements.

Mass Transport. Mass transport in fuel cells can be modeled by *Warburg* circuit elements. Time does not permit the derivation of Warburg elements here. However, they are based on (and can be derived from) diffusion processes. The impedance of an "infinite" Warburg element (used for an infinitely thick diffusion layer) is given by the equation

$$Z = \frac{\sigma_i}{\sqrt{w}}(1 - j) \qquad (7.19)$$

where σ_i in this equation is the *Warburg coefficient* for a species i (not the conductivity) and is defined as:

$$\sigma_i = \frac{RT}{(n_i F)^2 A \sqrt{2}}\left(\frac{1}{c_i^0 \sqrt{D_i}}\right) \qquad (7.20)$$

where A is the electrode area, c_i is the bulk concentration of species i, and D_i is the diffusion coefficient of species i. Thus, σ_i characterizes the effectiveness of transporting species i to or away from a reaction interface. If species i is abundant (c_i^0 is large) and diffusion is fast (D_i is high), then σ_i will be small and the impedance due to mass transport of species i will be negligible. On the other hand, if the species concentration is low and diffusion is slow, σ_i will be large and the impedance due to mass transport can become significant. Note from Equation 7.19 that the Warburg impedance also depends on the frequency of the potential perturbation. At high frequencies the Warburg impedance is small since diffusing reactants do not have to move very far. However, at low frequencies the reactants must diffuse farther, thereby increasing the Warburg impedance.

The equivalent circuit and corresponding Nyquist diagram of the infinite Warburg impedance element are given in Figure 7.10. Note that the infinite Warburg impedance shows a characteristic increasing linear response with decreasing ω. The infinite Warburg impedance appears as a diagonal line with a slope of 1.

The infinite Warburg impedance is only valid if the diffusion layer is infinitely thick. In fuel cells, this is rarely the case. As we learned in Chapter 5, convective mixing in fuel cell flow structures usually restricts the diffusion layer to the thickness of the electrode. For such situations, the impedance at lower frequencies no longer obeys the infinite Warburg

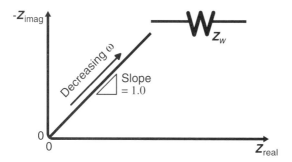

Figure 7.10. Circuit diagram and Nyquist plot for a Warburg element used to model diffusion processes. The impedance response is a diagonal line with a slope of 1. Impedance increases from left to right with decreasing frequency.

equation. In these cases, it is better to use a *porous bounded Warburg* model (also called the "O" diffusion element), which has the form

$$Z = \frac{\sigma_i}{\sqrt{w}}(1-j) \ \tanh\left(\delta\sqrt{\frac{jw}{D_i}}\right) \qquad (7.21)$$

where δ is the diffusion layer thickness. As shown in Figure 7.11, at high frequencies or cases where δ is large, the porous bounded Warburg impedance converges to the infinite Warburg behavior. However, at low frequencies or for small diffusion layers, the porous bounded Warburg impedance moves back toward the real axis.

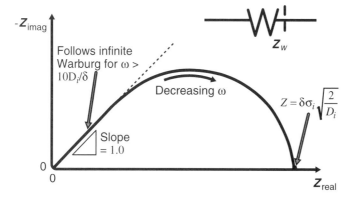

Figure 7.11. Circuit diagram and Nyquist plot for a porous bounded Warburg element, which is used to model finite diffusion processes (with diffusion occurring through a fixed diffusion layer thickness from an inexhaustible bulk supply of reactants). This situation is typical in fuel cell systems. At high frequency the porous bounded Warburg impedance response mirrors the behavior of an infinite Warburg; at low frequency it returns toward the real impedance axis. (This makes intuitive sense: A finite diffusion layer thickness should yield a finite real impedance.) The low-frequency real axis impedance intercept yields information about the diffusion layer thickness.

TABLE 7.1. Impedance Summary of Common Equivalent Circuit Elements

Circuit Element	Impedance
Resistor	R
Capacitor	$1/jwC$
Constant-phase element	$1/[Q(jw)^{\alpha}]$
Inductor	jwL
Infinite Warburg	$(\sigma_i/\sqrt{w})(1-j)$
Finite (porous bounded) Warburg	$(\sigma_i/\sqrt{w})(1-j)\ \tanh\left(\delta\sqrt{jw/D_i}\right)$
Series impedance elements	$Z_{\text{series}} = Z_1 + Z_2$
Parallel impedance elements	$1/Z_{\text{parallel}} = 1/Z_1 + 1/Z_2$

We have now assembled enough tools to describe basic fuel cell processes using equivalent circuit elements. The equivalent circuit elements that we have developed (as well as a few others) are summarized in Table 7.1.

Simple Equivalent Circuit Fuel Cell Model. We now construct a simple equivalent circuit model for a complete fuel cell using the elements described previously. We assume that our fuel cell suffers from the following loss processes:

1. Anode activation
2. Cathode activation
3. Cathode mass transfer
4. Ohmic loss

For simplicity, we assume that the cathode mass transfer process can be modeled with an infinite Warburg impedance element. Also, we assume that the anode kinetics are fast compared to the cathode activation kinetics. The physical picture, equivalent circuit model, and corresponding Nyquist plot for our fuel cell are shown in Figure 7.12. The Nyquist plot was generated using the equivalent circuit values given in Table 7.2. Note how the impedance response of this fuel cell model is given by a combination of the impedance behaviors from each individual element in our circuit! The Nyquist plot shows two semicircles followed by a diagonal line. The high-frequency, real-axis intercept corresponds to the ohmic resistance of our fuel cell model. The first loop corresponds to the RC model of the anode activation kinetics while the second loop corresponds to the RC model of the cathode activation kinetics. The diameter of the first loop gives R_f for the anode while the diameter of the second loop gives R_f for the cathode. Note how the cathode loop is significantly larger than the anode loop. This visually indicates that the cathode activation losses are significantly greater than the anode activation losses. From the R_f values, the kinetics of the anode and cathode reactions can be extracted using Equation 7.15. Fitting the C_{dl} values gives an indication of the effective surface area of the fuel cell electrodes.

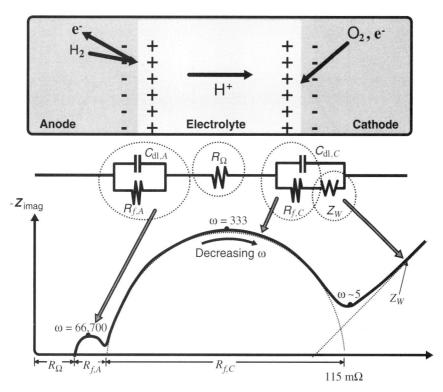

Figure 7.12. Physical picture, circuit diagram, and Nyquist plot for a simple fuel cell impedance model. The equivalent circuit for this fuel cell consists of two parallel RC elements to model the anode and cathode activation kinetics, an infinite Warburg element to simulate cathode mass transfer effects, and an ohmic resistor to simulate the ohmic losses. While schematically shown in the electrolyte region, the ohmic resistor models the ohmic losses arising from all parts of the fuel cell (electrolyte, electrodes, etc.). The impedance response shown in the Nyquist plot is based on the circuit element values given in Table 7.2. Each circuit element contributes to the shape of the Nyquist plot, as indicated in the diagram. The ohmic resistor determines the high-frequency impedance intercept. The small semicircle is due to the anode RC element while the large semicircle is due to the cathode RC element. The low-frequency diagonal line comes from the infinite Warburg element.

TABLE 7.2. Summary of Values Used to Generate Nyquist Plot in Figure 7.12

Fuel Cell Process	Circuit Element	Value
Ohmic resistance	R_Ω	10 mΩ
Anode Faradaic resistance	R_{fA}	5 mΩ
Anode double-layer capacitance	C_{dlA}	3 mF
Cathode Faradaic resistance	R_{CA}	100 mΩ
Cathode double-layer capacitance	C_{dlA}	30 mF
Cathode Warburg coefficient	σ	0.015

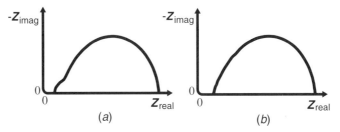

Figure 7.13. In H_2–O_2 fuel cells the cathode impedance is often significantly larger than the anode impedance. In these cases, the cathode impedance can mask the impedance of the anode, as shown to varying degrees in (*a*) and (*b*). This masking (or "merging") also occurs if the RC time constants for the anode and cathode reactions overlap. If R_f for the anode is extremely small, the RC time constant for the anode may correspond to frequencies that are beyond the limits of most impedance hardware. (EIS is usually limited to $f < 100$ kHz.) In these cases, the anode impedance may be unmeasurable.

The diagonal line at low frequencies is due to mass transport as modeled by the infinite Warburg impedance. From the frequency–impedance data of this line, the mass transport properties of the fuel cell can be extracted. If a porous bounded Warburg is used instead, a diffusion layer thickness could also be extracted.

For clarity in this example, we deliberately chose RC values for the anode and cathode that allowed the two semicircles to be distinguished from one another. In many real fuel cells, however, the RC loop for the cathode overwhelms the RC loop for the anode, as shown in Figure 7.13.

To fully understand fuel cell behavior, it is essential to measure the impedance response at several different points along a fuel cell's i–V curve. The impedance behavior of a fuel cell will change along the i–V curve, depending on which loss processes are dominant. Figure 7.14 gives several illustrative examples. At low currents, the activation kinetics dominate and R_f is large, while the mass transport effects can be neglected. In these situations, an impedance response similar to that shown in Figure 7.14*a* is typical. At higher currents (higher activation overvoltages), R_f decreases since the activation kinetics improve with increasing η_{act} (refer to Equation 7.15). Thus, the activation impedance loop decreases, as shown in Figure 7.14*b*. A decreasing impedance loop with increasing activation overvoltage is indicative of an activated electrochemical reaction. At high currents, mass transport effects occur and the impedance response may look something like Figure 7.14*c*.

While the power of EIS is considerable, the technique is complex and fraught with pitfalls. *Caution! There be dragons here!* Due to time and space limitations, this EIS overview is not comprehensive. Interested readers who plan to use EIS for fuel cell characterization are highly encouraged to consult the extensive literature on EIS beforehand [38]–[40].

Example 7.1. Assume point *a* on the i–V curve in Figure 7.14 corresponds to $i = 0.25$ A and $V = 0.77$ V. Assume that point *b* on the i–V curve corresponds to $i = 1.0$ A and $V = 0.65$ V. From the EIS data in Figure 7.14, calculate η_{ohmic} and η_{act} at points *a* and *b* on the fuel cell i–V curve. Assume that only ohmic and activation

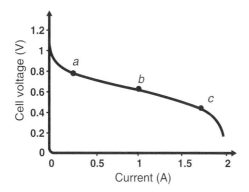

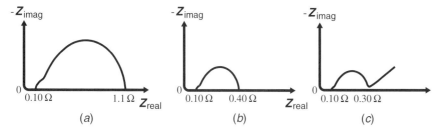

Figure 7.14. EIS characterization of a fuel cell requires impedance measurements at several different points along an i–V curve. The impedance response will change depending on the operating voltage. (a) At low current, the activation kinetics dominate and R_f is large, while the mass transport effects can be neglected. (b) At intermediate current (higher activation overvoltages), the activation loops decrease since R_f decreases with increasing η_{act}. (Refer to Equation 7.15.) (c) At high current, the activation loops may continue to decrease but the mass transport effects begin to intercede, resulting in the diagonal "warburg" response at low frequency.

losses contribute to fuel cell performance. If the activation losses are wholly due to the cathode, calculate i_0 and α for the cathode based on your η_{act} values ($T = 300$ K, $n = 2$, and $E_{\text{thermo}} = 1.2$ V).

Solution: At point a, $i = 0.25$ A, $R_{\text{ohmic}} = 0.10\ \Omega$, and $\eta_{\text{tot}} = 1.2$ V $- 0.77$ V $= 0.43$ V. Thus

$$\eta_{\text{ohmic}} = i R_{\text{ohmic}} = (0.25\ \text{A})(0.10\ \Omega) = 0.025\ \text{V}$$
$$\eta_{\text{act}} = \eta_{\text{tot}} - \eta_{\text{ohmic}} = 0.43\ \text{V} - 0.025\ \text{V} = 0.405\ \text{V} \tag{7.22}$$

Note: It is *not* appropriate to write $\eta_{\text{act}} = i R_f$, since R_f changes as a function of i. Thus, the best we can do is *infer* the activation loss by subtracting the ohmic loss from the total loss. At point b, $i = 1.0$ A, $R_{\text{ohmic}} = 0.10\ \Omega$, and $\eta_{\text{tot}} = 1.2$ V $- 0.65$ V $= 0.55$ V. Thus

$$\eta_{\text{ohmic}} = i R_{\text{ohmic}} = (1.0\ \text{A})(0.10\ \Omega) = 0.10\ \text{V}$$
$$\eta_{\text{act}} = \eta_{\text{tot}} - \eta_{\text{ohmic}} = 0.55\ \text{V} - 0.1\ \text{V} = 0.45\ \text{V} \tag{7.23}$$

Note that R_f decreases at point b, but the total activation loss still increases slightly (from 0.405 to 0.45 V). This is expected; the total activation loss increases with increasing current, but the "effective resistance" of the activation process decreases. We can fit the EIS data from a and b to Equation 7.15 to extract j_0 and α:

$$\text{For point } b: \qquad \eta_{\text{act}} = -\left(\frac{RT}{\alpha n F}\right) \ln i_0 + \left(\frac{RT}{\alpha n F}\right) \ln i$$

$$0.405 \text{ V} = -\left(\frac{RT}{\alpha n F}\right) \ln i_0 \tag{7.24}$$

Substitution into a similar equation for point a allows us to solve for α:

$$\text{For point } a: \qquad \eta_{\text{act}} = -\left(\frac{RT}{\alpha n F}\right) \ln i_0 + \left(\frac{RT}{\alpha n F}\right) \ln i$$

$$0.405 \text{ V} = 0.45 \text{ V} + \left(\frac{RT}{\alpha n F}\right) \ln 0.25 \tag{7.25}$$

$$\alpha = 0.398 \qquad \text{for } T = 300 \text{ K}, n = 2$$

Substituting α back into the equation for point b yields i_0:

$$0.45 \text{ V} = -\left(\frac{(8.314)(300)}{(0.398)(2)(96500)}\right) \ln i_0$$

$$i_0 = 9.5 \times 10^{-7} \text{ A} \tag{7.26}$$

If we knew the area of the fuel cell, we could then calculate the more fundamental properties $\text{ASR}_{\text{ohmic}}$ and j_0 from R_{ohmic} and i_0.

7.3.5 Current Interrupt Measurement

The current interrupt method can provide some of the same information provided by EIS. While not as accurate or as detailed as an impedance experiment, current interrupt has several major advantages compared to impedance:

- Current interrupt is extremely fast.
- Current interrupt generally requires simpler measurement hardware.
- Current interrupt can be implemented on high-power fuel cell systems. (Such systems are generally not amenable to EIS.)
- Current interrupt can be conducted in parallel with a j–V curve measurement.

For these reasons, current interrupt has found wide acceptance in the fuel cell research community, especially for characterization of large fuel cells (e.g. residential or vehicular fuel cell stacks).

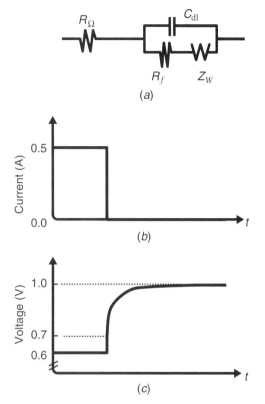

Figure 7.15. (*a*) Simplified equivalent circuit of a fuel cell system. The RC components from the anode and cathode have been consolidated into a single branch. (*b*) Hypothetical current interrupt profile applied to the circuit in (*a*). In this example, an original steady-state current load of 500 mA is abruptly zeroed. (*c*) Hypothetical time response of fuel cell voltage when the current interrupt in (*b*) is applied to the system. The instantaneous rebound in the voltage is associated with the pure ohmic losses in the system. The time-dependent voltage rebound is associated with the activation and mass transport losses in the system.

The basic idea behind the current interrupt technique is illustrated in Figure 7.15. When a constant-current load on a fuel cell system is abruptly interrupted, the resulting time-dependent voltage response will be representative of the capacitive and resistive behaviors of the various components in the fuel cell. The same equivalent circuit models that were used to analyze the impedance behavior of fuel cells may be used to understand the current interrupt behavior of fuel cells.

For example, consider the simple equivalent circuit fuel cell model shown in Figure 7.15*a*. If the current flowing through this cell is abruptly interrupted, as shown in Figure 7.15*b*, the corresponding voltage–time response will resemble Figure 7.15*c*. Interruption of the current causes an immediate rebound in the voltage, followed by an additional, time-dependent rebound in the voltage. The immediate voltage rebound is associated with the ohmic resistance of the fuel cell. The time-dependent rebound is associated with the much slower reaction and mass transport processes.

The voltage rebound process can be understood via the circuit diagram in Figure 7.15a. As the circuit diagram illustrates, the reaction and mass-transport processes are modeled by time-dependent RC and Warburg elements. Due to their capacitive nature, the voltage across these elements recovers over a period of time. The recovery time for the RC element can be approximated by its RC time constant. Because the voltage rebound across the resistor is immediate while the voltage rebound across the RC/Warburg element is time dependent, the voltage–time response can be used to separate the two contributions. Example 7.2 illustrates this technique.

Example 7.2. Calculate η_{ohmic} and R_{ohmic} from the current interrupt data in Figure 7.15.

Solution: In Figure 7.15, when the fuel cell is held under 500-mA current load, the steady-state voltage is 0.60 V. When the current is abruptly zeroed, the cell voltage instantaneously rises to 0.7 V. We associate this instantaneous rebound in the cell voltage with the ohmic processes in the fuel cell. Therefore, the fuel cell must have been experiencing an ohmic loss of 100 mV at the 500 mA current load point:

$$\eta_{\text{ohmic}} = 0.70 \text{ V} - 0.60 \text{ V} = 0.10 \text{ V} \quad (\text{at } i = 500 \text{ mA}) \tag{7.27}$$

The ohmic resistance may be calculated from η_{ohmic} and the current:

$$R_{\text{ohmic}} = \frac{\eta_{\text{ohmic}}}{i} = \frac{0.10 \text{ V}}{0.50 \text{ A}} = 0.02 \ \Omega \tag{7.28}$$

After a long relaxation time, the fuel cell's voltage recovers to a final value of around 1.0 V. Thus, the activation and concentration losses in this fuel cell must amount to about 0.30 V at a 500-mA current load (1.0 V − 0.70 V = 0.30 V).

To get accurate results from the current interrupt technique, the current should be interrupted sharply and cleanly (on the order of microseconds to milliseconds), and a fast oscilloscope should be employed to record the voltage response. Current interrupt is often implemented in parallel with $i–V$ curve measurements. It is especially useful for determining the ohmic component of fuel cell loss at each measurement point on the fuel cell $i–V$ curve. Typically, after a fuel cell $i–V$ data point is recorded, a current interrupt measurement is then made to determine R_Ω at that point. Then, the $i–V$ measurement procedure is stepped to the next current level and the voltage is allowed to equilibrate to the steady state. In this way, the $i–V$ curve information is collected along with detailed ohmic-loss information from each point. The ohmic-loss portion of the $i–V$ curve data can then be removed; such curves are called "iR-free" or "iR-corrected" $i–V$ curves. When fit to the Tafel equation, these "iR-corrected" curves allow the activation and concentration losses to be separated. The result is a nearly complete quantification of the ohmic, activation, and concentration losses associated with the fuel cell.

7.3.6 Cyclic Voltammetry

Cyclic voltammetry is typically used to characterize fuel cell catalyst activity in more detail. In a standard CV measurement, the potential of a system is swept back and forth between two voltage limits while the current response is measured. The voltage sweep is generally linear with time, and the plot of the resulting current versus voltage is called a cyclic voltammogram. An illustration of a typical CV waveform is provided in Figure 7.16.

In fuel cells, CV measurements can determine in situ catalyst activity by using a special "hydrogen-pump-mode" configuration. In this mode, argon gas is passed through the cathode instead of oxygen, while the anode is supplied with hydrogen. The CV measurement is performed by sweeping the voltage of the system between about 0 and 1 V with respect to the anode. An example of a hydrogen-pump-mode cyclic voltammogram from a fuel cell is shown in Figure 7.17. When the potential increases from 0 V, a current begins to flow. (See Figure 7.17.) There are two contributions to this current. One contribution is constant—a simple, capacitive charging current which flows in response to the linearly changing voltage. The second current response is nonlinear and corresponds to a hydrogen adsorption reaction occurring on the electrochemically active anode catalyst surface. As the voltage increases further, this reaction current reaches a peak and then falls off as the entire catalyst surface becomes fully saturated with hydrogen. The active catalyst surface area can be obtained by quantifying the total charge (Q_h) provided by hydrogen adsorption on the catalyst surface. The total charge essentially corresponds to the area under the hydrogen adsorption reaction peak in the CV after converting the potential axis to time and making sure to exclude the capacitive charging current contribution.

An active catalyst area coefficient A_c may be calculated which represents the ratio of the measured active catalyst surface area compared to the active surface area of an atomically smooth catalyst electrode of the same size:

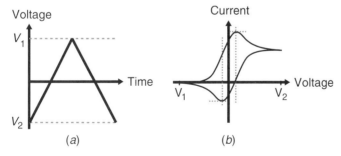

(a) (b)

Figure 7.16. Schematic of a (CV) waveform and typical resulting current response. (*a*) In a CV experiment, the voltage is swept linearly back and forth between two voltage limits (denoted V_1 and V_2 on the diagram). (*b*) The resulting current is plotted *as a function of voltage*. When the voltage sweeps past a potential corresponding to an active electrochemical reaction, the current response will spike. After this initial spike, the current will level off when most of the readily available reactants have been consumed. On the reverse voltage scan, the reverse electrochemical reaction (with a corresponding reverse current direction) may be observed. The shape and size of the peaks give information about the relative rates of reaction and diffusion in the system.

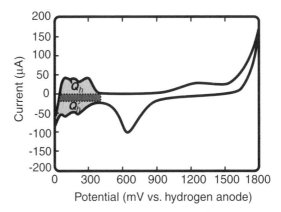

Figure 7.17. Fuel cell CV curve. The peaks marked Q_h and Q'_h represent the hydrogen adsorption and desorption peaks on the platinum fuel cell catalyst surface, respectively. The grey rectangular area between the two peaks denotes the approximate contribution from the capacitive charging current. The active catalyst surface area can be calculated from the area under the Q_h or Q'_h peak (recognizing that the voltage axis can be converted to a time axis if the scan rate of the experiment is known).

$$A_c = \frac{\text{measured active catalyst surface area}}{\text{geometric surface area}} = \frac{Q_h}{Q_m A_{\text{geometric}}} \qquad (7.29)$$

where Q_m is the adsorption charge for an atomically smooth catalyst surface, generally accepted to be 210 $\mu C/cm^2$ for a smooth platinum surface.

As noted before, a highly porous, well-made fuel cell electrode may have an active surface area that is orders of magnitude larger than its geometric area. This effect is expressed through A_c.

7.4   EX SITU CHARACTERIZATION TECHNIQUES

While the direct in situ electrical characterization techniques are the most popular methods used to study fuel cell behavior, indirect *ex situ* characterization techniques can provide enormous additional insight into fuel cell performance. Most *ex situ* techniques focus on evaluating the physical or chemical structure of fuel cell components in an effort to identify which elements most significantly impact fuel cell performance. Pore structure, catalyst surface area, electrode/electrolyte microstructure, and electrode/electrolyte chemistry are among the most important characteristics to evaluate.

7.4.1   Porosity Determination

The porosity ϕ of a material is defined as the ratio of void space to the total volume of the material. To be effective, fuel cell electrodes and catalyst layers must exhibit substantial porosity. Furthermore, this pore space should be interconnected and open to the surface.

Porosity determination is accomplished in several ways. First, if the density of a porous sample (ρ_s) can be determined by measuring its mass and volume, and the bulk density of the material used to make the sample is also known (ρ_b), then the porosity may be calculated as

$$\phi = 1 - \frac{\rho_s}{\rho_b} \qquad (7.30)$$

For fuel cells, however, *effective porosity* is more important than total porosity. Effective porosity counts only the pore space which is interconnected and open to the surface. (In other words dead pores are ignored.) Effective porosity can be determined using volume infiltration techniques. For example, the total volume of a porous sample is first determined by immersing the sample in a liquid which does not enter the pores. For example, at low pressure, mercury will not infiltrate pore spaces due to surface tension effects. Then, the sample may be inserted into a container of known volume which contains an inert gas. The gas pressure in the container is noted; then a second evacuated chamber of known volume is connected to the system and the new system pressure is noted. Using the ideal gas law, the volume of open pores in the sample may be obtained and thus the effective porosity.

Pore size distributions may be obtained from mercury porosimetry. In this method, the porous sample is placed into a chamber which is then evacuated. Mercury is then injected into the porous sample, first at extremely low pressure and then at steadily increasing pressures. The volume of mercury taken up at each pressure is noted. Mercury will enter a pore of radius r only when the pressure p in the chamber is

$$p \geq \frac{2\gamma}{r} \cos \theta \qquad (7.31)$$

where γ is the surface tension of mercury and θ is the contact angle of mercury. Fitting this equation to the experimental mercury uptake–pressure data allows approximate pore size distribution curves to be calculated.

7.4.2 BET Surface Area Determination

As discussed many times, the most effective fuel cell catalyst layers have extremely high real surface areas. Surface area determination, therefore, represents an important characterization tool. As we learned for CV, the electrochemically active surface area can be determined from specialized in situ electrochemical measurements. Additionally, the double-layer capacitance C_{dl} in impedance measurements may be used to roughly estimate surface areas based on the fact that a smooth reaction interface should have a capacitance of about 30 μF/cm^2. However, for the most accurate surface area determination, an ex situ technique known as the Brunauer–Emmett–Teller method is employed.

The BET method makes use of the fact that a fine layer of an inert gas like nitrogen, argon, or krypton will absorb on a sample surface at extremely low temperatures. In a typical experiment, a dry sample is evacuated of all gas and cooled to 77 K, the temperature of liquid nitrogen. A layer of inert gas will physically adhere to the sample surface, lowering the pressure in the analysis chamber. From the measured absorption isotherm of the experiment, the surface area of the sample can be calculated.

7.4.3 Gas Permeability

High surface area and high porosity accomplish nothing if the fuel cell electrode and catalyst structure exhibit low permeability. Permeability measures the ease with which gases move through a material. Even highly porous materials can have low permeability if most of their pores are closed or fail to interconnect. Fuel cell electrodes and catalyst layers should have high permeabilities. On the other hand, fuel cell electrolytes need to be gas tight. Permeability K is determined by measuring the volume of gas (ΔV) that passes through a sample in a given period of time (Δt) when driven by a given pressure drop ($\Delta p = p_1 - p_2$):

$$K = \frac{I}{\Delta p} - \frac{\Delta V}{\Delta t} \frac{2p_2}{(p_1 + p_2)\,\Delta p} \tag{7.32}$$

where I is a constant.

7.4.4 Structure Determinations

Significant information about microstructure, porosity, pore size distribution, and interconnectedness is gleaned from microscopy. Optical microscopy (OM), scanning electron microscopy (SEM), transmission electron microscopy (TEM), and atomic force microscopy (AFM) are invaluable characterization techniques. Specific quantitative structural information can be provided from X-ray diffraction (XRD) measurements, which provide crystal structure, orientation, and chemical compound information. This information is extremely important when developing new electrode, catalyst, or electrolyte materials. Furthermore, XRD peak broadening measurements can provide information about particle size (in catalyst powder samples) or grain size (in bulk crystalline samples). Combined with TEM, XRD allows structural, chemical, and powder size distribution determinations for catalyst particles as small as 10 Å.

7.4.5 Chemical Determinations

When developing new catalyst, electrode, or electrolyte materials, it is always important to know what you have. Therefore chemical determinations of composition, phase, bonding, or spatial distribution are just as important as structural determinations. For chemical determinations, TEM and XRD prove invaluable. In addition, other techniques like auger electron spectroscopy (AES), X-ray photoelectron spectroscopy (XPS), and secondary-ion mass spectrometry (SIMS) can provide useful information. While it is beyond the scope of this book to describe the advantages and disadvantages of these techniques, the interested reader is invited to consult the literature available on the subject.

CHAPTER SUMMARY

This chapter discussed many of the major techniques used to characterize fuel cells. We have seen that fuel cell characterization has two major goals: (1) to quantitatively separate good fuel cell designs from bad fuel cell designs, and (2) to understand *why* fuel cell designs are good or bad.

- In situ electrical characterization techniques make use of the three fundamental electrochemical variables (voltage, current, and time) to probe fuel cell behavior.

- Ex situ characterization techniques focus on correlating the structure (porosity, grain size, morphology, surface area, etc.) or the chemistry (composition, phase, spatial distribution) of fuel cell components to fuel cell performance.

- The major in situ electrical characterization techniques are (1) $j-V$ curve measurement, (2) electrochemical impedance spectroscopy (EIS), (3) current interrupt, and (4) cyclic voltammetry (CV).

- A careful $j-V$ curve measurement yields the steady-state performance of a fuel cell under well-documented conditions. A fuel cell's $j-V$ performance is sensitive to the measurement procedure and test conditions. Fuel cell $j-V$ curves can only be fairly compared if they are acquired using similar measurement procedures and testing conditions.

- Current interrupt, EIS, and CV measurements utilize the non-steady-state (*dynamic*) behavior of fuel cells to distinguish between the major processes that contribute to fuel cell performance.

- Current interrupt distinguishes ohmic and nonohmic fuel cell processes. The immediate voltage rise after an abrupt current interruption is associated with ohmic processes, while the time-dependent voltage rise is associated with activation and mass transport processes. Current interrupt is fast and relatively easy to implement. It is especially attractive for high-power systems.

- In EIS, the impedance of a fuel cell system is measured over many orders of magnitude in frequency. A Nyquist plot of the resulting impedance data can be fit to an equivalent circuit model of the fuel cell. From this fit, the ohmic, activation, and mass transport losses in the fuel cell can often be resolved separately. Electrochemical impedance spectroscopy can be slow and requires sophisticated hardware. It is difficult to implement for high-power systems.

- While the subject of impedance is complex (no pun intended), you should become familiar with the equivalent circuit models of common fuel cell components and the resulting impedance responses that these models produce.

- In a standard CV measurement, the potential of a system is swept back and forth between two voltage limits while the current response is measured. In general, CV measurements are used to determine in situ catalyst activity, although they may also be used for detailed reaction kinetics analysis.

- Some of the more popular *ex situ* characterization techniques include porosity analysis, surface area determination, permeability measurement, inspection microscopy (OM, SEM, TEM, AFM), and chemical analysis (XRD, AES, XPS, SIMS).

CHAPTER EXERCISES

Review Questions

7.1   What are the two main goals of fuel cell characterization?

7.2   List at least three major operation variables that can affect fuel cell performance. For each, provide what you believe is the most important equation that describes how fuel cell performance is affected by the variable in question.

7.3   Discuss the relative advantages and disadvantages of EIS versus current interrupt measurement.

7.4   A fuel cell's $j-V$ curve is acquired at two different scan rates: 1 and 100 mA/s.

    (a)   Which scan rate will result in better apparent performance? (Assume the scans were acquired with increasing current starting at zero current.)

    (b)   Which portion of the $j-V$ curve (low current density, moderate current density, high current density) will be most affected by the change in scan rate and why?

7.5   **(a)**   Draw a schematic EIS curve for a fuel cell with one blocking electrode (represented by a series RC) and one activated electrode (represented by a parallel RC). Assume that the RC product for the parallel RC is *much smaller* than the RC product for the series RC.

    (b)   Draw a schematic EIS curve for the scenario above if the RC product for the parallel RC is *much greater* than the RC product for the series RC.

    (c)   Draw a schematic EIS curve for a fuel cell modeled by two parallel RC elements, an ohmic resistance component, and a porous bounded Warburg element. Assume the time constants of the two parallel RC elements are separated by at least two orders of magnitude.

7.6   Sketch an example material structure that has high porosity but low permeability.

Calculations

7.7   In Example 7.1 we calculated η_{ohmic}, η_{act}, i_0, and α from the $i-V$ and EIS data in points a and b of Figure 7.14. In this problem, calculate η_{ohmic}, η_{act}, i_0, and α from the $i-V$ and EIS data in points b and c of Figure 7.14. Assume that point c on the $i-V$ curve corresponds to $i = 1.75$ A and $V = 0.45$ V. Assume that the activation losses are wholly due to the cathode and $T = 300$ K and $n = 2$.

7.8   Calculate the approximate active platinum catalyst area coefficient from the CV curve in Figure 7.17 assuming that it was acquired from a 0.1×0.1-cm square test electrode at a scan rate of 10 mV/s.

PART II

FUEL CELL TECHNOLOGY

CHAPTER 8

OVERVIEW OF FUEL CELL TYPES

8.1 INTRODUCTION

As described in the first chapter of this book, there are five major types of fuel cells, differentiated from one another on the basis of their electrolyte:

1. Phosphoric acid fuel cell
2. Polymer electrolyte membrane fuel cell
3. Alkaline fuel cell
4. Molten carbonate fuel cell
5. Solid oxide fuel cell

Many of the discussions and examples in the first part of this book focused on the PEMFC and the SOFC. Of all the fuel cell types, the PEMFC and the SOFC appear well positioned to deliver on the promise of the technology. Still, the other fuel cell classes have unique advantages, properties, and histories which make a succinct overview worthwhile. In the following sections we briefly discuss each of the five major fuel cell types. We conclude the chapter with a summary of the relative merits of each type.

8.2 PHOSPHORIC ACID FUEL CELL

In the PAFC, liquid H_3PO_4 electrolyte (either pure or highly concentrated) is contained in a thin SiC matrix between two porous graphite electrodes coated with a platinum catalyst.

Hydrogen is used as the fuel and air or oxygen may be used as the oxidant. The anode and cathode reactions are

$$\text{Anode:} \qquad\qquad\qquad H_2 \rightarrow 2H^+ + 2e^-$$

$$\text{Cathode:} \qquad \frac{1}{2}O_2 + 2H^+ + 2e^- \rightarrow H_2O$$

(8.1)

A schematic of a PAFC is provided in Figure 8.1. Figure 8.2 gives a photograph of a 200 kW stationary power commercial PAFC system. Pure phosphoric acid solidifies at

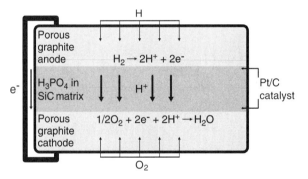

Figure 8.1. Schematic of H_2–O_2 PAFC. The phosphoric acid electrolyte is immobilized within a porous SiC matrix. Porous graphitic electrodes coated with a Pt catalyst mixture are used for both the anode and the cathode. Water is produced at the anode.

Figure 8.2. Photograph of PureCell^{TM} 200 power system, a commercial 200-kW PAFC. The unit includes a reformer, which processes natural gas into H_2 for fuel. This system provides clean, reliable power at a range of locations from a New York City police station to a major postal facility in Alaska to a credit-card processing system facility in Nebraska to a science center in Japan. It also can provide heat for the building. (Photograph courtesy of UTC Fuel Cells, Inc.)

42°C. Therefore, PAFCs must be operated above this temperature. Because freeze–thaw cycles can cause serious stress issues, commissioned PAFCs are usually maintained at operating temperature. Optimal performance occurs at temperatures of 180–210°C. Above 210°C, H_3PO_4 undergoes an unfavorable phase transition which renders it unsuitable as an electrolyte. The SiC matrix provides mechanical strength to the electrolyte, keeps the two electrodes separated, and minimizes reactant gas crossover. During operation, H_3PO_4 must be continually replenished because it gradually evaporates to the environment (especially during higher temperature operation). Electrical efficiencies of PAFC units are $\approx 40\%$ with combined heat and power units achieving $\approx 70\%$.

Because PAFCs employ platinum catalysts, they are susceptible to carbon monoxide and sulfur poisoning at the anode. This is not an issue when running on pure hydrogen but can be important when running on reformed or impure feedstocks. Susceptibility depends on temperature; because the PAFC operates at higher temperatures than the PEMFC, it exhibits somewhat greater tolerance. Carbon monoxide tolerance at the anode can be as high as 0.5–1.5%, depending on the exact conditions. Sulfur tolerance in the anode, where it is typically present as H_2S, is around 50 ppm (parts per million).

PAFC Advantages

- Mature technology
- Excellent reliability/long-term performance
- Electrolyte is relatively low-cost

PAFC Disadvantages

- Expensive platinum catalyst
- Susceptible to CO and S poisoning
- Electrolyte is a corrosive liquid that must be replenished during operation

8.3 POLYMER ELECTROLYTE MEMBRANE FUEL CELL

The PEMFC is constructed from a proton-conducting polymer electrolyte membrane, usually a perfluorinated sulfonic acid polymer. Because the polymer membrane is a proton conductor, the anode and cathode reactions in the PEMFC (like the PAFC) are

$$\text{Anode:} \qquad H_2 \rightarrow 2H^+ + 2e^-$$
$$\text{Cathode:} \qquad \frac{1}{2}O_2 + 2H^+ + 2e^- \rightarrow H_2O$$

$$(8.2)$$

A schematic diagram of a PEMFC is provided in Figure 8.3. Figure 8.4 gives a photograph of the power train layout of a Honda FCX fuel cell vehicle powered by PEMFCs.

The polymer membrane employed in PEMFCs is thin (20–200 μm), flexible, and transparent. It is coated on either side with a thin layer of platinum-based catalyst and porous carbon electrode support material. This electrode–catalyst–membrane–catalyst–electrode

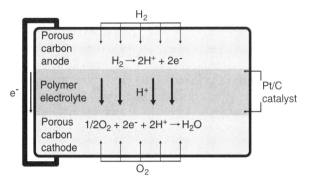

Figure 8.3. Schematic of H_2–O_2 PEMFC. Porous carbon electrodes (often made from carbon paper or carbon cloth) are used for both the anode and the cathode. The electrodes are coated with a Pt catalyst mixture. Water is produced at the anode.

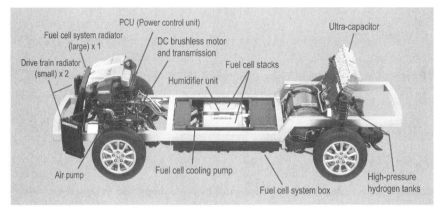

Figure 8.4. Photograph of Honda FCX fuel cell car power train. Two PEMFC stacks generate 86 kW of electricity. The capacitor delivers a high rate of electrical energy to the motor during startup and acceleration and stores electricity recovered during braking. The drive train consists of a motor, transmission, and drive shaft, with the motor producing 80 kW maximum power and 272 N · m maximum torque. Two high-pressure hydrogen storage tanks have a capacity of 156.6 L of hydrogen at 350 atm. The power control unit (PCU) manages the electrical power output of the stacks, capacitor, and motor. The humidifier unit recycles product water from the fuel cell stacks for use in humidifying incoming reactant air to the stack. The cooling system uses three radiators to cool the system and stack. (Photograph courtesy of Honda Motor Co., Ltd.)

sandwich structure is referred to as a membrane electrode assembly (MEA). The entire MEA is less than 1mm thick. Because the polymer membrane must be hydrated with liquid water to maintain adequate conductivity (see Chapter 4), the operating temperature of the PEMFC is limited to 90°C or lower. Due to the low operating temperature, platinum-based materials are the only practical catalysts currently available. While H_2 is the fuel of choice, for low-power (< 1-kW) portable applications, liquid fuels such as methanol and

DIRECT METHANOL FUEL CELL

Methanol, a liquid fuel, is an attractive candidate for portable fuel cell applications due to its high-energy density. The methanol electro-oxidation reaction in acidic electrolytes, such as the PEMFC environment, is

$$CH_3OH + H_2O \rightarrow CO_2 + 6H^+ + 6e^- \tag{8.3}$$

A fuel cell running on methanol requires water as an additional reactant at the anode. It produces CO_2 at the anode as a waste product.

As with the H_2–O_2 PEMFC, the best catalysts for low-temperature methanol fuel cells are Pt based. Unfortunately, the j_0 values for the methanol reaction are quite low, *resulting in large activation overvoltage losses at the anode in addition to the cathode*. The low exchange current density reflects the complexity of the methanol oxidation reaction. The reaction occurs by many individual steps, several of which can lead to the formation of undesirable intermediates, including CO, which acts as a poison.

Carbon monoxide tolerance is provided by alloying the Pt catalyst with a secondary component such as Ru, Sn, W, or Re. Ruthenium is considered to be most effective at providing tolerance. It creates an adsorption site capable of forming OH_{ads} species. These OH_{ads} species react with the bound CO species to produce CO_2, thereby removing the poison. Current DMFCs exhibit power densities of about 30–100 mW/cm^2. (Compare this to H_2 PEMFC power densities of 300–1000 mW/cm^2.) In addition to the considerable activation overvoltage losses at the anode, DMFCs suffer from significant methanol crossover through the electrolyte.

formic acid are also being considered. One such liquid fuel solution, the direct methanol fuel cell (DMFC), is a PEMFC that directly oxidizes methanol (CH_3OH) to provide electricity. The DMFC is under extensive investigation at this time (2005). Some researchers assign these alternative-fuel PEMFCs their own fuel cell class. The text box above provides additional information on the DMFC.

The PEMFC currently exhibits the highest power density of all the fuel cell types (300–1000 mW/cm^2). It also provides the best fast-start and on–off cycling characteristics. For these reasons, it is well suited for portable power and transport applications. Fuel cell development at most of the major car companies is almost exclusively focused on the PEMFC.

PEMFC Advantages

- Highest power density of all the fuel cell classes
- Good start–stop capabilities
- Low-temperature operation makes it suitable for portable applications

PEMFC Disadvantages

- Uses expensive platinum catalyst

- Polymer membrane and ancillary components are expensive
- Active water management is often required
- Very poor CO and S tolerance

8.4 ALKALINE FUEL CELL

The AFC employs an aqueous potassium hydroxide electrolyte. In contrast to acidic fuel cells where H^+ is transmitted from the anode to the cathode, in an alkaline fuel cell OH^- is conducted from the cathode to the anode. The anode and cathode reactions are therefore

$$\text{Anode:} \qquad\qquad H_2 + 2OH^- \rightarrow 2H_2O + 2e^-$$
$$(8.4)$$
$$\text{Cathode:} \qquad \frac{1}{2}O_2 + 2e^- + H_2O \rightarrow 2OH^-$$

Thus water is consumed at the cathode of an AFC while it is produced (twice as fast) at the anode. If the excess water is not removed from the system, it can dilute the KOH electrolyte, leading to performance degradation. A schematic diagram of an AFC is provided in Figure 8.5. Figure 8.6 gives a photograph of an AFC fuel cell unit that was used on the NASA Apollo missions.

For reasons that are still poorly understood, the cathode activation overvoltage in an AFC is significantly less than in an acidic fuel cell of similar temperature. Under some conditions, nickel (rather than platinum) catalysts can even be used at the cathode. Because the ORR kinetics proceed much more rapidly in an alkaline media compared to an acidic media, AFCs can achieve operating voltages as high as 0.875 V. Remember that a high operating voltage leads to high efficiency—an important point if fuel is at a premium.

Depending upon the concentration of KOH in the electrolyte, the AFC can operate at temperatures between 60 and 250°C. Alkaline fuel cells require pure hydrogen and pure oxygen as fuel and oxidant because they cannot tolerate even atmospheric levels of carbon

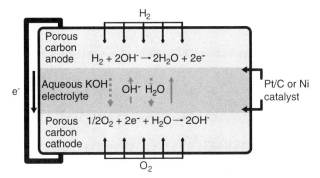

Figure 8.5. Schematic of a H_2–O_2 AFC. Porous carbon or nickel electrodes are used for both the anode and the cathode. Either Pt or non–precious metal catalyst alternatives can be used. Water is produced at the anode and consumed at the cathode; therefore the water must be extracted from the anode waste stream or recycled through the electrolyte using electrolyte recirculation.

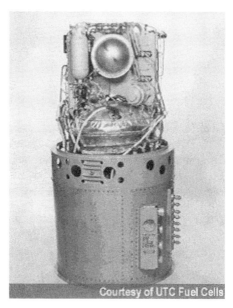

Figure 8.6. Photograph of UTC AFC. These fuel cell units supplied the primary electric power for the Apollo space missions. The units were rated to 1.5 kW with a peak power capability of 2.2 kW, weighed 250 lb, and were fueled with cryogenic H_2 and O_2. Fuel cell performance during the Apollo missions was exemplary. Over 10,000 h of operation were accumulated in 18 missions without an in-flight incident. (Photograph courtesy of UTC Fuel Cells, Inc.)

dioxide. The presence of CO_2 in an AFC degrades the KOH electrolyte as follows:

$$2OH^- + CO_2 \rightarrow CO_3^{2-} + H_2O \tag{8.5}$$

Over time, the concentration of OH^- in the electrolyte declines. Additionally, K_2CO_3 can begin to precipitate out of the electrolyte (due to its lower solubility), leading to significant problems. These issues can be partially mitigated by the use of CO_2 scrubbers and the continual resupply of fresh KOH electrolyte. However, both solutions entail significant additional cost and equipment.

Due to these limitations, the AFC is not economically viable for most terrestrial power applications. However, the AFC demonstrates impressively high efficiencies and power densities, leading to an established application in the aerospace industry. Alkaline fuel cells were employed on the Apollo missions as well as on the Space Shuttle Orbiters.

AFC Advantages

- Improved cathode performance
- Potential for non–precious metal catalysts
- Low materials costs, extremely low cost electrolyte

AFC Disadvantages

- Must use pure H_2–O_2
- KOH electrolyte may need occasional replenishment
- Must remove water from anode

8.5 MOLTEN CARBONATE FUEL CELL

The electrolyte in the MCFC is a molten mixture of alkali carbonates, Li_2CO_3 and K_2CO_3, immobilized in a $LiOAlO_2$ matrix. The carbonate ion, CO_3^{2-}, acts as the mobile charge carrier in the MCFC. The anode and cathode reactions are therefore

$$\text{Anode:} \qquad H_2 + CO_3^{2-} \rightarrow CO_2 + H_2O + 2e^-$$
$$\text{Cathode:} \qquad \tfrac{1}{2}O_2 + CO_2 + 2e^- \rightarrow CO_3^{2-}$$

$$(8.6)$$

In the MCFC, CO_2 is produced at the anode and consumed at the cathode. Therefore MCFC systems must extract the CO_2 from the anode and recirculate it to the cathode. (This situation contrasts with the AFC, where CO_2 must be excluded from the cathode.) The CO_2 recycling process is actually less complicated than one might suppose. Typically, the waste stream from the anode is fed to a burner, where the excess fuel combusts. The resulting mixture of steam and CO_2 is then mixed with fresh air and supplied to the cathode. The heat released at the combustor preheats the reactant air, thus improving the efficiency and maintaining the operating temperature of the MCFC. A schematic diagram of a MCFC is provided in Figure 8.7. Figure 8.8 gives a photograph of a 25-kW pressurized MCFC system.

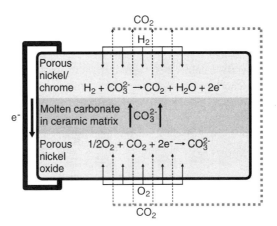

Figure 8.7. Schematic of H_2–O_2 MCFC. The molten carbonate electrolyte is immobilized in a ceramic matrix. Nickel-based electrodes provide corrosion resistance, electrical conductivity, and catalytic activity. The CO_2 must be recycled from the anode to the cathode to sustain MCFC operation since CO_3^{2-} ions are otherwise depleted. Water is produced at the anode.

(*a*)

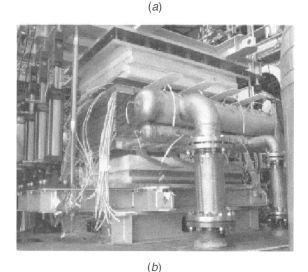

(*b*)

Figure 8.8. Photograph of a 25-kW pressurized MCFC system. (*a*) The individual single cell measures 120 cm $\times$ 81.4 cm $\times$ 0.65 cm (6000 cm^2 of active cell area) and produces 625 watts. (*b*) The stack system is composed of 40 single cells to produce 25 kW in total. During operation, the entire system is enclosed in a pressurized vessel at 3 atm to improve performance. (Photograph courtesy of Korea Institute of Science and Technology/Korea Electric Power Research Institute.)

The electrodes in a typical MCFC are nickel based; the anode usually consists of a nickel/chromium alloy while the cathode consists of a lithiated nickel oxide. At both electrodes, the nickel provides catalytic activity and conductivity. At the anode, the chromium additions maintain the high porosity and surface area of the electrode structure. At the cathode, the lithiated nickel oxide minimizes nickel dissolution, which could otherwise adversely affect fuel cell performance.

The relatively high operating temperature (650°C) of the MCFC provides fuel flexibility. The MCFC can run on hydrogen, simple hydrocarbons (like methane), and simple alcohols. Carbon monoxide tolerance is not an issue for MCFCs; rather than acting as a poison, CO acts as a fuel!

Due to stresses created by the freeze–thaw cycle of the electrolyte during start-up/shutdown cycles, the MCFC is best suited for stationary, continuous power applications. The electrical efficiency of a typical MCFC unit is near 50%. In combined heat and power applications efficiencies could reach close to 90%.

MCFC Advantages

- Fuel flexibility
- Non–precious metal catalyst
- High-quality waste heat for cogeneration applications

AFC Disadvantages:

- Must implement CO_2 recycling
- Corrosive, molten electrolyte
- Degradation/lifetime issues
- Relatively expensive materials

8.6 SOLID-OXIDE FUEL CELL

The SOFC employs a solid ceramic electrolyte. The most popular SOFC electrolyte material is yttria-stabilized zirconia (YSZ), which is an oxygen ion (oxygen vacancy) conductor. Since O^{2-} is the mobile conductor in this case, the anode and cathode reactions are

$$\text{Anode:} \qquad H_2 + O^{2-} \rightarrow H_2O + 2e^-$$
$$\text{Cathode:} \qquad \tfrac{1}{2}O_2 + 2e^- \rightarrow O^{2-} \tag{8.7}$$

In a SOFC, water is produced at the anode, rather than at the cathode, as in a PEMFC. A schematic of a SOFC is provided in Figure 8.9. Figure 8.10 is a photograph of an SOFC prototype.

The anode and cathode materials in a SOFC are different. The fuel electrode must be able to withstand the highly reducing high-temperature environment of the anode, while the air electrode must be able to withstand the highly oxidizing high-temperature environment of the cathode. The most common material for the anode electrode in the SOFC is a nickel–YSZ cermet (a cermet is a mixture of ceramic and metal). Nickel provides conductivity and catalytic activity. The YSZ adds ion conductivity, thermal expansion compatibility, and mechanical stability and maintains the high porosity and surface area of the anode structure. The cathode electrode is usually a mixed ion-conducting and electronically conducting ceramic material. Typical cathode materials include strontium-doped lanthanum manganite

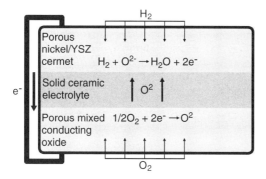

Figure 8.9. Schematic of H_2–O_2 SOFC. The ceramic electrolyte is solid state. A nickel–YSZ cermet anode and a mixed conducting ceramic cathode provide the required thermal, mechanical, and catalytic properties at high SOFC operating temperatures. Water is produced at the anode.

Figure 8.10. Photograph of Siemens-Westinghouse 220-kW hybrid SOFC/micro gas-turbine system. This system was delivered to Southern California Edison in May 2000. (Courtesy of Siemens-Westinghouse.)

(LSM), lanthanum–strontium ferrite (LSF), lanthanum–strontium cobaltite (LSC), and lanthanum strontium cobaltite ferrite (LSCF). These materials show good oxidation resistance and high catalytic activity in the cathode environment.

The operating temperature of the SOFC is currently between 600 and 1000°C. The high operating temperature provides both challenges and advantages. The challenges include stack hardware, sealing, and cell interconnect issues. High temperature makes the materials requirements, mechanical issues, reliability concerns, and thermal expansion matching tasks more difficult. Advantages include fuel flexibility, high efficiency, and the ability to employ cogeneration schemes using the high quality waste heat that is generated. The electrical efficiency of the SOFC is about 50–60%; in combined heat and power applications, efficiencies could reach 90%.

An intermediate-temperature (400–700°C) SOFC design could remove most of the disadvantages associated with high-temperature operation while maintaining the most significant SOFC benefits. Such SOFCs could employ much cheaper sealing technologies and robust, inexpensive metal (rather than ceramic) stack components. At the same time, these SOFCs could still provide reasonably high efficiency and fuel flexibility. However, there are still many fundamental problems that need to be solved before the routine operation of lower temperature SOFCs can be achieved.

SOFC Advantages

- Fuel flexibility
- Non–precious metal catalyst
- High quality waste heat for cogeneration applications
- Solid electrolyte
- Relatively high power density

SOFC Disadvantages

- Significant high-temperature materials issues
- Sealing issues
- Relatively expensive components/fabrication

8.7 SUMMARY COMPARISON

Currently, none of the fuel cell types are ready for commercial application. Until significant cost, power density, reliability, and durability improvements are made, fuel cells will remain a niche technology. Of the five fuel cell types we have discussed, PEMFCs and SOFCs offer the best prospects for continued improvement and eventual application. While PAFCs and AFCs benefited from early historical development, the other fuel cell types have caught up and offer further advantages that will make them more attractive in the long run. Due to their high energy/power density and low operating temperature, the PEMFC and the DMFC appear uniquely suited for portable power applications. Both the PEMFC and the SOFC can be applied to residential power and other small-scale stationary power applications. High-power applications (above 250 kW or so) are best served by SOFC and combined-cycle (SOFC–turbine) technology. High-temperature fuel cells offer attractive efficiency and fuel flexibility advantages. They also generate higher quality waste heat which can be used in combined applications. While all fuel cells operate best on hydrogen, those operating at higher temperatures offer improved impurity tolerance and the possibility of internal reforming of hydrocarbon fuels to yield hydrogen. Table 8.1 summarizes the major benifits and characteristics of the five fuel cell classes discussed in this chapter.

CHAPTER SUMMARY

This chapter briefly covered the five major fuel cell types. Different electrolytes lead to differences in reaction chemistry, operating temperature, cell materials, and cell design. These

TABLE 8.1. Comparison Summary of the Five Major Fuel Cell Types

Fuel Cell Type	Electrical Efficiency (%)	Power Density (mW/cm^2)	Power Range (kW)	Internal Reforming	CO Tolerance	Balance of Plant
PAFC	40	150–300	50–1000	No	Poison (<1%)	Moderate
PEMFC	40–50	300–1000	0.001–1000	No	Poison (<50 ppm)	Low–moderate
AFC	50	150–400	1–100	No	Poison (<50 ppm)	Moderate
MCFC	45–55	100–300	100–100,000	Yes	Fuel	Complex
SOFC	50–60	250–350	10–100,000	Yes	Fuel	Moderate

247

differences lead to important distinctions between the relative advantages, disadvantages, and characteristics of the five fuel cell types.

- The five major fuel cell types are phosphoric acid fuel cell (PAFC), polymer electrolyte membrane fuel cell (PEMFC), alkaline fuel cell (AFC), molten carbonate fuel cell (MCFC), and solid-oxide fuel cell (SOFC). They differ from one another on the basis of their electrolyte.

- You should be able to identify and discuss the important differences in reaction chemistry, operating temperature, cell design, catalyst, and electrode material for each of the five major fuel cell types.

- You should be able to write the H_2–O_2 anode and cathode half reactions for each of the five fuel cell classes.

- PAFC advantages include technological maturity, reliability, and low electrolyte cost. Disadvantages include the requirement for expensive platinum catalyst, poisoning susceptibility, and corrosive liquid electrolyte.

- PEMFC advantages include high power density, low operating temperature, and good start–stop cycling durability. Disadvantages include the requirement for expensive platinum catalyst, high-cost membrane and cell components, poor poison tolerance, and water management issues.

- AFC advantages include improved cathode performance, non–precious metal catalyst potential, and inexpensive electrolyte/cell materials. Disadvantages include system complexities introduced by the requirement for water removal at the cathode, occasional replenishment of the KOH electrolyte, and the requirement for pure H_2 and O_2 gas. (The AFC cannot tolerate even atmospheric levels of CO_2.)

- MCFC advantages include fuel flexibility, non–precious metal catalyst, and the production of high-quality waste heat for cogeneration applications. Disadvantages include system complexity introduced by the requirement for CO_2 recycling, corrosive molten electrolyte, and relatively expensive cell materials.

- SOFC advantages include fuel flexibility, non–precious metal catalyst, completely solid-state electrolyte, and the production of high-quality waste heat for cogeneration applications. Disadvantages include system complexity introduced by the high operating temperature, high-temperature cell sealing difficulties (especially under thermal cycling), and relatively expensive cell components/fabrication.

- While all fuel cells run best on H_2 gas, the high-temperature fuel cells can also run on simple hydrocarbon fuels or CO via direct electrooxidation or internal reforming.

- Historically, the PAFC and the AFC benefited from extensive research and development. Today, the PEMFC and the SOFC appear poised to best meet potential applications. PEMFCs are especially suited for portable and small stationary applications while SOFCs appear suited for distributed-power and utility-scale power applications.

CHAPTER EXERCISES

Review Questions

8.1   (a) Why is nickel used in many high-temperature fuel cells? (b) In SOFC anodes, why is YSZ mixed with the nickel? (c) In MCFC anodes, why is chromium added to the nickel?

8.2   What do you think is the single most significant advantage of high-temperature fuel cells compared to low-temperature fuel cells? Defend your answer.

8.3   Draw a diagram similar to the one in Figure 8.9 for a SOFC operating on CO fuel. Show both the anode and cathode half reactions clearly, as well as the reactants, products, and ionic species.

Calculations

8.4   **(a)**   Given the information in the caption of Figure 8.8, calculate the power density (per unit area) for the individual single cell of the MCFC system.

      (b)   Based on the information, estimate the volumetric power density of the MCFC stack system.

8.5   The fuel cell car shown in Figure 8.4 can travel 430 km at 100 km/h with a full tank of hydrogen. Given the information in the caption of Figure 8.4, estimate the power output from the fuel cell during the travel. Assume hydrogen is an ideal gas, and the fuel cell efficiency is 55%.

8.6   Consider a SOFC system with an electrical efficiency of 55%. Suppose the SOFC rejects heat at 800°C.

      (a)   If a heat engine takes this input heat from the fuel cell and rejects it at 100°C, what is the Carnot efficiency of this heat engine?

      (b)   Assume that the practical efficiency of the heat engine is 60% of the Carnot efficiency. In this case, if the heat engine and fuel cell are combined, what would be the net electrical efficiency of the combined system?

CHAPTER 9

OVERVIEW OF FUEL CELL SYSTEMS

In this chapter, we move beyond the single-fuel-cell unit to the complete fuel cell system. The ultimate goal of any fuel cell system is to deliver the right amount of power to the right place at the right time. To meet that goal, a fuel cell system generally includes a *set* of fuel cells in combination with a suite of additional components. A set of fuel cells are required since a single fuel cell provides only about 0.6–0.7 V at operational current levels. Other components besides the fuel cells themselves are needed to keep the cells running. These components include devices that provide the fuel supply, cooling, power regulation, and system monitoring, to name a few. Often, these devices can take up more room (and cost) than the fuel cell unit itself. Those that draw electrical power from the fuel cell are called ancillaries, or parasitic power devices.

The target application strongly dictates fuel cell system design. In portable fuel cell systems, where mobility and energy density are at a premium, there is a strong incentive to minimize system components. In utility-scale, stationary power generation, where reliability and energy efficiency are at a premium, there is a strong incentive to include beneficial system components. The two example fuel cell system designs shown in Figure 9.1 compare these opposing designs.

This chapter covers the major subsystems included in a typical fuel cell design. These subsystems, some of which are illustrated in Figure 9.1, include the following:

- The fuel cell stack (referred to as the *fuel cell subsystem*)
- The thermal management subsystem
- The fuel delivery/processing subsystem
- The power electronics subsystem

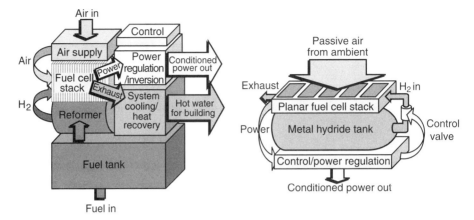

Figure 9.1. Schematic of two fuel cell systems: (*a*) stationary residential-scale fuel cell system; (*b*) portable fuel cell system.

In addition to detailing these subsystems, this chapter also discusses other relevant system design issues such as system pressurization, humidification, and portable fuel cell sizing.

9.1 FUEL CELL STACK (FUEL CELL SUBSYSTEM)

As we have learned, the voltage of a single fuel cell is limited to about 1 V. Furthermore, we recognize that, under load, the output voltage of a single hydrogen fuel cell is typically 0.6–0.7 V. This range generally corresponds to an operational "sweet spot" where the electrical efficiency of the fuel cell is still tolerable (around 45%) and the power density output of the fuel cell is near its maximum. However, most real-world applications require electricity at several, tens, or even hundreds of volts. How do we get 0.6-V fuel cells to supply the high-voltage requirements of real-world applications? One option is to interconnect multiple fuel cells in series. Connected in series, fuel cell voltages sum. This technique, known as fuel cell "stacking," permits fuel cell systems to meet any arbitrary voltage requirement.

In addition to building voltage, a stack design should also meet the following requirements:

- Simple and inexpensive to fabricate
- Low-loss electrical interconnects between cells
- Efficient manifolding scheme (to provide reactant gas distribution)
- Efficient cooling scheme (for high-power stacks)
- Reliable sealing arrangements between cells

Figure 9.2 illustrates the most common form of fuel cell interconnection, referred to as *vertical* or *bipolar plate* stacking. In this configuration, a single conductive flow plate is

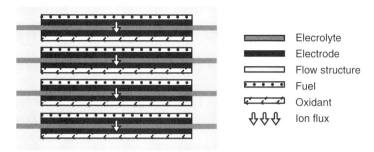

Figure 9.2. Vertical stack interconnection. Fuel cells are serially interconnected via bipolar plates. A bipolar plate simultaneously acts as the anode of one cell and cathode of the neighboring cell. In this diagram, the flow structures, which must be conductive, act as bipolar plates.

in contact with both the fuel electrode of one cell and the oxidant electrode of the next, connecting the two fuel cells in series. The plate serves as the anode in one cell and the cathode in the next cell, hence the name *bipolar plate*. Bipolar stacking is similar to how batteries are stacked on top of one other in a flashlight. Bipolar stacks have the advantage of straightforward electrical connection between cells and exhibit extremely low ohmic loss due to the huge electrical contact area between cells. The bipolar plate design leads to fuel cell stacks which are robust. Most conventional PEMFC stacks adapt this configuration.

Bipolar configurations can be hard to seal. Consider the fuel cell assembly shown in Figure 9.3. It should be apparent from this 3D view that gas will leak out the edges of the porous and gas-permeable electrodes unless edge seals are provided around every cell in

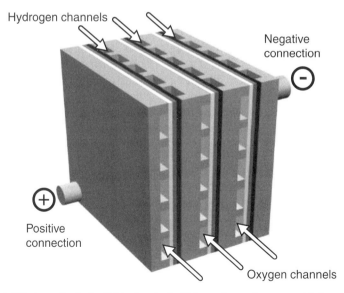

Figure 9.3. A 3D view of a fuel cell bipolar stack. Unless edge seals are provided around each cell, it is clear that this stack will leak.

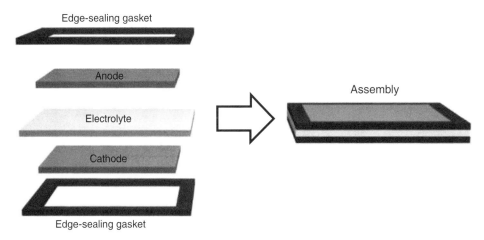

Figure 9.4. An example sealing method which incorporates gaskets around the edges of each cell.

the stack. A common way to provide edge seals is to make the electrolyte slightly larger than the porous electrodes and then fit a sealing gasket around both sides. This technique is illustrated in Figure 9.4. Under compression, the edge gaskets create a gas-tight seal around each cell.

Planar interconnection designs have recently been explored as alternatives to vertical stacking. In planar configurations, cells are connected *laterally* rather than *vertically*. While planar designs are less amenable to large-scale power systems due to their increased electrical resistance losses, the format yields form factor advantages for certain portable applications such as laptop computers or personal digital assistants. Figure 9.5 illustrates two possible planar interconnection configurations. The upper diagram presents the so-called *banded membrane* design, in which the cathode of one cell is electrically connected to the anode of another cell across (or around) the membrane. Such construction can yield better volumetric packaging compared to conventional vertical stacks in low-power applications. However, the most critical disadvantage of this configuration is that interconnections must ultimately cross from one side of the membrane to the other. These cross-membrane inter-

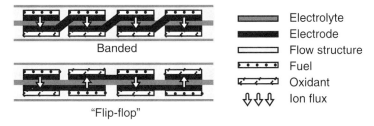

Figure 9.5. Planar series interconnection. Two planar interconnection schemes are shown, the banded and flip-flop designs. In contrast to the banded configuration, the flip-flop scheme has single-level interconnects that never cross the membrane plane.

connections are made at the outer perimeter of a cell array by "edge tabs," or by routing breaches through the central area of the membrane. Interconnection at the perimeter limits design flexibility and may require longer conductor lengths, thereby increasing resistive losses. Breach interconnection through the membrane presents an extremely difficult challenge with respect to local sealing, and the problem is particularly severe for polymer electrolytes that may deform grossly according to humidity level. To overcome the challenges associated with the banded membrane design, the planar *flip-flop* configuration has been proposed. The lower diagram in Figure 9.5 illustrates such a configuration. The most prominent feature of the flip-flop design is the interconnection of electrodes from two different cells on the same side of the membrane.

For SOFCs, sealing issues make the planar and vertical stacking arrangements shown in Figures 9.2 and 9.5 less desirable. Although these designs have been successfully implemented for SOFCs, a stacking arrangement which minimizes the number of seals is preferred. One highly successful method to minimize seals is to employ a tubular geometry, as shown in Figure 9.6. Tubular geometries are especially useful for high-temperature fuel cells, which encounter large temperature gradients. Over large temperature gradients, the thermal expansion coefficients of different materials are more difficult to match and therefore sealing is more challenging. The SOFC systems from Siemens-Westinghouse use this technique. A photograph of a Siemens-Westinghouse tubular fuel cell stack is shown in Figure 9.7.

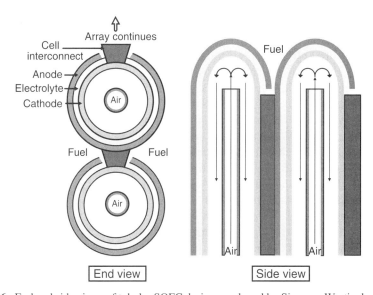

Figure 9.6. End and side views of tubular SOFC design employed by Siemens-Westinghouse. Air is fed through the inside of the tubes while the fuel stream is fed along the outside of the tubes. Series stacking is accomplished by the continuation of more cells in the vertical direction, while parallel stacking can be accomplished by the addition of cells in the horizontal direction.

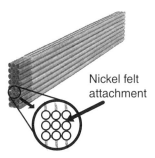

Figure 9.7. Photograph and end-on detail of a small (24 cell) stack of Siemens-Westinghouse tubular SOFCs. Each tube is 150 cm long with a diameter of 2.2 cm. (Courtesy of Siemens-Westinghouse.)

9.2   THE THERMAL MANAGEMENT SUBSYSTEM

As we know, fuel cells are usually only about 30–60% electrically efficient at typical operating power densities. Energy not converted into electrical power is dissipated as heat. If the rate of heat generation is too high, the fuel cell stack can overheat. If stack cooling is not sufficient, the stack may exceed its recommended operating temperature limits, or thermal gradients may arise within the stack. Thermal gradients within the stack can have a negative effect on cell performance by causing cells to operate at different voltages. In such cases, the fuel cell needs active cooling to maintain its optimal operating temperature and avoid thermal gradients within the stack. Cooling requirements strongly depend on fuel cell type and size. Small, low-temperature fuel cells (such as PEMFCs) can frequently get by with "passive" cooling (cooling via natural convection), while high-temperature fuel cells (such as SOFCs and MCFCs) and larger low-temperature fuel cells (PEMFCs and PAFCs) require "active" cooling (cooling via forced convection). High-power-density automotive stacks require active liquid cooling.

As mentioned above, low-power portable PEMFC systems (<100 W) often need no cooling other than what is naturally provided by the ambient. As the fuel cell size decreases, surface-to-volume ratios increase. Natural cooling from the flowing reactants and products and heat transfer from the stack's walls to the surrounding air via natural convection is sufficient to counterbalance heat production from within the fuel cell. In fact, small PEMFC systems can actually benefit from *self-heating* effects. Because PEMFCs work best at 60–80°C, small PEMFC systems can be designed to heat themselves to this temperature range by careful heat transfer design for a range of electric power output levels.

Larger portable systems (>100 W) generally require cooling by forced convection of fluids. Figure 9.8 shows an example bipolar-plate design which includes additional channels for active air cooling. An actively cooled stack will also need an ancillary device like a fan, blower, or pump to circulate fluid through the cooling channels. Unfortunately, this ancillary device will consume some of the power generated by the fuel cell, referred to as parasitic power. The choice of fan, blower, or pump depends on the required cooling rate and the resulting pressure drop that must be overcome in the coolant channels. Fans and blowers consume little power but work best for low-pressure situations. Pumps are needed when higher fluid air flow rates (and consequently greater pressure drops) are required.

Additional internal channels for cooling

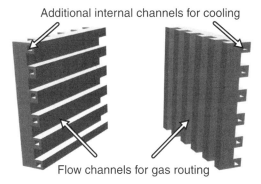

Flow channels for gas routing

Figure 9.8. Examples of fuel cell bipolar plates with additional internal channels provided for integrated cooling of fuel cell stack.

The effectiveness of a cooling system can be evaluated by considering the amount of heat removal it accomplishes compared to the electrical power it consumes:

$$\text{Effectiveness} = \frac{\text{heat removal rate}}{\text{electrical power consumed by fan, blower, or pump}} \qquad (9.1)$$

Effectiveness ratios of 20–40 are generally attainable for well-designed cooling systems.

High-power-density stacks often employ active liquid cooling instead of active cooling with a gas (such as air). Active liquid cooling is used when the volume of the fuel cell stack is constrained (such as in automotive applications). Because the heat capacities of liquids are much greater than the heat capacities of gases, small liquid-cooled channels will be able to remove far more heat than their gas-cooled equivalents. In a liquid-based cooling system, the cooling liquid typically needs to be recycled, as only a limited onboard supply is available (unlike air cooling). This may add complexity to the system. If the cooling liquid is water, it must be deionized so that it cannot carry an electric current. Most automotive fuel cell stacks (50–90 kW) are liquid cooled; some use a water–glycol mixture.

By contrast, high-temperature fuel cells like MCFCs and SOFCs require high temperatures to operate, and therefore employ different cooling designs. The cells are also less susceptible to overheating. In fact, the generated heat is often used beneficially. Heat released by the electrochemical oxidation at these high-temperature cells is used to (1) provide heat for the high-temperature electrochemical reactions at the cells themselves, (2) preheat inlet gases, and (3) provide heat for necessary upstream chemical reactions, which break down hydrocarbons (known as internal reforming, discussed in the following sections). During normal operation, excess flows of process air are often sufficient to cool these stacks.

The above sections focused on the importance of cooling fuel cell stacks to avoid (1) stack overheating and (2) large temperature gradients within the stack. Cooling fuel cell stacks is also important from another perspective: Heat released by the stack can often be used for a useful purpose. Heat released by the stack can be recovered for (1) internal system heating and (2) external heating. Examples of internal heating include (1) preheating the inlet gases to the fuel cell stack and (2) vaporizing water to humidify inlet gases for the stack. Examples of external heating include (1) using an automotive fuel cell system to provide space heating for the passengers in a vehicle and (2) using a stationary fuel cell system to provide space heating and hot water for a building. Heat recovery for both internal system heating and external heating is discussed in detail in Chapter 10. Heat can be recovered not only from the fuel cell stack but also from other system components, as discussed in Chapter 10.

> **Example 9.1.** The fuel cell system shown on the left of Figure 9.1 is a MCFC that produces 200 kW of electric power with an electrical efficiency of 52% based on the heating value of natural gas fuel consumed by it. (1) Calculate the quantity of heat released by the fuel cell. Assume any energy not produced as electric power from the fuel cell stack is released as heat. (2) You would like to use the heat released by the fuel cell to heat a building. Assume that you can recover 70% of the available heat for this purpose, with 30% of the available heat lost to the surroundings. Calculate the amount of heat recovered and the amount lost to the environment.
>
> *Solution:*
>
> 1. Based on Chapter 2, the real electrical efficiency of the fuel cell stack is described by
>
> $$\epsilon_R = \frac{P_e}{\Delta \dot{H}_{(HHV),fuel}} \qquad (9.2)$$
>
> where P_e is the electrical power output of the fuel cell stack. We assume any energy that is not produced as electric power from the stack is produced as heat. This assumes that the parasitic power draw from pumps, compressors, and other components is negligible. The amount of heat released by the fuel cell is the maximum quantity of recoverable heat ($d H_{MAX}$). The maximum heat recovery efficiency ($\epsilon_{H,MAX}$) is
>
> $$\epsilon_{H,MAX} = 1 - \epsilon_R = 1 - 0.52 = 0.48 = 48\% \qquad (9.3)$$
>
> The amount of heat released by the fuel cell is
>
> $$d\dot{H}_{MAX} = \frac{(1 - \epsilon_R)P_e}{\epsilon_R} = \frac{(1 - 0.52)200 \text{ kW}}{0.52} = 185 \text{ kW} \qquad (9.4)$$
>
> 2. The amount of heat recovered is 0.70×185 kW $= 130$ kW and the amount of heat lost to the environment is 0.30×185 kW $= 55$ kW.

9.3 FUEL DELIVERY/PROCESSING SUBSYSTEM

Providing fuel for a fuel cell is often the most difficult task that a system designer faces. Almost all practical fuel cells today use hydrogen or compounds containing hydrogen as fuel. As a result, there are effectively two main options for fueling a fuel cell:

1. Use hydrogen directly or
2. Use a hydrogen carrier

A *hydrogen carrier* is a convenient chemical species which is used to convey hydrogen to a fuel cell. For example, methane, CH_4, is a convenient hydrogen carrier because it is far more readily available than hydrogen. If hydrogen is used directly, it must be created first, via one of several processes we will learn more about in Chapter 10. It must also be stored before use.

For stationary fuel cell systems, availability is one of the most important criteria affecting the choice of fuel. By contrast, for portable fuel cells, the storage effectiveness of the fuel is critical. Storage effectiveness can be measured using (1) gravimetric energy density and (2) volumetric energy density:

$$\text{Gravimetric energy density} = \frac{\text{stored enthalpy of fuel}}{\text{total system mass}} \tag{9.5}$$

$$\text{Volumetric energy density} = \frac{\text{stored enthalpy of fuel}}{\text{total system volume}} \tag{9.6}$$

These metrics express the energy content stored by a fuel system relative to the fuel system size. These metrics can be used regardless of whether a direct H_2 storage system or a H_2 carrier system is employed.

The major options for fueling are now discussed in more detail.

9.3.1 H_2 Storage

In a H_2 storage system, the fuel cell is supplied directly with H_2 gas. There are several major advantages to direct hydrogen supply:

- Most types of fuel cells run best on pure H_2.
- Impurity/contaminant concerns are eliminated.
- The fuel cell system is simplified
- Hydrogen has a long storage "shelf-life" (except for liquid H_2).

Unfortunately, H_2 is not a widely available fuel. Furthermore, effective high-density storage systems for H_2 do not exist. The three most common ways to store hydrogen are as follows:

1. As a compressed gas
2. As a liquid
3. In a metal hydride

Each of these storage options is briefly discussed below. Table 9.1 summarizes typical characteristics of each of the three direct H_2 storage methods.

TABLE 9.1. Comparison of Various Direct H_2 Storage Systems

Storage System	Mass Storage Efficiency (% H_2/kg)	Volumetric Storage Density (kg H_2/L)	Gravimetric Storage Energy Density (kWh/kg)	Volumetric Storage Energy Density (kWh/L)
Compressed H_2, 300 bars	3.1	0.014	1.2	0.55
Compressed H_2, 700 bars	4.8	0.033	1.9	1.30
Cryogenic liquid H_2	14.2	0.043	5.57	1.68
Metal hydride (conservative)	0.65	0.028	0.26	1.12
Metal hydride (optimistic)	2.0	0.085	0.80	3.40

Note: The mass and volume of the entire storage system (tank, valves, regulators, tubing) is taken into account in these data.

HYDROGEN STORAGE EFFICIENCY

The effectiveness of a direct hydrogen storage system can also be measured by (1) hydrogen mass storage efficiency and (2) hydrogen volume storage density. These two parameters describe the amount of hydrogen that can be stored by a direct storage system relative to the storage system size:

$$\text{Mass storage efficiency} = \frac{\text{mass of H}_2 \text{ stored}}{\text{total system mass}} \times 100\% \tag{9.7}$$

$$\text{Volume storage density} = \frac{\text{mass of H}_2 \text{ stored}}{\text{total system volume}} \tag{9.8}$$

- **Compressed** H_2. This is the most straightforward way to store hydrogen. The H_2 is compressed to very high pressures inside specially designed gas cylinders. Storage efficiencies are rather modest but generally improve with cylinder size and increased pressurization. Current cylinder technology permits storage pressures as high as 700 bars. However, high-pressure storage can introduce significant safety problems. Additionally, the act of pressurizing the H_2 is energy intensive. Approximately 10% of the energy content of H_2 gas must be expended to pressurize it to 300 bars. Fortunately, as the storage pressure increases still further, the losses do not increase at the same rate. The additional energy expended to further compress the H_2 is balanced by the fact that more H_2 is stored.

- **Liquid** H_2. If hydrogen gas is cooled to 22 K, it will condense into a liquid. Liquefaction permits H_2 storage at low pressure. Liquid hydrogen has the highest mass storage density of the direct H_2 storage options, about 0.071 g/cm^3. The storage container must be a thick, double-walled reinforced vacuum insulator to maintain the cryogenic temperatures. Therefore, although volumetric storage efficiencies are still modest, mass storage efficiencies can be impressive. (For this reason, liquid H_2 is used for the Space Shuttle, where gravimetric energy density is especially important.) Perhaps most problematically, H_2 liquefaction is extremely energy intensive; the energy required to liquify H_2 is approximately 30% of the energy content of the H_2 fuel itself.

- **Metal Hydride.** Common metal hydride materials include iron, titanium, manganese, nickel, and chromium alloys. Ground into extremely fine powders and placed into a container, these metal alloys work like "sponges" which can absorb large quantities of H_2 gas usually by dissociating the H_2 into H atoms, which are then absorbed within the alloy. Upon slight heating, the hydrides will release their stored H_2 gas. Metal hydrides can absorb incredibly large quantities of H_2. In fact, H gas atoms can be packed inside some metal hydrides in a manner that achieves a higher volumetric energy density than liquid hydrogen! Unfortunately, the hydride materials themselves are quite heavy, so gravimetric energy density is modest. Furthermore, the materials are expensive. Metal hydride storage may be most attractive for certain portable applications.

9.3.2   Using a H_2 Carrier

Using a H_2 carrier instead of hydrogen gas can permit significantly higher gravimetric and volumetric energy storage densities. These H_2 carriers are especially attractive for portable applications. Typical H_2 carriers include methane (CH_4), methanol (CH_3OH), sodium borohydride ($NaBH_4$), formic acid (HCO_2H), and gasoline ($C_nH_{1.87n}$).

Hydrogen carriers are also attractive for stationary applications. Because H_2 gas does not occur naturally on its own, it must be derived from another hydrogen-containing compound. Unlike natural gas or oil, we cannot "drill" for hydrogen. Thus, most stationary fuel cells operate on more widely available fuels like methane or biogas. Using these carrier fuels, fuel cells can still offer high electrical efficiency, modularity, and low emissions compared to existing power plant options.

Unfortunately, most H_2 carriers are not directly usable in a fuel cell. Instead, they must be chemically processed to produce H_2 gas, which is then fed to the fuel cell. A few H_2 carriers *are* directly usable. These include methane used in SOFCs and MCFCs and methanol used in direct methanol fuel cells (DMFCs).

To compare the "effectiveness" of H_2 carriers in providing fuel for a fuel cell, it is important to consider how much of the energy stored in the original carrier is actually usable by the fuel cell. For example, the energy density of methanol is considerably greater than that of compressed hydrogen, but a fuel cell may only be able to convert 20% of methanol's energy into electricity, whereas it could convert 50% of the compressed hydrogen's energy into electricity. In this case, the effectiveness of the methanol fuel compared to hydrogen is only 0.40. A H_2 carrier system's effectiveness is defined as the percentage of a carrier's energy that can be converted into electricity in a fuel cell compared to the percentage of the energy in hydrogen gas that can be converted into electricity:

$$\text{Carrier system effectiveness} = \frac{\% \text{ conversion of carrier to electricity}}{\% \text{ conversion of } H_2 \text{ to electricity}} \tag{9.9}$$

Adjustment by this effectiveness value permits a fair comparison between the storage energy density of a direct H_2 system and a H_2 carrier system for portable fuel cells.

Returning to our methanol example, methanol reforming requires a 50% molar mixture of methanol and water, according to the reaction

$$CH_3OH + H_2O \rightarrow CO_2 + 3H_2 \tag{9.10}$$

If a hypothetical methanol fueling system consists of a 1-L 50% methanol–50% water fuel reservoir plus an additional 1-L reformer, the net volumetric energy density for the fueling system would be 1.71 kWh/L. If we assume that the effectiveness ratio for utilizing the energy content carried in this fuel–water mixture is 0.7, then this methanol fuel system would be equivalent to a direct hydrogen system which has a volume storage energy density of 1.2 kWh/L. On a gravimetric basis, this methanol fuel system might be equivalent to a direct hydrogen system with a gravimetric energy density of 1.4 kWh/kg. The storage metrics and effectiveness of several carrier fuel storage systems are detailed in Table 9.2.

As alluded to earlier in this section, there are two major ways to utilize hydrogen carriers. They can be electro-oxidized directly in a fuel cell to generate electricity (but only if they are relatively simple, easily reacted species) or they can be reformed (chemically

TABLE 9.2. Comparison of Various Carrier H$_2$ Storage Systems

Storage System	Gravimetric Storage Energy Density (kWh/kg)	Volumetric Storage Energy Density (kWh/L)	Carrier Effectiveness
Direct methanol (50% molar mix with H$_2$O)	4	3.4	0.40
Reformed Methanol (50% molar mix with H$_2$O)	2	1.7	0.70
Reformed NaBH$_4$ (30% molar mix with H$_2$O)	1.5	1.5	0.90

Note: The mass and volume of the entire storage system (tank, valves, reformer, etc.) is taken into account in these data.

processed) into hydrogen gas, which is then used by the fuel cell to produce electricity. Reforming can be further subdivided according to whether (1) it occurs in a chemical reactor outside the fuel cell (external reforming) or (2) it occurs at the catalyst's surface inside the fuel cell itself (internal reforming). These three options are now briefly discussed.

- **Direct Electro-Oxidation.** Direct electro-oxidation is attractive primarily because it is simple. No additional external chemical reactors or other components are required compared to a normal H$_2$–O$_2$ fuel cell, although different catalysts, electrolytes, and electrode materials may need to be used. Example fuels that can be directly electro-oxidized in a fuel cell include methanol, ethanol, and formic acid. In direct electro-oxidation, electrons are directly stripped from a fuel molecule. The extra steps required to first reform the fuel into hydrogen are thus avoided. As an example, the reaction chemistry of the direct methanol fuel cell was presented in Chapter 8. Unfortunately, fuel cells operating directly on non–hydrogen fuels suffer significant power density and electrical energy efficiency reductions due to kinetic complications. Because of these complications, a fuel cell operating directly on a non–hydrogen fuel needs to be much larger than a fuel cell operating on hydrogen to provide the same power. Often it must be larger by a factor of 10. This need can greatly offset the energy density gains produced by switching to a carrier fuel in the first place. A careful examination of the balance between fuel reservoir size, fuel cell size, and fuel efficiency is required to determine whether direct electro-oxidation of a carrier fuel makes sense.

- **External Reforming.** Fuel processors use heat, often in combination with catalysts and steam, to break down H$_2$ carrier fuels to H$_2$. During a fuel reforming process, additional species such as CO and CO$_2$ may also be produced. At best, these side products dilute the H$_2$ gas fed to the fuel cell, slightly lowering performance. At worst, they can act as poisons to the fuel cell, severely reducing performance. In such cases, additional processing steps are required to increase the H$_2$ content of the gas and remove the poisons before the reformate (reformed H$_2$ gas mixture) is fed to the fuel cell. This processing is especially important for low-temperature fuel cells. Some of these chemical processes release heat (exothermic) while others require heat to be supplied (endothermic). For high-temperature fuel cells, required heat may be supplied by the fuel cell itself. For low-temperature fuel cells, some of the fuel may be burned to provide this heat. The size and complexity of an external fuel processor depend on the type of fuel reformed, whether impurities or poisons need to be removed, and how much reformate needs to be produced. Figure 9.9 shows a few examples of external fuel processors. Chapter 10 discusses in detail the design of fuel processor subsystems.

- **Internal Reforming.** In internal reforming, the reforming process occurs inside the fuel cell stack itself, at the surface of the anode's catalysts. Internal reforming only works with high-temperature fuel cells with certain fuels. In these cells, the high-operating-temperature catalysts work not only for generating electricity but also for carrying out the fuel reforming reactions. In a typical internal reforming scheme, the H_2 carrier gas is mixed with steam before being fed to the fuel cell anode. The gas and steam react over the anode catalyst surface to produce H_2, CO, and CO_2. Carbon monoxide generation in these high-temperature fuel cells is not a problem, as it can be used directly as fuel. Alternatively, CO can react with more steam to produce further H_2. Compared to external reforming, internal reforming presents several major advantages. These include reduced system complexity (the need for an external chemical reactor is eliminated), reduced system cost, higher system efficiency, higher conversion efficiency, and direct heat transfer between endothermic reforming reactions and exothermic electrochemical reations.

(a)   (b)

Figure 9.9. Two example external reformers. (*a*) A Honda Home Energy Station that generates hydrogen from natural gas for use in fuel cell vehicles while supplying electricity and hot water to the home through fuel cell cogeneration functions. This unit, located in New York, is a second-generation model (developed in collaboration with Plug Power Inc.) which unifies a natural gas reformer and pressurizing units into one compact component to reduce the volume. The unit can produce up to 2 standard cubic meters of hydrogen per hour. (Courtesy of Honda Motors Co., Ltd.) (*b*) A Pacific Northwest National Laboratory microfuel processor that converts methanol into hydrogen and carbon dioxide. The system includes a catalytic combustor, a steam reformer, two vaporizers, and a recuperative heat exchanger embedded in a device no larger than a dime! It is the smallest integrated catalytic fuel processor in the world. (Courtesy of Pacific Northwest National Laboratory.)

Direct electro-oxidation is best applied in portable applications, where simple systems, minimal ancillaries, low power, and long run time are needed. Fuel reforming is best applied in stationary applications, where fuel flexibility is important and the excess heat can also be used either by the system or by sources of heat demand outside the system. Currently, fuel reforming technology appears unlikely for portable and transportation applications. In 2004, the U.S. Department of Energy decided to discontinue on-board fuel processor R&D for fuel cell vehicles. It outlined the main reasons as follows:

- Current fuel processing technologies do not meet technical and economic targets.
- There is no clear path forward to meet the criteria necessary for full implementation/integration in fuel cell vehicles.
- Only marginal improvement is expected in efficiency and emissions between a hybrid gasoline–electric vehicle and a fuel cell vehicle operating on gasoline that is reformed on-board the vehicle.

9.3.3 Fuel Delivery/Processing Subsystem Summary

Fuel cell type and application ultimately determine the best fuel delivery subsystem for a given situation. For stationary applications like distributed generation, fuel processing subsystems may operate on locally available fuels such as methane or biogas. For transportation systems, compressed gas H_2 storage is currently a leading candidate. For small portable fuel cells, metal hydride storage and direct electro-oxidation of liquid fuels (especially direct methanol) are leading candidates. While direct H_2 fuel delivery subsystems are relatively simple, carrier-gas–based fuel processing subsystems can be quite complex. Because of their complexity, fuel processing subsystems will be discussed in greater detail in Chapter 10.

Table 9.3 summarizes the relative storage energy densities, advantages, disadvantages, and applications of the major fuel delivery/processing subsystems. Note that these tendencies were extrapolated from real-world subsystems. Storage densities vary considerably depending on the details of the system design, size, and intended application.

9.4 POWER ELECTRONICS SUBSYSTEM

The power electronics subsystem consists of (1) power regulation, (2) power inversion, (3) monitoring and control, and (4) power supply management. These four tasks of the power electronics subsystem will be discussed in detail in the following four sections.

Fuel cell power conditioning generally involves two tasks: (1) *power regulation* and (2) *power inversion*. Regulation means providing power at an exact voltage and maintaining that voltage constant over time, even as the current load changes. Inversion means converting the DC power provided by a fuel cell to AC power, which most electronic devices consume. For almost all fuel cell applications, power regulation is essential. For most stationary and automotive fuel cell systems, inversion is also essential. Stationary systems need to supply electricity to the surrounding AC electric grid and to power AC devices in buildings.

TABLE 9.3. Qualitative Summary of Various Fuel/Fuel System Choices for Mobile and Stationary Fuel Cell Applications

Fuel System	Gravimetric Storage Energy Density	Volumetric Storage Energy Density	Fuel Availability	Fuel Suitability for Fuel Cell	Comments
		Fuel Systems for Mobile Applications			
Compressed H_2	Moderate	Moderate	Low	High	For transportation
Cryogenic H_2	Moderate–high	Moderate	Low	High	Liquefaction is energy intensive
Metal hydride	Low	High	Low	High	Expensive, heavy
Direct methanol	High	High	Moderate	Low–Moderate	For portable applications
Reformed methanol	Moderate–high	Moderate–high	Moderate	Moderate	For transportation applications
Reformed gasoline	Low	Low	High	Low	Expensive, hard to reform
		Fuels for Stationary Generation Applications			
Neat hydrogen	Low	Low	Low	High	Must have H_2 source!
Methane	Moderate	Moderate	High	Moderate	Best for high-temperature fuel cells
Biogas	Low	Low	Low	Moderate	Best for high-temperature fuel cells

Automotive systems often need to invert DC power to AC power for an AC electric motor, which tends to be more efficient than a DC motor. Inversion is unnecessary for some portable fuel cell applications; for example, a fuel cell laptop uses DC power directly. Unfortunately, power conditioning comes at a price, in terms of both economics and efficiency. Power conditioning will typically add about 10–15% to the cost of a fuel cell system. Also, power conditioning reduces the efficiency of a fuel cell system by about 5–20%. Careful selection of the optimal power conditioning solution for a given application is essential. Power regulation and power inversion are discussed next.

9.4.1   Power Regulation

Most applications require power that is delivered at a specific voltage level that is stable over time. Unfortunately, the power provided by a fuel cell is not perfectly stable; a fuel cell's voltage is highly dependent on temperature, pressure, humidity, and flow rate of reactant gases. Cell voltage changes dramatically depending on the current load. For example, looking at the polarization curve of a single cell as shown in Figure 1.9, we see that voltage can experience a 2-to-1 decline with current draw. Also, even if multiple fuel cells are carefully stacked together in series, the voltage of the system will often not be exactly what is desired for a given application. For these reasons, fuel cell power is generally regulated using DC/DC converters. A DC/DC converter takes a fluctuating DC fuel cell voltage as input and converts it to fixed, stable, specified DC voltage output.

There are two major types of DC/DC converters: step-up converters and step-down converters. In a step-up converter, the input voltage from a fuel cell is stepped up to a higher fixed output voltage. In a step-down converter, the input voltage from a fuel cell is stepped down to a lower fixed output voltage. In either case, regardless of the value of the input voltage (and even if it changes in time), it will be stepped to the converter's specified output voltage, within certain limits. While a step-down converter sounds reasonable, a step-up voltage seems impossible. Are we getting something for nothing? The answer is no! In either case, total power must be conserved, minus some losses. For example, a typical step-up converter might step a fuel cell stack's input from 10 V and 20 A to an output of 20 V and 9 A. Although the voltage has been boosted by a factor of 2, the current has been cut by slightly more than one-half. You could calculate the efficiency of this converter by comparing the output power to the input power:

$$\text{Efficiency} = \frac{\text{output power}}{\text{input power}} = \frac{20 \text{ V} \times 9 \text{ A}}{10 \text{ V} \times 20 \text{ A}} = 0.90 \qquad (9.11)$$

This step-up converter is 90% efficient. DC/DC converters are generally 85–98% efficient. Step-down converters are typically more efficient than step-up converters, and converter efficiency improves as the input voltage increases. For this reason, fuel cell stacking is important. While theoretically possible, it would be extremely inefficient to take a single fuel cell at 0.5 V and step it up to 120 V. Figure 9.10 illustrates several examples of the voltage and current relationships for step-up and step-down converters. In a fuel cell, a step-

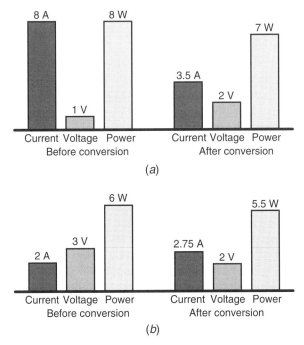

Figure 9.10. Example current–voltage–power relationships for (*a*) a step-up converter and (*b*) a step-down converter.

up converter can be used to maintain a constant voltage, regardless of the load. This idea is shown schematically in Figure 9.11. Keep in mind that, as we just discussed, stepping up the voltage lowers the current output commensurately. Thus, as shown by the arrows, point X on the fuel cell $j–V$ curve corresponds to point X' on the step-up converter curve, while point Y on the fuel cell $j–V$ curve corresponds to point Y' on the step-up converter curve.

9.4.2 Power Inversion

In most stationary applications, such as utility or residential power, the fuel cell will be connected to the surrounding electricity grid or must meet the needs of common household appliances. In these cases, AC rather than DC power is required. Depending on the exact application, either one-phase or three-phase AC power will be required. Utilities and large industrial customers require three-phase power, whereas most residences and businesses need only single-phase AC power. Fortunately, both single-phase and three-phase power inversion technologies are well developed and highly efficient. Similar to DC/DC converters, DC/AC inverters are typically 85–97% efficient.

Figure 9.12 introduces a typical single-phase inverter solution, known as pulse-width modulation. In pulse-width modulation, a series of switches trigger periodic DC voltage pulses through a regulator circuit. By varying the width of these pulses (starting with a

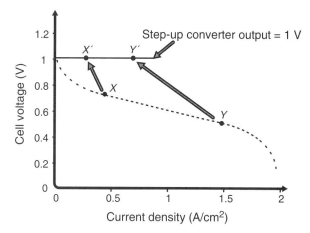

Figure 9.11. A DC/DC converter may be used to transform a fuel cell's variable j–V curve behavior into a constant-voltage output. Up conversion to the higher fixed-voltage output of the converter is accompanied by a commensurate reduction in current, as shown by points X vs. X' and Y vs. Y'.

few short pulses and then increasing the pulse widths before decreasing them again), a reasonable approximation to a sine wave can be created in the resulting current response.

9.4.3 Monitoring and Control System

A large fuel cell system is essentially a complex *electrochemical processing plant*. During operation, many variables such as stack temperature, gas flow rates, power output, cooling, and reforming need monitoring and control. A fuel cell control system generally consists of three separate aspects: a system-monitoring aspect (gauges, sensors, etc., that monitor

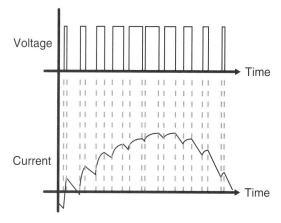

Figure 9.12. Pulse-width voltage modulation allows DC to be transformed into an approximately sinusoidal current waveform.

the conditions of the fuel cell), a system actuation aspect (valves, pumps, switches, etc., that can be regulated to impose changes on the system), and a central control unit, which mediates the interaction between the monitoring sensors and the control actuators. The objective of the central control unit is usually to keep the fuel cell operating at a stable, specified condition. The central control unit can be regarded as the "brains" of the fuel cell system. Most control systems use feedback algorithms to maintain the fuel cell at a stable operating point. For example, a feedback loop might be implemented between a fuel cell stack temperature sensor and the thermal management subsystem. In such a feedback loop, if the control unit senses that the temperature of the fuel cell stack is increasing, it might increase the flow rate of cooling air through the stack. On the other hand, if the fuel cell stack temperature decreases, the control system might reduce the cooling air flow rate. A schematic fuel cell control system is shown in Figure 9.13.

9.4.4 Power Supply Management

Power supply management is the part of the power electronics subsystem used to match the fuel cell system's electrical output with that demanded by the load. Fuel cells have a slower dynamic response than other electronic devices, such as batteries and capacitors, due to lag times in system components such as pumps, compressors, and fuel reformers. Fuel cell systems can operate with or without energy buffers such as batteries or capacitors. Without any energy buffers, the response of fuel cell systems may be on the order of seconds to hundreds of seconds. With energy buffers, the system's response time can be reduced to milliseconds. Power supply management also incorporates a strategy for serving a changing electric load. A midsize car consumes 25 kW of electrical power on average but up to 120 kW at its peak. A fuel cell system's power supply must be designed and controlled to supply power even under large fluctuations in load. In distributed generation applications, power supply management also incorporates a strategy for the fuel cell system to interact with the local grid and to respond to changes in electrical demand from the buildings it

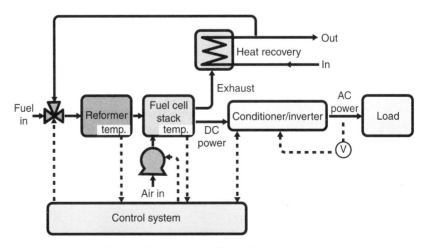

Figure 9.13. Schematic fuel cell control system.

serves. For example, during a power outage, a fuel cell system must either shut down or disconnect from the grid completely because if it continues to feed electrical power into the grid, it could electrocute engineers repairing the outage.

9.5 CASE STUDY OF FUEL CELL SYSTEM DESIGN: SIZING A PORTABLE FUEL CELL

Portable fuel cell systems are subject to several important constraints not faced by stationary fuel cell systems. When designing portable power systems, two critical constraints are the power and lifetime energy requirements of the application. For example, a laptop computer might draw 10 W of power (power requirement) and need to run for 3 h (energy requirement). Given fuel cell power density information, it is relatively straightforward to size a fuel cell system that will produce 10 W of power. Given fuel energy density information, it is also straightforward to size a fuel reservoir that will supply the system sufficiently for 3 hours of use. However, a more difficult task is to determine the *optimal ratio* between the fuel cell size and fuel reservoir size such that the power and energy requirements of the applications are met with minimum possible volume or weight. This optimization is an exercise in fuel cell sizing and illustrates the complex trade-offs between energy density and power density in fuel cell systems.

As an example of the subtleties of system sizing, consider a hypothetical fuel cell system consisting of a 99-L fuel reservoir and a 1-L fuel cell. Suppose that this fuel cell system must deliver 100 W of power. The 1-L fuel cell must therefore obtain a power density of 100 W/L to provide the required power. At 100 W/L, we will assume that the fuel cell is 40% efficient. Thus, the 99-L fuel reservoir, when used at 40% efficiency, effectively provides 39.6 L of extractable fuel energy.

Now, suppose that we resize the system such that the fuel reservoir is 98 L and the fuel cell is 2 L. To deliver 100 W of power, the fuel cell must now obtain a power density of 50 W/L. At this reduced power density, the efficiency of the fuel cell will be greater. (This is because it can run at a lower current density and a higher voltage point and still meet the reduced power density requirements.) Assume that the fuel cell is 50% efficient at a power density of 50 W/L. In this case, the 98-L fuel reservoir used at 50% efficiency effectively provides 49 L of extractable fuel energy. By changing the size of the fuel cell relative to the fuel reservoir, we have greatly extended the lifetime of this system without increasing its total volume! Essentially, we have sacrificed a small amount of the fuel reservoir volume to provide room for a larger fuel cell, but this sacrifice is more than compensated for by the fact that we are using the remaining fuel more efficiently (due to the reduced power density demands on the fuel cell).

Continuing the above example, if we sacrifice even more of the fuel reservoir to further increase the efficiency of the fuel cell, we can generate still greater system lifetimes. At some point, however, an optimum will be reached. How can we determine this optimum? Essentially, given a fixed system volume and a fixed power requirement, we want to maximize the "in-use" time of the system. The following text box describes how this optimum can be calculated given the properties of the fuel cell, fuel reservoir, and volume and power requirements of the system. By calculating this optimum over a range of system sizes and power requirements, a *Ragone* plot may be generated.

OPTIMIZING A PORTABLE FUEL CELL SYSTEM

Optimizing a portable fuel cell essentially involves the following problem: For a given system volume and power requirement, what is the best ratio between the volume of the fuel cell and the volume of the fuel reservoir to maximize the lifetime of the system? (This optimization exercise can also be worked out on a gravimetric basis.) Figure 9.14 illustrates the key terms:

$$p_{FC} = \text{power density of the fuel cell unit}$$

$$x = \text{volume fraction taken up by the fuel cell unit}$$

$$e_F = \text{energy density of the fuel reservoir}$$

$$1 - x = \text{volume fraction taken up by the fuel reservoir}$$

$$V = \text{total volume of the system}$$

$$P = \text{total system power requirement}$$

$$E = \text{total extractable energy from the fuel reservoir}$$

Maximizing the system's in-use time means maximizing E, the total extractable energy from the fuel reservoir. The system power P and the total system volume V are the constraints on the maximization. The power density of the fuel cell unit (p_{FC}) and the energy density of the fuel reservoir (e_F) are the knowns, and the volume fraction taken up by the fuel cell unit relative to the fuel reservoir (x) is the unknown.

This problem can be solved in the following manner. First, construct an expression for the total extractable energy from the fuel reservoir (E), since this is what we are trying to maximize:

$$E = (1 - x)Ve_F\varepsilon \tag{9.12}$$

(continued)

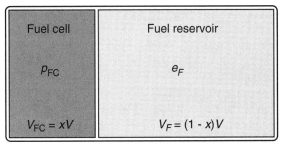

Entire system: V, P, E
$P = xVp_{FC}, E = (1 - x)Ve_F\varepsilon$

Figure 9.14. Optimizing a fuel cell system involves finding the best ratio between fuel cell size and fuel reservoir size so that the system provides the required power for the longest possible time.

In this expression, ε gives the efficiency at which the fuel contained in the fuel reservoir is utilized by the fuel cell and will be a function of the power density of the fuel cell (p_{FC}). In other words, $\varepsilon = \varepsilon(p_{FC})$. At high power densities, the fuel cell will be less efficient at utilizing fuel; at low power densities, the fuel cell will be more efficient. The functional dependence between the fuel cell power density and efficiency must be guessed or determined. (It can be calculated from the fuel cell's i–V curve and stack volume information.) After explicitly acknowledging the functional dependence of ε, Equation 9.12 becomes

$$E = (1 - x)Ve_F\varepsilon(p_{FC}) \tag{9.13}$$

The system must attain a total power given by P. This constrains p_{FC} such that $xVp_{FC} = P$. Introducing this constraint into our optimization equation gives

$$E = (1 - x)Ve_F\varepsilon\left(\frac{P}{xV}\right) \tag{9.14}$$

The volume fraction x which maximizes E can then be determined by setting the derivative of this expression with respect to x equal to zero and solving for x. Inserting x back into Equation 9.14 determines the optimal value of E.

A Ragone plot nicely summarizes the trade-offs between energy density and power density, allowing a designer to compare the maximum design limits for a set of different power systems. NASA engineers designing a portable power source for a space mission (where weight is critical) might pore over a gravimetric Ragone plot like the one shown in Figure 9.15. This plot displays the relationship between gravimetric power density and gravimetric energy density for a variety of portable power systems. A Ragone plot for volumetric power and energy density would likely look similar. A curve on the Ragone plot represents the locus of power density/energy density design points available to a designer using a particular technology. For example, consider the design of a 10-kg portable fuel cell system that needs to deliver 100 W of power (net system power density 10 W/kg). A glance at Figure 9.15 indicates that such a system will provide an energy density of around 250 Wh/kg, and we can thus expect its lifetime to be about 25 h. If the system instead needs a power of 200 W (increasing the net system power density to 20 W/kg), then the energy density of the system will fall to about 150 Wh/kg and we can expect its lifetime to fall to about 8 h. This trade-off occurs because in order to increase the power of the fuel cell system, we have to devote more of the system mass to the fuel cell itself. This restructuring leaves less mass available for fuel. In the extreme, we could imagine designing a fuel cell system where 100% of the system weight is taken by the fuel cell (leaving 0% available for fuel). The power density of such a system would simply correspond to the power density of the fuel cell itself. The energy density of the system would be zero. This design point corresponds to the power axis intercept of the fuel cell Ragone curve. At the other extreme, a fuel cell system that is 100% fuel would have a power density of zero and an energy density which corresponds to the energy density of

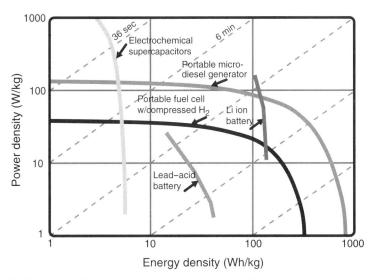

Figure 9.15. Gravimetric Ragone plots for a variety of portable power solutions showing trade-offs between system power density and system energy density. The dashed diagonal lines indicate contours of constant lifetime for various power density/energy density ratios.

the fuel itself. This design point corresponds to the energy density axis intercept of the fuel cell Ragone curve.

Fuel cell systems and combustion systems are fully scalable; their Ragone curves extend fully across the energy density/power density space. In batteries and capacitors, power and capacity are convoluted; their Ragone curves cannot extend over the full energy density/power density space.

CHAPTER SUMMARY

- A fuel cell system generally consists of a set of fuel cells combined with a suite of other system components. A set of fuel cells is required to meet the voltage requirements of real-world applications. The suite of system components typically includes devices to provide cooling, fueling, monitoring, power conditioning, and control for the fuel cell device.

- Fuel cell system design is strongly application dependent. For example, in portable applications, where mobility and energy density are at a premium, there is an incentive to minimize system ancillaries.

- Fuel cell stacking refers to the combination of multiple fuel cells in series to build voltage. The most common stacking arrangements include the vertical (bipolar) configuration, the planar banded configuration, the planar flip-flop configuration, and the tubular configuration.

- As stack size and power density increase, stack cooling becomes more and more essential. Internal air or water cooling channels can be integrated into fuel cell stack designs to provide effective cooling.

- Stack cooling is used to prevent (1) overheating and (2) thermal gradients within the stack.

- Heat released by the stack can be recovered for (1) internal system heating and (2) external heating of source of thermal demand (such as a building's heating loop).

- A cooling system's effectiveness can be computed by comparing the rate of cooling accomplished versus the power consumed by the cooling system. Good designs attain effectiveness ratios of 20–40.

- Fuel candidates for stationary power applications should be evaluated primarily on their availability for fuel cell use. Fuel system candidates for mobile applications should be additionally evaluated on their gravimetric and volumetric storage energy density metrics.

- There are two primary fueling options for fuel cells: direct hydrogen or a hydrogen carrier.

- Advantages of direct hydrogen include high performance, simplicity, and the elimination of impurity concerns. Unfortunately, hydrogen is not a widely available fuel and current hydrogen storage solutions are suboptimal.

- The major direct hydrogen storage solutions include compressed gas storage, cryogenic liquid storage, and reversible metal hydride storage.

- Hydrogen carriers are often far more widely available than hydrogen gas fuel and can greatly facilitate storage.

- Hydrogen carriers can either be directly electro-oxidized in the fuel cell to produce electricity or reformed to produce H_2 gas, which is then electro-oxidized by the fuel cell to produce electricity.

- Other than H_2, only a few simple fuels can be directly electro-oxidized. Direct electro-oxidation assures a simple fuel cell system but often dramatically lowers fuel cell performance.

- Fuel reforming processes produce hydrogen from the carrier stream. Impurities and poisons may also be generated. Depending on the fuel cell, these contaminants may need to be removed from the fuel prior to use. In high-temperature fuel cells, the reforming process can occur inside the fuel cell (internal reforming) rather than in a separate chemical reactor (external reforming).

- For portable applications, direct or reformed methanol fuel systems may provide energy density improvements compared to direct hydrogen storage solutions.

- For stationary applications, reformed methane gas and biogas are the leading fuel solutions due to their wide availability and low cost compared to hydrogen.

- The power delivered by a fuel cell must be conditioned to ensure a stable, reliable electrical output.

- Power conditioning includes power regulation and power inversion. Power regulation uses DC/DC converters to step up or step down the variable voltage of a fuel cell stack to a predetermined, fixed output. Power inversion is used to transform the DC power provided by a fuel cell into AC power. (Power inversion is not needed in all cases.)

- In both power regulation and power inversion, total power is conserved (minus some losses). DC/DC converters and DC/AC inverters are typically 85–98% efficient.
- The fuel cell control unit is the "brains" of the fuel cell system. Control units use feedback loops between system-monitoring elements (sensors) and system actuation elements (valves, switches, fans) to maintain operation within a desired range.
- Power supply management matches the fuel cell system's electrical output with that demanded by the load through the use of energy buffers and special controls.
- Portable fuel cell sizing involves trade-offs between the size of the fuel cell unit and the size of the fuel reservoir unit. Correctly evaluating this sizing trade-off requires a careful optimization.
- These trade-offs can be expressed via Ragone plots, which allow the power density/energy density limitations of multiple portable power solutions to compared against one another.

CHAPTER EXERCISES

Review Questions

9.1 Imagine a combination of the vertical and tubular stacking configurations. Draw a possible stacking arrangement involving a series of stacked donut-shaped cells where H_2 is provided to the stack up the central tubelike core and air is provided around the outside. Do not forget about sealing!

9.2 Why does the U.S government believe fuel cell vehicles with on-board gasoline reforming are unlikely? Do you agree or disagree? Why?

Calculations

9.3 **(a)** Assuming STP conditions, what is the rate of heat generation from a 1000-W hydrogen-fueled PEM running at 0.7 V?

     (b) The fuel cell in part (a) is equipped with a cooling system that has an effectiveness rating of 25. To maintain a steady-state operating temperature, assuming no other sources of cooling, what is the parasitic power consumption of the cooling system?

9.4 In Section 9.3.2, it was stated that a fuel system consisting of a 1-L reformer plus a 1-L fuel reservoir containing a 50:50 molar mix of methanol and water had a net energy density of 1.71 kWh/L (in terms of the heating value of the fuel). Derive this value. Assume STP and use the HHV enthalpy for methanol. Assume that the density of water is 1.0 g/cm^3 and that the density of methanol is 0.79 g/cm^3. Clearly show all steps.

9.5 We would like to compute the carrier system effectiveness of a fuel cell operating on reformed natural gas. Since the reforming process is not perfectly efficient, in this example we assume that the enthalpy content of H_2 provided to the fuel cell amounts to only 75% of the original enthalpy content of the natural gas. Furthermore, we recognize that the H_2 supplied by the reformer will be diluted with CO_2, other inert gases, and perhaps even some poisons. We assume that these diluents lower the efficiency of the fuel cell by 20% compared to operation on pure H_2. What is the total net effectiveness of this reformed natural gas system?

9.6 Assume that the functional relationship between the power density of a fuel cell unit and the electrical efficiency of fuel utilization can be described as

$$\varepsilon(p_{FC}) = A - Bp_{FC} \tag{9.15}$$

In this equation, as the volumetric power density (p_{FC}) of the fuel cell goes up, the energy efficiency ε goes down (for A and B positive). Using the procedure outlined in the optimization text box, derive the expression for the optimal value of X (the volume fraction occupied by the fuel cell unit) given a system volume of V and a power requirement of P. (b) Calculate X if $V = 100$ L, $P = 500$ W, $A = 0.7$ and $B = 0.003$ L/W. Check to make sure that the fuel cell power density required by your solution is reasonable.

CHAPTER 10

FUEL CELL SYSTEM INTEGRATION AND SUBSYSTEM DESIGN

Having introduced fuel cell system components in Chapter 9, we now look in greater detail at the subsystems surrounding the fuel cell, especially the ones needed to deliver fuel and to manage heat. Through the context of a stationary fuel cell system example, we will explore the details of subsystem design. In particular:

1. We will gain a broad overview of how the four primary fuel cell subsystems fit together in the context of a stationary fuel cell example.
2. We will learn in detail about one of the most important subsystems, the fuel processing subsystem.
3. We will learn in detail about another crucial subsystem, the heat management subsystem.

The chapter is split into these three main sections.

The *fuel processing subsystem* is a miniature chemical plant. Its primary purpose is to chemically convert a readily available fuel such as a hydrocarbon (HC) fuel or liquid hydrogen into a hydrogen-rich fluid that can be oxidized at the fuel cell's anode. It also serves to convert fuel or oxidant not consumed at the fuel cell's anode and cathode into useful energy. The *heat management subsystem* is a controlled system of heat exchangers that heat or cool system components, channeling heat from one system component to another and delivering excess heat to an external heat sink (such as a building's heating system) where it can be used. In this chapter, we will learn about (1) the integration of a fuel cell system's four primary subsystems, (2) the chemical engineering building blocks of the fuel processing subsystem and (3) a methodology for managing heat within the system that maximizes energy efficiency.

10.1   INTEGRATED OVERVIEW OF FOUR PRIMARY SUBSYSTEMS

A fuel cell system can convert the chemical energy in a fuel into both electrical power *and* useful heat. Figure 10.1 shows a diagram of a stationary fuel cell system, showing the primary chemical reactors, mass flows, and heat flows (a process diagram). This particular fuel cell system uses a hydrogen fuel cell stack and consumes natural gas fuel. This fuel cell system provides both electricity and heat for a building, an application known as combined heat and power (CHP).

The fuel cell system illustrated in Figure 10.1 contains all four primary subsystems previously introduced in chapter 9: (1) the fuel processing subsystem, (2) the fuel cell subsystem, (3) the power electronics subsystem, and (4) the thermal management subsystem. The fuel processing subsystem consists of the streams of flowing gases (illustrated by arrows) and the series of chemical reactors (illustrated by cylinders). The fuel cell subsystem is shown by the fuel cell stack, the pump and compressors, and the stack's coolant loop. The power electronics subsystem incorporates the thin, dark-shaded electricity lines and connecting boxes in the upper right-hand corner. The thermal management subsystem is represented by dashed heat stream lines with arrows, and physically includes a network of heat exchangers, flowing fluids, and pumps.

COMBINED HEAT AND POWER

Combined heat and power, or cogeneration, is the simultaneous production of electricity and heat from the same energy source. A CHP power plant produces both electric power and heat. This heat can be recovered for a useful purpose, such as warming a building space or water or for a useful industrial process (like brewing beer!). For CHP plants, it is useful to define the term *overall efficiency* (ε_O). The overall efficiency is the sum of the electrical efficiency (ε_R) of the power plant and its heat recovery efficiency (ε_H):

$$\varepsilon_O = \varepsilon_R + \varepsilon_H < 100\% \tag{10.1}$$

where ε_O cannot exceed 100%. Combined heat and power fuel cell systems have achieved $\varepsilon_R = 50\%$ and $\varepsilon_H = 20\%$ for $\varepsilon_O = 70\%$ [41]. Another important term for CHP power plants is the heat-to-power ratio (H/P). The H/P is the ratio of retrievable heat ($d\dot{H}$) to net system electrical power ($P_{e,\text{SYS}}$):

$$\frac{H}{P} = \frac{d\dot{H}}{P_{e,\text{SYS}}} \tag{10.2}$$

For the CHP fuel cell system above, $H/P = \varepsilon_H/\varepsilon_R = 0.20/0.50 = 0.40$. The H/P varies for different types of power plant designs, usually between 0.25 and 2. As another example, your university may use a CHP natural gas power plant to provide electricity and heat to your campus. For such a plant, typical values are $\varepsilon_R = 40\%$ and $\varepsilon_H = 20\%$, $\varepsilon_O = 60\%$, and $H/P = 0.50$.

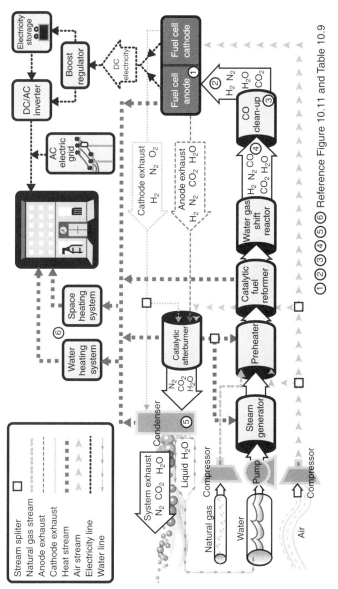

Figure 10.1. Process diagram of CHP fuel cell system.

NATURAL GAS FUEL

Natural gas is one of the most common fuels for heating buildings and for fueling power plants. Natural gas is primarily composed of methane (CH_4). A sample composition of desulfurized natural gas fuel is shown in Table 10.1. Actual natural gas composition varies by region according to the source of the gas (the gas field from which it is extracted) and regulations regarding its purity. Actual natural gas will also at a minimum contain trace quantities of sulfur compounds. Sulfur compounds occur naturally in gas fields but are also added by gas supplier companies as an odorant.

The four subsystems shown in Figure 10.1 perform several functions:

1. The *fuel processing subsystem* chemically converts a HC fuel such as natural gas into a hydrogen- (H_2-) rich gas. This subsystem also purifies the gas to remove or reduce any poisons such as carbon monoxide (CO) or sulfur compounds. This purified gas can then be tolerated by sensitive catalysts (such as platinum) at the fuel cell's electrode and within the fuel processor's chemical reactors. For example, in Figure 10.1, the reactor labeled 3, "CO clean-up," purifies the stream of CO. Finally, this subsystem takes any excess fuel and oxidant not consumed by the fuel cell and recycles them within the system.

2. The *fuel cell subsystem* consists primarily of a fuel cell stack (labeled 1 in Figure 10.1) that converts a H_2-rich gas and oxidant into DC electricity and heat, along with pumps and compressors that convey reactants and products, and the heating and cooling loops required for the stack and these streams.

3. The *power electronics subsystem*, shown in Figure 10.1 by the electricity lines, converts the fuel cell's DC electrical power to AC power used in the building. The power electronics subsystem also balances the building's electrical demand with the electricity supplied by the fuel cell system by using an energy storage device such as a battery or capacitor or by relying on the surrounding AC electrical grid.

TABLE 10.1. Composition of Natural Desulfurized Gas Fuel

CH_4	0.9674	N_2	0.0045
C_2H_6	0.0164	H_2O	0
C_3H_8	0.0019	CO	0
C_4H_{10}	0.0005	CO_2	0.0091
C_5H_{12}	0.0002	H_2	0
O_2	0		

Note: Natural gas is typically greater than 90% methane (CH_4), but compositions vary by region. It typically contains a small percentage of more complex hydrocarbons (HC), including ethane (C_2H_6), propane (C_3H_8), butane (C_4H_{10}), and pentane (C_5H_{12}). Actual natrual gas will also contain trace sulfur compounds.

4. The *thermal management subsystem*, shown in Figure 10.1 by the dashed heat streams, captures heat released by the fuel cell and by the fuel processing subsystem. This heat is used either to warm other system components (such as a steam generator) or to heat the building. Excess heat is rejected to the atmosphere.

The following sections briefly discuss this stationary fuel cell system's four primary subsystems to give a better understanding of their design.

10.1.1 Fuel Processor Subsystem

The details of the fuel processor subsystem are shown in Figure 10.2. The main purpose of this fuel processor is to convert a HC fuel (such as CH_4) into a H_2-rich gas. The system consists of a series of catalytic chemical reactors, heat management devices, reactant and product delivery streams, and extraction equipment. First, liquid water is heated and converted to steam in a steam generator (labeled 1). Steam could be needed for several downstream processes, including humidifying the fuel cell's inlet gases and providing a reactant for the fuel processor. Second, compressed natural gas fuel is combined with compressed air and/or steam and warmed in a preheater (labeled 2). Third, the fuel mixture enters a fuel reformer (labeled 3), where it reacts at high temperature ($>600°C$), often in the presence of a catalyst, producing a H_2-rich stream (referred to as the *reformate stream*). Fourth, the reformate stream enters a water gas shift reactor (labeled 4), which increases the quantity of H_2 in the stream and decreases the CO content. Fifth, in the CO clean-up reactor (labeled 5), the reformate is stripped of CO via either chemical reaction or physical separation, so that the CO will not poison the fuel cell. Sixth, in the afterburner section (labeled 6), exhaust exiting from the fuel cell anode and cathode is combusted catalytically to recover heat for other fuel processing stages or to provide heat to a source of thermal demand outside the fuel cell system. Depending on the H_2-utilization of the fuel cell, a large quantity of H_2 may be available at the fuel cell's exhaust outlet, between 15 and 45% of incoming fuel energy. Also, combustion of H_2 in the afterburner produces water, which can be reused in other parts of the system. Finally, as shown in Figure 10.2, after the catalytic afterburner, a condenser converts steam back to liquid water by cooling this stream. The condenser can be used to capture the latent heat of condensation. In a fuel cell system, a condenser is important for both recapturing heat and recovering liquid water to achieve *neutral system water balance*. Regardless of the source of fuel, almost all fuel processor subsystem designs will incorporate (1) an afterburner, (2) a steam generator, and (3) a condenser to achieve a higher overall efficiency.

The fuel reformer's efficiency (ε_{FR}) is often described in terms of the higher heating value (HHV) of H_2 in the reformate that is subsequently sent to the fuel cell ($\Delta H_{(HHV),H_2}$) compared with the HHV of the fuel input ($\Delta H_{(HHV),fuel}$), including any fuel that must be combusted to provide energy for the reformer itself:

$$\varepsilon_{FR} = \frac{\Delta H_{(HHV),H_2}}{\Delta H_{(HHV),fuel}} \tag{10.3}$$

(For a discussion of HHV, see Chapter 2.) A control volume analysis of the fuel reformer encapsulates chemical reactor 3 in Figure 10.2. The fuel processor's efficiency (ε_{FP}) is

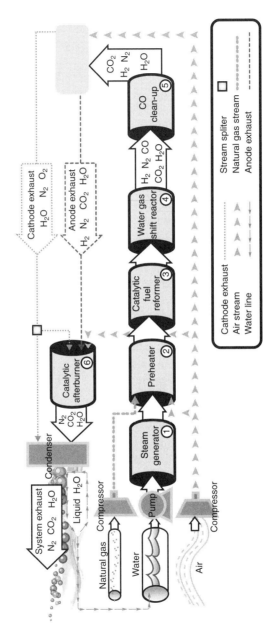

Figure 10.2. Fuel processor subsystem.

NEUTRAL SYSTEM WATER BALANCE

Neutral water balance is achieved when all of the water that is consumed by system components is produced by other components internal to the system. In other words, no additional water needs to be added from an external source. For example, some parts of the fuel cell system may consume liquid water (such as the fuel processor) and other parts of the system may produce it (such as the fuel cell and the condenser). To achieve neutral water balance, water vapor in the fuel cell's exhaust stream should be condensed. A fuel cell system can achieve neutral water balance if

$$\sum \dot{m}_p - \sum \dot{m}_c \geq 0 \tag{10.4}$$

where $\sum \dot{m}_p$ is the sum of the mass flow rates of produced water and $\sum \dot{m}_c$ is the sum of the mass flow rates of consumed water. To achieve neutral water balance, the system needs the sum of condensed water, $\sum \dot{m}_{CD}$, to equal $\sum \dot{m}_c$, or

$$\sum \dot{m}_{CD} = \sum \dot{m}_c \tag{10.5}$$

where

$$\sum \dot{m}_p = \sum \dot{m}_{CD} + \sum \dot{m}_{NCD} \tag{10.6}$$

and $\sum \dot{m}_{NCD}$ is the sum of the mass flow rates of noncondensed water, that is, water that leaves the system as a vapor. The quantity of noncondensed water ($\dot{m}_{NCD}$) depends primarily on the outlet temperature of the condenser or gas stream. In some cases, the inlet air stream contains water vapor from natural humidity that must be accounted for in the system water balance.

described in similar terms, where ε_{FP} is the ratio of the HHV of H_2 in the reformate ($\Delta H_{(HHV),H_2}$) compared with the HHV of the fuel input ($\Delta H_{(HHV),fuel}$), including any fuel that must be combusted to provide energy for the fuel processor itself:

$$\varepsilon_{FP} = \frac{\Delta H_{(HHV),H_2}}{\Delta H_{(HHV),fuel}} \tag{10.7}$$

A control volume analysis of the fuel processor may encapsulate all chemical reactors (nos. 1–6) in Figure 10.2. In both cases, the denominator should incorporate all energy inputs to the fuel reforming and/or fuel processing stages. A realistic ε_{FP} for a natural gas fuel processor is 85%.

10.1.2 Fuel Cell Subsystem

The fuel cell subsystem converts the H_2-rich fuel stream to DC electrical power. As shown in Figure 10.3, a H_2-rich fuel stream and water are fed to the fuel cell's anode. This stream is often intentionally humidified for PEMFC systems so as to maintain electrolyte hydra-

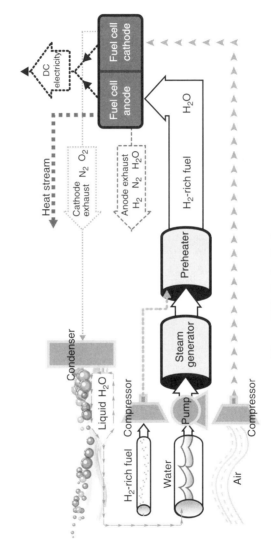

Figure 10.3. Fuel cell subsystem.

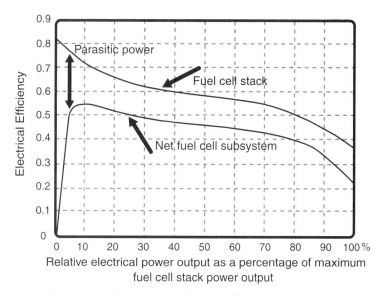

Figure 10.4. Gross and net efficiency of a fuel cell subsystem.

tion.t Simultaneously, compressed air is fed to the fuel cell's cathode. As discussed in previous chapters, in a hydrogen fuel cell, H_2 and O_2 undergo oxidation–reduction reactions at the electrodes and produce electricity, heat, and water [42].

Figure 10.4 shows that the gross fuel cell stack electrical efficiency differs from the net fuel cell subsystem electrical efficiency. The difference between these efficiencies is due to the parasitic power required to run pumps, compressors, and other system devices. This *parasitic power* is drawn from the fuel cell stack itself, thus reducing the net electrical power truly available from the system. Figure 10.4 shows that, for a fuel cell stack, the maximum efficiency occurs at minimum power. By contrast, for a fuel cell subsystem, the minimum efficiency occurs at minimum power. The fuel cell subsystem's net electrical efficiency ($\varepsilon_{R,\text{SUB}}$) is described in terms of the net electrical power of a fuel cell subsystem ($P_{e,\text{SUB}}$) and the HHV of H_2 in the inlet gas ($\Delta \dot{H}_{(\text{HHV}),H_2}$),

$$\varepsilon_{R,\text{SUB}} = \frac{P_{e,\text{SUB}}}{\Delta \dot{H}_{(\text{HHV}),H_2}} \tag{10.8}$$

A realistic $\varepsilon_{R,\text{SUB}}$ is 42%.

> ***Example 10.1.*** The net electrical power of a fuel cell subsystem ($P_{e,\text{SUB}}$) can be expressed as
>
> $$P_{e,\text{SUB}} = P_e - P_{e,P} \tag{10.9}$$
>
> where P_e is the gross electrical power output of the stack and $P_{e,P}$ is the electrical parasitic power. Based on Figure 10.4, develop an equation to approximate the behavior of $P_{e,P}$.

Solution: One possible solution is an equation of the form $P_{e,P} = \alpha + \beta P_e$, where α represents a fixed parasitic power load (such as 1 kW) and βP_e is the parasitic power that is a percentage (such as $\beta = 0.10$) of load. Here, α represents the "upfront energy cost" of operating the system. The term α is likely to refer to the minimum power draw required to turn on components like pumps and compressors.

10.1.3 Power Electronics Subsystem

The power electronics subsystem, detailed in Figure 10.5, incorporates both (1) power conditioning (discussed in Chapter 9) and (2) supply management.

1. *Power conditioning* devices convert a fuel cell's low-voltage DC power to high-quality DC or AC power (normally 120 V and 60 Hz single phase for U.S. domestic applications and three phase for commercial and industrial applications). A fuel cell subsystem produces DC electricity at a voltage that varies with power output level. As we learned in Chapter 1, a single fuel cell's voltage declines at higher currents, potentially by as much as a factor of 2. A fuel cell stack's voltage follows the same pattern and may also deteriorate with time. To compensate for these changes in fuel cell stack voltage, a boost regulator may be used, as shown in Figure 10.5. The boost regulator matches the fuel cell stack's output voltage with the inverter's input voltage by compensating for voltage fluctuations. The inverter then converts the fuel cell stack's DC power into AC power, which may be filtered to enhance its quality.

2. *Supply management* matches the instantaneous supply of electricity with that demanded through electrical storage buffers and/or power from the surrounding utility grid (the network of electricity lines that provide our buildings with power today). To ensure that the electricity demanded by the load can be supplied, a fuel cell system may rely on an electricity storage device such as a battery or capacitor for back-up power. The fuel cell system may charge the storage device, as shown in Figure 10.5, when electricity demand is low. Alternatively, the fuel cell system may rely on the surrounding AC electricity grid to

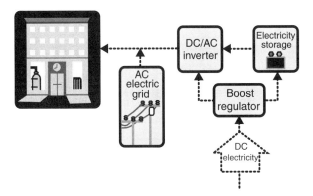

Figure 10.5. Power electronics subsystem.

make up for any additional power needed, as shown in Figure 10.5. Also, a fuel cell system may sell excess electricity back to the surrounding grid.

The power electronics subsystem net electrical efficiency ($\varepsilon_{R,\mathrm{PE}}$) compares the net electrical power of the fuel cell subsystem ($P_{e,\mathrm{SUB}}$) with that of the fuel cell system ($P_{e,\mathrm{SYS}}$):

$$\varepsilon_{R,\mathrm{PE}} = \frac{P_{e,\mathrm{SYS}}}{P_{e,\mathrm{SUB}}} \tag{10.10}$$

If the power electronics subsystem is simplified to include only a boost regulator (a type of DC–DC converter) in series with a DC–AC converter, $\varepsilon_{R,\mathrm{PE}}$ is also

$$\varepsilon_{R,\mathrm{PE}} = \varepsilon_{R,\mathrm{DC-DC}} \times \varepsilon_{R,\mathrm{DC-AC}} \tag{10.11}$$

where $\varepsilon_{R,\mathrm{DC-DC}}$ is the electrical efficiency of the DC–DC converter and $\varepsilon_{R,\mathrm{DC-AC}}$ is the electrical efficiency of the DC–AC converter. If $\varepsilon_{R,\mathrm{DC-DC}} = \varepsilon_{R,\mathrm{DC-AC}} = 96\%$, a realistic $\varepsilon_{R,\mathrm{PE}}$ is 92%.

10.1.4 Thermal Management Subsystem

The thermal management subsystem, shown in Figure 10.6, recovers waste heat from the system for both internal system use and external use, such as for heating a building's air space and hot water. The thermal management subsystem manages heat flows from both the fuel processing subsystem and the fuel cell subsystem. For the thermal management subsystem shown in Figure 10.6, heat is recovered from (1) the catalytic fuel reformer (if

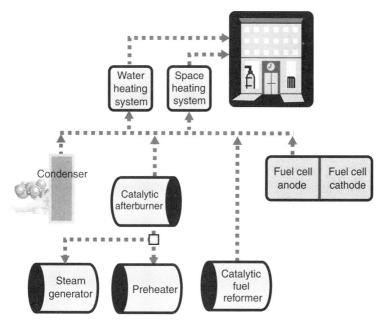

Figure 10.6. Thermal management subsystem.

it operates *exothermically*), (2) the fuel cell stack, (3) the catalytic afterburner, and (4) the condenser. Heat is delivered to (1) the steam generator, (2) the preheater, (3) the building's hot-water heating system, and (4) the building's space heating system. All of these streams are shown in Figure 10.6. Heat can be transferred within the system via both direct and indirect heat transfer. For example, in some fuel processor designs, upstream exothermic processes directly supply heat to downstream endothermic processes. This is the case when heat output from the catalytic afterburner is used to heat the stream generator, as shown in Figure 10.6. The heat recovery efficiency of the thermal management system depends on the design and control of the heat exchangers.

The heat recovery subsystem efficiency can be described in terms of the heat recovery efficiency of the fuel processor subsystem ($\varepsilon_{FP,H}$) and the heat recovery efficiency of the fuel cell subsystem ($\varepsilon_{SUB,H}$), according to

$$\varepsilon_{FP,H} = \varepsilon_{TM}(1 - \varepsilon_{FP}) \tag{10.12}$$

$$\varepsilon_{SUB,H} = \varepsilon_{TM}(1 - \varepsilon_{R,SUB}) \tag{10.13}$$

where ε_{TM} is the thermal management subsystem efficiency, the percentage of heat successfully recovered for a useful purpose compared with the heat available. Well-designed systems of heat exchangers may be able to capture 80% of available heat ($\varepsilon_{TM} = 80\%$).

EXOTHERMIC AND ENDOTHERMIC REACTORS

Some of the chemical reactors in the fuel cell system produce heat; their reactions are *exothermic*. Other reactors are *endothermic*; their reactions require heat to be added. Endothermic reactors are heat sinks and require heat to be conveyed to them from exothermic reactors or other heat sources.

10.1.5   Net Electrical and Heat Recovery Efficiencies

CHP fuel cell systems can achieve high overall efficiencies (ε_O), where

$$\varepsilon_O = \varepsilon_R + \varepsilon_H \tag{10.14}$$

The fuel cell system's electrical efficiency (ε_R) compares the net electrical output of the system with the HHV of the fuel input:

$$\varepsilon_R = \frac{P_{e,SYS}}{\Delta \dot{H}_{(HHV),fuel}} \tag{10.15}$$

where

$$\varepsilon_R = \varepsilon_{FP} \times \varepsilon_{R,SUB} \times \varepsilon_{R,PE} \tag{10.16}$$

$$= \frac{\Delta \dot{H}_{(HHV),H_2}}{\Delta \dot{H}_{(HHV),fuel}} \frac{P_{e,SUB}}{\Delta \dot{H}_{(HHV),H_2}} \frac{P_{e,SYS}}{P_{e,SUB}} \tag{10.17}$$

The fuel cell system's heat recovery efficiency ε_H is the sum of the heat recovery efficiency of the fuel cell system in terms of the original fuel input ($\varepsilon_{SUB,H,fuel}$) and the heat recovery efficiency of the fuel processor ($\varepsilon_{FP,H}$). This can be expressed by

$$\varepsilon_{SUB,H,fuel} = \varepsilon_{FP} \times \varepsilon_{TM} \times (1 - \varepsilon_{R,SUB}) \qquad (10.18)$$

and

$$\varepsilon_H = \varepsilon_{SUB,H,fuel} + \varepsilon_{FP,H} \qquad (10.19)$$

Example 10.2. The text above gives realistic efficiency values for the various subsystems of the fuel cell system shown in Figure 10.1. Based on these efficiencies, (1) calculate the fuel cell system's electrical efficiency, (2) the system's heat recovery efficiency, and (3) the system's overall efficiency and (4) report the H/P.

Solution

1. For the four subsystems discussed above, Table 10.2 summarizes the efficiencies for the four individual subsystems, along with the system's net electrical efficiency (ε_R), calculated as

$$\varepsilon_R = \varepsilon_{FP} \times \varepsilon_{R,SUB} \times \varepsilon_{R,PE} = 0.85 \times 0.42 \times 0.92 = 0.328 \qquad (10.20)$$

$$= 33\% \qquad (10.21)$$

2. Table 10.2 also summarizes the thermal recovery efficiencies for subsystems along with the overall system heat recovery efficiency (ε_H). A thermal management system that is 80% efficient can recover 80% of available heat from the fuel processor subsystem and the fuel cell subsystem, according to

$$\varepsilon_{FP,H} = \varepsilon_{TM} \times (1 - \varepsilon_{FP}) = 0.80(1 - 0.85) = 0.12$$

$$\varepsilon_{SUB,H} = \varepsilon_{TM} \times (1 - \varepsilon_{R,SUB}) = 0.80(1 - 0.42) = 0.46$$

$$\varepsilon_{SUB,H,fuel} = \varepsilon_{FP} \times \varepsilon_{TM} \times (1 - \varepsilon_{R,SUB}) = 0.85[0.80(1 - 0.42)] = 0.39$$

$$\varepsilon_H = \varepsilon_{SUB,H,fuel} + \varepsilon_{FP,H} = 0.12 + 0.39 = 0.51$$

$$= 51\% \qquad (10.22)$$

TABLE 10.2. Electrical Efficiency and Heat Recovery Efficiency for Four Main Subsystems

	Fuel Processing Subsystem	Fuel Cell Subsystem	Power Electronics Subsystem	Thermal Management Subsystem	Overall System
Electrical efficiency	85%	42%	92%	NA	33%
Heat recovery efficiency	12%	46%	NA	80%	51%

3. $\varepsilon_O = \varepsilon_R + \varepsilon_H = 0.33 + 0.51 = 84\%$.
4. $H/P = \varepsilon_H/\varepsilon_R = 0.51/0.33 = 1.55$.

10.2 EXTERNAL REFORMING: FUEL PROCESSING SUBSYSTEMS

As discussed in Chapter 9, H_2 gas can be produced from any substance containing hydrogen atoms, such as water, acids or bases, or hydrocarbon (HC) fuels. One method for producing hydrogen is to convert a HC fuel into a hydrogen-rich gas. This method is known as fuel processing. Most buildings in developed countries are connected to distribution networks for natural gas (NG) or have access to liquid petroleum gas (LPG). Either of these two HC fuels can be chemically processed into a hydrogen-rich gas for use in a fuel cell. A fuel processing subsystem is a miniature chemical plant. The complexity of the fuel processing subsystem depends on the type of fuel cell it serves and the type of fuel it is processing. A fuel processing subsystem consists of a series of catalytic chemical reactors that convert natural gas (or LPG) into a low-impurity, high-hydrogen-content gas. Both the PEMFC and PAFC are sensitive to impurities in their feed gases which might otherwise poison (block) catalyst sites for reaction. Therefore, they require extensive fuel processing systems that employ multiple stages. By contrast, MCFCs and SOFCs generally operate at temperatures high enough to achieve internal reforming, such that the fuel and steam mixture can often be fed directly into the fuel cell stack. Since low-temperature fuel cells have the most stringent fuel processing requirements, we will take a look at a typical fuel processing subsystem for a PEMFC or PAFC. Such a subsystem will probably consist of at least three primary reactor processes (see Figure 10.2):

- Fuel reforming (labeled no. 3)
- Water gas shift reaction (labeled no. 4)
- Carbon monoxide clean-up (labeled no. 5)

Although outside the scope of our discussion, the sulfur in the fuel must also be removed in an upstream processing step. For now, let's examine the three main fuel processing stages.

10.2.1 Fuel Reforming Overview

The overall goal of fuel reforming is to convert a HC fuel into a hydrogen-rich gas. The primary conversion may be accomplished with or without a catalyst via one of three major types of fuel reforming processes:

- Steam reforming (SR)
- Partial oxidation (POX) reforming
- Autothermal reforming (AR)

To compare the effectiveness of various fuel reforming processes, we introduce the concept of H_2 yield (y_{H_2}), which represents the molar percentage of H_2 in the reformate stream at the outlet of the fuel reformer:

$$y_{H_2} = \frac{n_{H_2}}{n} \qquad (10.23)$$

In this equation, n_{H_2} is the number of moles of H_2 produced by the fuel reformer and n is the total number of moles of all gases at the outlet. In a similar manner, we introduce the concept of a steam-to-carbon ratio (S/C), which represents the ratio of the number of moles of molecular water (n_{H_2O}) to the moles of atomic carbon (n_c) in a fuel (such as methane CH_4) in a chemical stream:

$$\frac{S}{C} = \frac{n_{H_2O}}{n_c} \qquad (10.24)$$

Each of the three main reforming processes produce varying H_2 yields, require different steam-to-carbon ratios, and possess unique advantages and disadvantages. The major characteristics of these three reforming processes are described in Tables 10.3 and 10.4. In the following sections, each will be discussed in greater detail.

10.2.2 Steam Reforming

Steam reforming is an endothermic reaction that combines a HC fuel with steam over a catalyst at high temperature, according to

$$C_x H_y + x H_2 O_{(g)} \leftrightarrow x CO + \left(\tfrac{1}{2} y + x\right) H_2 \Rightarrow CO, CO_2, H_2, H_2O \qquad (10.25)$$

The SR of natural gas typically has a H_2 yield of 76% on a dry basis (i.e., no water vapor is in the gas stream) [43]. Because no oxygen in air is involved in the reaction, the outlet stream is not diluted by N_2 in air, and therefore the H_2 yield is the highest of the three reforming types. To increase the H_2 yield, Le Chatelier's principle tells us that operating the reaction with excess water vapor would help shift the reaction's equilibrium to favor H_2 production. To further increase the H_2 yield, the CO in the outlet of the SR reactor can be "shifted" to H_2 via a second reaction, the water–gas shift (WGS) reaction:

$$CO + H_2 O_{(g)} \leftrightarrow CO_2 + H_2 \qquad (10.26)$$

The WGS reaction can increase the H_2 yield by about 5%. The primary SR reactions for methane are summarized in Table 10.5. (In this table and throughout the chapter, enthalpies of reaction are reported at STP. We will use these STP values for back-of-the-envelope calculations in the chapter.)

A steam reformer must be designed to capture heat to sustain its endothermic reaction. A common steam reformer design is the tubular reformer. A tubular reformer consists of a furnace that contains tubes filled with catalysts through which the SR reactants pass. Most SR catalysts are gradually poisoned by sulfur compounds so input fuel must be cleaned to low sulfur levels (10–15 ppm). The endothermic SR reaction takes place inside the tubes. Often, the tubes are heated by the combustion of some of the input fuel. Alternatively,

TABLE 10.3. Comparison of Chemical Reaction Characteristics of Three Primary Fuel Reforming Reactions

Type	Chemical Reaction	Temperature Range (°C)	Hydrogen Output Gas Composition (with Natural Gas Fuel)					Exothermic or Endothermic?
			H_2	CO	CO_2	N_2	Other	
Steam reforming	$C_xH_y + xH_2O_{(g)} \leftrightarrow xCO + (\frac{1}{2}y + x)H_2$ $\Rightarrow CO, CO_2, H_2, H_2O$	700–1000	76%	9%	15%	0%	Trace NH_3, CH_4, SO_x	Endothermic
Partial oxidation	$C_xH_y + \frac{1}{2}xO_2 \leftrightarrow xCO + \frac{1}{2}yH_2$	>1000	41%	19%	1%	39%	Some NH_3, CH_4, SO_x, HC	Exothermic
Autothermal reforming	$C_xH_y + zH_2O_{(g)} + (x - \frac{1}{2}z)O_2$ $\leftrightarrow xCO + (z + \frac{1}{2}y)H_2$ $\Rightarrow CO, CO_2, H_2, H_2O$	600–900	47%	3%	15%	34%	Trace NH_3, CH_4, SO_x, HC	Neutral

Note: The steam reforming reaction produces the highest H_2 yield and the cleanest exhaust. The low H_2 yield for the partial oxidation and autothermal reforming reactions is a result of their intake of air; the O_2 in air partially oxidizes the fuel while the N_2 in air dilutes the H_2 composition in the outlet gas. For all three reactions, the H_2 yield can be increased by downstream use of the water gas shift reaction. In the chemical reaction for steam reforming, the first line shows the typical reactants and products in their correct molar ratios. The second line below this shows the full range of products for an actual reactor, which may include not only CO and H_2 but also CO_2 and H_2O. The chemical reaction for autothermal reforming is shown in a similar manner. Concentrations are noted on a dry basis (i.e., no water vapor in gas stream).

TABLE 10.4. Advantages and Disadvantages of Three Primary H$_2$ Production Methods

Type	Advantages	Disadvantages
Steam reforming	Highest H$_2$ yield	Requires careful thermal management to provide heat for reaction, especially for (a) start-up and (b) dynamic response
		Only works on certain fuels
Partial oxidation	Quick to start and respond because reaction is exothermic	Lowest H$_2$ yield
	Quick dynamic response	Highest pollutant emissions (HCs, CO)
	Less careful thermal management required	
	Works on many fuels	
Autothermal reforming	Simplification of thermal management by combining exothermic and endothermic reactions in same process	Low H$_2$ yield
	Compact due to reduction in heat exchangers	Requires careful control-system design to balance exothermic and endothermic processes during load changes and start-up.
	Quick to start	

Note: Autothermal reforming combines steam reforming and partial oxidation to achieve some of the benefits of both, including simple heat management and quick response. Partial oxidation provides the greatest fuel-type flexibility.

TABLE 10.5. Steam Reforming Reactions

Reaction Number	Reaction Type	Stoichiometric Formula	$\Delta \hat{h}_{rxn}^0$ (kJ/mol)
1	Steam reforming	$CH_4 + 2H_2O_{(g)} \rightarrow CO_2 + 4H_2$	+165.2
2	Water–gas shift reaction	$CO + H_2O_{(g)} \rightarrow CO_2 + H_2$	−41.2
3	Evaporation	$H_2O_{(l)} \rightarrow H_2O_{(g)}$	+44.1

Note: The main steam reforming reaction is endothermic. Vaporized water (steam) is a reactant. The water gas shift reaction increases H_2 yield.

within a fuel cell system, the heat for the endothermic SR reaction could be provided by combusting the anode exhaust gas (the unconsumed fuel exiting the fuel cell's anode) in a catalytic afterburner, such as the one labeled 6 in Figure 10.2.

> ***Example 10.3.*** (1) For an idealized reformer consuming methane (CH_4) fuel and operating with combined SR and WGS reactions, what is the maximum H_2 yield? (2) What is the steam-to-carbon ratio for the combined reactions? (3) In a real fuel reformer, why might you want to operate the reactor with a higher steam-to-carbon ratio? (4) What quantity of heat is consumed by the reaction, assuming, for simplicity, that the reactants and products enter and leave the reactor at STP?
>
> *Solution*
>
> 1. For SR of CH_4, we have
>
> $$CH_4 + H_2O_{(g)} \leftrightarrow CO + 3H_2 \qquad (10.27)$$
>
> and for the WGS reaction, we have
>
> $$CO + H_2O_{(g)} \leftrightarrow CO_2 + H_2 \qquad (10.28)$$
>
> For the two combined reactions, we have
>
> $$CH_4 + 2H_2O_{(g)} \leftrightarrow CO_2 + 4H_2 \qquad (10.29)$$
>
> which is the first reaction (1) shown in Table 10.5. This combined reaction has a hydrogen yield of
>
> $$y_{H_2} = \frac{4 \text{ mol } H_2}{4 \text{ mol } H_2 + 1 \text{ mol } CO_2} = 0.80 \qquad (10.30)$$
>
> or 80%.
>
> 2. The steam-to-carbon ratio is
>
> $$\frac{S}{C} = \frac{n_{H_2O}}{n_c} = 2 \qquad (10.31)$$

3. You might want to operate with a higher steam-to-carbon ratio to reduce carbon deposition and to increase the H_2 yield, according to Le Chatelier's principle. Carbon deposition occurs due to reaction 3 (thermal decomposition) from Table 10.6. Typically, a S/C ratio of 3.5 to 4.0 can prevent carbon formation.

4. According to Table 10.5, 165.2 kJ/mol of CH_4 must be provided to the endothermic reaction if water is in a vapor state. If water enters in a liquid state, an additional 44.1 kJ/mol of H_2O or an additional 88.2 kJ/mol of CH_4 is required. In total, if water enters in a liquid state, 253.4 kJ/mol of CH_4 must be provided to the endothermic reaction if it were to take place at STP.

10.2.3 Partial Oxidation Reforming

Partial oxidation reforming is an exothermic reaction that combines a HC fuel with some oxygen to *partially oxidize* (or partially combust) the fuel into a mixture of CO and H_2, usually in the presence of a catalyst. In *complete* combustion, a HC fuel combines with sufficient oxygen (O_2) to completely oxidize all products to CO_2 and H_2O. In complete combustion, the product stream contains no H_2, CO, O_2, or fuel. For example, the complete combustion of propane (C_3H_8) is

$$C_3H_8 + xO_2 \leftrightarrow yCO_2 + zH_2O \tag{10.32}$$

No H_2, CO, O_2, or C_3H_8 is produced. According to the conservation of mass, the number of moles of H, C, and O must be equal on both sides of the equation. Then, we obtain

$$C_3H_8 + 5O_2 \leftrightarrow 3CO_2 + 4H_2O \tag{10.33}$$

The minimum quantity of O_2 required is 5 mol O_2/ mol C_3H_8. This minimum quantity of O_2 required for complete combustion is called the *stoichiometric* amount of O_2.

In POX (or *partial* combustion), a HC fuel combines with less than the stoichiometric amount of O_2, such that the *incomplete* combustion products CO and H_2 are formed. For example, the incomplete combustion of propane (C_3H_8) is

$$C_3H_8 + xO_2 \leftrightarrow yCO + zH_2 \tag{10.34}$$

According to the conservation of mass, we then obtain

$$C_3H_8 + 1.5O_2 \leftrightarrow 3CO + 4H_2 \tag{10.35}$$

The quantity of O_2 required is 1.5 mol O_2/mol C_3H_8, far less than the stoichiometric amount. Operating with less than the stoichiometric amount of O_2 is also called operating *fuel rich* or O_2 *deficient*. More generally, for any HC fuel, POX is defined as

$$C_xH_y + \tfrac{1}{2}xO_2 \leftrightarrow xCO + \tfrac{1}{2}yH_2 \tag{10.36}$$

As with SR, the H_2 yield can then be further increased by shifting the CO in the outlet to H_2 via the WGS reaction:

TABLE 10.6. Partial Oxidation Reactions

Reaction Number	Reaction Type	Stoichiometric Formula	$\Delta \hat{h}_{rxn}^{0}$ (kJ/mol)
1	Partial oxidation	$CH_4 + \frac{1}{2}O_2 \rightarrow CO + 2H_2$	−35.7
2	Partial oxidation	$CH_4 + O_2 \rightarrow CO_2 + 2H_2$	−319.1
3	Thermal decomposition	$CH_4 \rightarrow C + 2H_2$	+75.0
4	Methane combustion	$CH_4 + 2O_2 \rightarrow CO_2 + 2H_2O_{(l)}$	−880
5	CO combustion	$CO + \frac{1}{2}O_2 \rightarrow CO_2$	−283.4
6	Hydrogen combustion	$H_2 + \frac{1}{2}O_2 \rightarrow H_2O_{(l)}$	−284

Note: Autothermal reforming reactions include these and the steam reforming reactions in Table 10.5.

$$CO + H_2O_{(g)} \leftrightarrow CO_2 + H_2 \qquad (10.37)$$

The primary reactions of the POX reforming process for methane gas are listed in Table 10.6.

> **Example 10.4.** An idealized POX fuel reformer consumes methane (CH_4) and air. (1) What is the maximum H_2 yield? (2) What quantity of heat is released by the reaction if it were to take place at STP? (3) What is the reformer efficiency? The HHV of methane is 55.5 MJ/kg (880 MJ/kmol) and the HHV of H_2 is 142 MJ/kg (284 MJ/kmol) at STP.

Solution

1. Operating on air, for every mole of O_2 we have 3.76 mol N_2, such that

$$C_xH_y + \frac{1}{2}x(O_2 + 3.76N_2) \leftrightarrow xCO + \frac{1}{2}yH_2 + 1.88xN_2 \qquad (10.38)$$

Then, for methane,

$$CH_4 + \frac{1}{2}(O_2 + 3.76N_2) \leftrightarrow CO + 2H_2 + 1.88N_2 \qquad (10.39)$$

Then, the reaction has a hydrogen yield of

$$y_{H_2} = \frac{2 \text{ mol } H_2}{2 \text{ mol } H_2 + 1 \text{ mol } CO + 1.88 \text{ mol } N_2} = 0.41 \qquad (10.40)$$

or 41%. Because O_2 in air is involved in the reaction, the outlet stream is diluted by N_2 in air, and therefore the H_2 yield is the lowest of the three reforming types.

2. According to Table 10.6, 35.7 kJ/mol CH_4 is released by the exothermic reaction at STP.

3. The fuel reformer efficiency in terms of HHV is

$$\varepsilon_{FR} = \frac{\Delta H_{(HHV),H_2}}{\Delta H_{(HHV),fuel}} = \frac{2 \text{ kmol } H_2 (284 \text{ MJ/kmolH}_2)}{1 \text{ kmol } CH_4 (880 \text{ MJ/kmol } CH_4)} = 65\% \quad (10.41)$$

10.2.4 Autothermal Reforming (AR)

Autothermal reforming combines (1) the SR reaction, (2) the POX reaction, and (3) the WGS reaction in a single process. Autothermal reforming combines these reactions such that (1) they proceed in the same chemical reactor and (2) the heat required by the endothermic SR reaction and the WGS reaction is exactly provided by the exothermic POX reaction. Autothermal reforming incorporates SR by including steam as a reactant. Similarly, it incorporates POX by including a substoichiometric amount of O_2 as a reactant. The AR reaction is

$$C_x H_y + z H_2 O_{(l)} + (x - \tfrac{1}{2}z) O_2 \leftrightarrow x CO_2 + \left(z + \tfrac{1}{2}y\right) H_2 \quad (10.42)$$

$$\Rightarrow CO, CO_2, H_2, H_2O \quad (10.43)$$

The value for the steam-to-carbon ratio, here shown as z/x, should be chosen such that the reaction is energy neutral, neither exothermic nor endothermic.

> **Example 10.5.** (1) For methane (CH_4), estimate the steam-to-carbon ratio that enables the AR reaction to be energy neutral. Assume H_2O enters as a liquid and the only products are CO_2 and H_2. For simplicity, assume that the reactants and products enter and leave the reactor at STP. (2) What is the H_2 yield? (3) What is the reformer efficiency?
>
> *Solution*
>
> 1. For the endothermic SR reaction, we have
>
> $$CH_4 + 2H_2O_{(l)} \leftrightarrow CO_2 + 4H_2 + 253.4 \text{ kJ/mol } CH_4 \quad (10.44)$$
>
> For the exothermic POX reaction, we have
>
> $$CH_4 + \tfrac{1}{2}O_2 \leftrightarrow CO + 2H_2 - 35.7 \text{ kJ/mol } CH_4 \quad (10.45)$$
>
> For the products of these combined reactions to produce only CO_2 and H_2, the CO in the POX reaction must be shifted to H_2 via the WGS reaction. Table 10.7 shows the solution to this problem. Table 10.7 shows the SR (1), POX (2), and WGS (3) reactions and the heat of reaction for each. By adding reaction 2 (POX) to reaction 3 (WGS), we get reaction 4, in which the CO is removed so that only CO_2 and H_2 are products. The enthalpy of reaction for each reaction also adds. We calculate that reaction 4 would have to take place 7.73 times for the energy it releases to equal

TABLE 10.7. Solution for Example 10.5

Reaction Number	Reaction Type	Chemical Formula	$\Delta \hat{h}^0_{rxn}$
1	SR	$1CH_4 + 2H_2O$ liquid $\rightarrow 1CO_2 + 4H_2$	$+253.4$
2	POX	$1CH_4 + 0.5O_2 \rightarrow 2H_2 + 1CO$	-35.7
3	WGS	$1H_2O$ liquid $+ 1CO \rightarrow 1CO_2 + 1H_2$	$+2.9$
4	POX + WGS	$1CH_4 + 1H_2O$ liquid $+ 0.5O_2 \rightarrow 1CO_2 + 3H_2$	-32.8
5	(POX + WGS) × 7.73	$7.73CH_4 + 7.73H_2O$ liquid $+ 3.86O_2 \rightarrow 7.73CO_2 + 23.2H_2$	-253.4
6	(POX + WGS) × 7.73 + SR	$8.73CH_4 + 9.73H_2O$ liquid $+ 3.86O_2 \rightarrow 8.73CO_2 + 27.2H_2$	0.0
7	[(POX + WGS) × 7.73 + SR]/8.73	$1CH_4 + 1.115H_2O$ liquid $+ 0.44O_2 \rightarrow 1CO_2 + 3.11H_2$	0.0

Note: Calculation of the appropriate steam-to-carbon ratio for autothermal reforming of methane. Autothermal reforming combines steam reforming (SR), partial oxidation (POX), and the water–gas shift (WGS) reactions to achieve neutral energy balance.

the energy consumed by reaction 1. This is shown as reaction 5. We add reactions 5 and 1 to attain reaction 6, which has an enthalpy of reaction of zero. We normalize reaction 6 by dividing by the number of moles of CH_4 to attain reaction 7. According to reaction 7, the steam-to-carbon ratio is

$$\frac{S}{C} = \frac{n_{H_2O}}{n_c} = 1.115 \tag{10.46}$$

and

$$z = 1.115. \tag{10.47}$$

2. What is the H_2 yield? Operating on air, for every mole of O_2, we have 3.76 mol N_2. For the 0.44 mol O_2 at the intake, we must also have 1.66 mol N_2. Then, the reaction has a H_2 yield of

$$y_{H_2} = \frac{3.11 \text{ mol } H_2}{3.11 \text{ mol } H_2 + 1 \text{ mol } CO_2 + 1.66 \text{ mol } N_2} = 0.54 \tag{10.48}$$

Because oxygen in air is involved in the reaction, the outlet stream is diluted by N_2 in air. This decreases the H_2 yield. However, the presence of water vapor as a reactant increases the H_2 yield. As a result, the H_2 yield is lower than for SR but higher than for POX.

3. The fuel reformer efficiency in terms of HHV is

$$\varepsilon_{FR} = \frac{\Delta H_{(HHV),H_2}}{\Delta H_{(HHV),fuel}} = \frac{3.11 \text{ kmol } H_2(284 \text{ MJ/kmol } H_2)}{1 \text{ kmol } CH_4(880 \text{ MJ/kmol } CH_4)} = 100\% \tag{10.49}$$

Example 10.6. You are designing a hydrogen generator to supply fuel cell vehicles with gaseous hydrogen. You want to use methane from nearby pipelines and liquid water from the utility as inputs. You choose the SR reaction as your primary fuel reforming method because of its high hydrogen yield. However, the endothermic SR reaction requires heat. To supply this heat, you design your steam reformer to burn some methane fuel. (1) Perform a back-of-the-envelope calculation to estimate the minimum quantity of methane fuel you must burn to provide enough heat for the steam reformer. Assume that heat transfer between your methane burner and the steam reformer is 100% efficient. Assume that the SR reactions achieve maximum H_2 yield, as in Example 10.3. Assume complete combustion of CH_4 with the stoichiometric quantity of O_2. For simplicity, we assume the reactions take place at STP. The HHV of methane is 55.5 MJ/kg (880 MJ/kmol) and the HHV of H_2 is 142 MJ/kg (284 MJ/kmol) at STP. (2) Calculate the reformer efficiency (ε_{FR}) in terms of HHV.

Solution

1. Assuming perfect heat transfer, based on the conservation of energy, the heat released by the exothermic reaction (Q_{out}) will equal the heat absorbed by the

endothermic reaction (Q_{in}),

$$Q_{in} = Q_{out} \tag{10.50}$$

whereby

$$n_{CH_4,SR}(\Delta\hat{h}^0_{rxn})_{SR} = n_{CH_4,C}(\Delta\hat{h}^0_{rxn})_C \tag{10.51}$$

where $n_{CH_4,SR}$ is the number of moles of CH_4 consumed by the SR reaction, $(\Delta\hat{h}^0_{rxn})_{SR}$ is the heat of reaction for the SR reaction, $n_{CH_4,C}$ is the moles of CH_4 consumed by the combustion reaction, $(\Delta\hat{h}^0_{rxn})_{SR}$ is the heat of reaction for the combustion of CH_4. Then, the ratio of $n_{CH_4,C}$ to $n_{CH_4,SR}$ is

$$\frac{n_{CH_4,C}}{n_{CH_4,SR}} = \frac{(\Delta\hat{h}^0_{rxn})_{SR}}{(\Delta\hat{h}^0_{rxn})_C} \tag{10.52}$$

Therefore, the ratio of the masses depends on the heats of reaction. According to Table 10.5, the SR reaction

$$CH_4 + 2H_2O_{(g)} \leftrightarrow CO_2 + 4H_2 \tag{10.53}$$

requires 165.2 kJ energy/mol CH_4 at STP. However, this reaction also assumes that H_2O is in vapor form [as indicated by the (g) for gas].

Because we are obtaining our H_2O in liquid form, we need to raise liquid H_2O to steam, according to the phase change reaction

$$H_2O_{(l)} \rightarrow H_2O_{(g)} \tag{10.54}$$

which requires +44.1 kJ energy/mol H_2O. Therefore, in total, for every mole of CH_4 reformed, we need to supply

$$(\Delta\hat{h}^0_{rxn})_{SR} = 165.2 \text{ kJ/mol } CH_4$$
$$+ 44.1 \text{ kJ/mol } H_2O \times 2 \text{ mol } H_2O/ \text{ mol } CH_4$$
$$= 253.4 \text{ kJ/mol } CH_4 \tag{10.55}$$

for the reaction. According to Table 10.6, the combustion of CH_4 is

$$CH_4 + 2O_2 \leftrightarrow CO_2 + 2H_2O \tag{10.56}$$

which releases -803.5 kJ/mol $CH_4 = (\Delta\hat{h}^0_{rxn})_C$ if water is produced in the vapor state. Therefore,

$$\frac{n_{CH_4,C}}{n_{CH_4,SR}} = \frac{253.4 \text{ kJ/mol } CH_4}{-803.5 \text{ kJ/mol } CH_4} = 0.315 \tag{10.57}$$

The moles, mass, or volume of CH_4 needed for combustion is at a minimum about 31.5% of the moles, mass, or volume of CH_4 consumed by the steam reformer.

2. The fuel reformer efficiency in terms of HHV is

$$\varepsilon_{FR} = \frac{\Delta H_{(HHV),H_2}}{\Delta H_{(HHV),fuel}} = \frac{4 \text{ mol } H_2(284 \text{ kJ/mol } H_2)}{1.315 \text{ kmol } CH_4(880 \text{ kJ/mol } CH_4)} = 98\% \qquad (10.58)$$

10.2.5 Water–Gas Shift Reactors

After bulk conversion of H_2 in the fuel reforming stage, the reformate is then usually sent through a WGS reactor. For example, in the fuel processor subsystem design shown in Figure 10.2, after the catalytic fuel reformer (labeled 3), the reformate enters a WGS reactor (4). The overall goals of the WGS reactor are to (1) increase the H_2 yield in the reformate stream and (2) decrease the CO yield. [Even small CO levels can damage certain types of fuel cells, such as PEMFCs, which tolerate less than 10 parts per million (ppm) of CO.] We have already seen how WGS can increase H_2 yields. We now examine the WGS reaction in more detail and discuss how it can also lower the CO yield in the reformate stream.

The WGS reaction reduces the CO yield in the reformate stream by the same percentage that it increases the H_2 yield. The CO yield (y_{CO}) is the molar percentage of CO in the

CATALYST DEACTIVATION

Catalysts can deactivate via two methods, *sintering* and *poisoning*, both of which are a concern in WGS reactors.

1. Sintering is a process in which the surface area of a catalyst decreases under the influence of high temperatures. Exposed to high temperatures, catalyst particles try to achieve a lower energy state by merging together to reduce their surface area. Over time, the reactor's catalyst loses its activity. For example, a WGS reactor may use a copper and zinc oxide catalyst supported on alumina. The zinc oxide molecules create a physical barrier that impedes the copper molecules from merging together. However, if the temperature is too high, the copper molecules can merge anyway. Thus, even a single high-temperature event can inactivate a reactor. For example, exposed to operating temperatures of 700°C, a catalyst's active surface area can decrease by a factor of 20 within the first few days of operation. Lower temperature operation reduces sintering because the copper molecules are less mobile.

2. Poisoning is essentially the chemical deactivation of a catalyst surface. For example, chemical impurities like sulfur can aggregate onto catalyst particles and deactivate them by blocking reaction sites. Poisoning reduces the activity of the catalysts at the front of the reactor first. The WGS reactor is particularly susceptible to sulfur poisoning.

reformate stream,

$$y_{CO} = \frac{n_{CO}}{n} \qquad (10.59)$$

where n_{CO} is the number of moles of CO in the reformate stream. The WGS reaction can reduce the CO yield to a range of 0.2–1.0%, typically in the presence of a catalyst.

If water enters as a vapor, the WGS reaction is slightly exothermic:

$$CO + H_2O_{(g)} \leftrightarrow CO_2 + H_2 \qquad \Delta \hat{h}_r(25°C) = -41.2 \text{ kJ/mol} \qquad (10.60)$$

According to Le Chatelier's principle, because the WGS reaction is exothermic, at high temperatures, it produces more reactants (CO and H_2O). At low temperatures, it produces more products (CO_2 and H_2). Therefore, at low temperatures, the reaction increases its H_2 yield. However, at high temperatures, the reaction rate is higher. To achieve the benefits of both a high H_2 yield at equilibrium and fast kinetics, the WGS process often proceeds in two stages. First, the WGS reaction proceeds at high temperature in one reactor to achieve a high reaction rate. Second, in a second reactor downstream of the first, the WGS reaction proceeds at low temperatures to increase the H_2 yield. Also according to Le Chatelier's principle, excess water vapor in the inlet shifts the reaction equilibrium to favor a higher H_2 yield.

10.2.6 Carbon Monoxide Clean-Up

Even after high- and low-temperature WGS processing, the amount of CO in the reformate stream is too high for most low-temperature fuel cells. For example, the most advanced PEMFC catalysts can withstand a CO yield of only 100 ppm or less. As a result, in fuel processor subsystem designs like the one shown in Figure 10.2, the reformate often passes through a "CO clean-up reactor" (labeled 5). The overall goal of this CO clean-up process is to reduce the CO yield to extremely low levels. This goal can be achieved by either (1) chemical reaction or (2) physical separation. In chemical reaction processes, another species reacts with CO to remove it. Two such processes are

1. selective methanation of CO and
2. selective oxidation of CO.

In both cases, the term *selective* means that a catalyst is used to promote one reaction that removes CO and to suppress another reaction that would otherwise consume H_2. In physical separation processes, either CO or H_2 is physically removed from the gas stream by selective adsorption or selective diffusion. Two such processes are

1. pressure-swing absorption and
2. palladium membrane separation.

These four CO clean-up processes are explained in the next four sections.

10.2.7   Selective Methanation of Carbon Monoxide to Methane

In selective methanation, a catalyst selectively promotes one reaction that removes CO over another that might otherwise consume H_2. Selective methanation promotes the CO methanation reaction,

$$CO + 3H_2 \leftrightarrow CH_4 + H_2O \qquad \Delta \hat{h}_r(25°C) = -206.1 \text{ kJ/mol} \qquad (10.61)$$

over the CO_2 methanation reaction,

$$CO_2 + 4H_2 \leftrightarrow CH_4 + 2H_2O \qquad \Delta \hat{h}_r(25°C) = -165.2 \text{ kJ/mol} \qquad (10.62)$$

The first reaction reduces the CO yield and the H_2 yield. The second reaction consumes even more H_2 while not reducing the CO yield. Therefore, a selective methane catalyst tries to promote the first reaction while suppressing the second. This relationship is summarized in Table 10.8. Selective methanation is an option when the CO concentration in the reformate stream is low, because even the promoted reaction consumes H_2.

10.2.8   Selective Oxidation of Carbon Monoxide to Carbon Dioxide

In selective oxidation, a catalyst selectively promotes a reaction that removes CO over another that consumes H_2. Selective oxidation promotes the CO oxidation reaction

$$CO + 0.5O_2 \leftrightarrow CO_2 \qquad \Delta \hat{h}_r(25°C) = -285 \text{ kJ/mol} \qquad (10.63)$$

over the H_2 oxidation reaction

$$H_2 + 0.5O_2 \leftrightarrow H_2O \qquad \Delta \hat{h}_r(25°C) = -284 \text{ kJ/mol} \qquad (10.64)$$

The first reaction decreases the CO yield while the second decreases the H_2 yield.

The change in Gibbs free energy (ΔG_{rxn}) for the CO reaction is increasingly more negative at lower temperatures, indicating a stronger driving force for that reaction at lower temperatures. Consequently, at lower temperatures, a higher percentage of CO adsorbs onto the catalyst surface. There, the CO blocks H_2 adsorption and oxidation. According

TABLE 10.8. Chemical Removal of CO from Reformate Stream

Reaction Type	Chemical Reaction	$\Delta \hat{h}_{rxn}^0$ (kJ/mol)	Catalyst Promotes ($\checkmark$) or Suppresses (x) Reaction?
1. Selective methanation	$CO + 3H_2 \leftrightarrow CH_4 + H_2O$	−206.1	$\checkmark$
	$CO_2 + 4H_2 \leftrightarrow CH_4 + 2H_2O$	−165.2	x
2. Selective oxidation	$CO + 0.5O_2 \leftrightarrow CO_2$	−285.0	$\checkmark$
	$H_2 + 0.5O_2 \leftrightarrow H_2O$	−284.0	x

Note: Catalysts selectively promote the consumption of CO over the consumption of H_2.

to Le Chatelier's principle, more CO adsorbs at higher CO concentrations. As a result, CO is typically removed via a series of consecutive selective oxidation catalyst beds, each of which operates at increasingly lower temperatures and lower CO concentrations. The decrease in CO adsorption due to lower concentrations in the later catalytic reactors is offset by the increasing effectiveness of CO adsorption from lower temperature operation.

> ***Example 10.7.*** You need to remove 0.2% CO from your reformate stream. (1) You decide to use the methanation process. You have developed a catalyst that is 100% selective for the methanation of CO reaction. How much H_2 is consumed? (2) To remove the same CO, you decide to try the selective oxidation process and can use a catalyst that is 100% selective for the oxidation reaction of CO. How much H_2 is consumed?
>
> *Solution*
>
> 1. For a catalyst with selectivity for CO of 100%, the removal of each molecule of CO still consumes three H_2 molecules, a process that wastes desired hydrogen. For the 0.2% of CO removed from the stream, the H_2 that is also removed is 0.6% of the total mixture.
> 2. For a catalyst with selectivity for CO of 100%, all 0.2% of CO can be removed while no H_2 is removed.

10.2.9 Pressure Swing Adsorption

One physical separation process for CO, pressure swing adsorption (PSA), removes not only CO but also all other species except H_2. It can produce a 99.99% pure H_2 stream. In a PSA system, all of the non-H_2 species in the reformate stream (such as HCs, CO, CO_2, and N_2) preferentially adsorb onto an adsorbent bed composed of zeolites, carbons, or silicas. The heat of adsorption characterizes the strength of surface–solute interactions, which are driven in part by the molecular weights of the adsorbing species. Only hydrogen passes through the bed unadsorbed due to its low molecular weight compared with all other species; the molecular weight of H_2 is 2.016 g/mol whereas all other molecules have a higher molecular weight. As a result, these beds adsorb most other species compared with H_2. Secondary determinants of adsorption include the molecule's polarity and shape.

A PSA unit operates with at least two such adsorption beds. Each adsorption process is a batch process. As a result, to have a continuous flow of reformate purified, at least two beds must operate in parallel: While one adsorbs impurities, the other desorbs. After one bed is saturated with all non-H_2 species, this saturated bed is isolated from fresh reformate by closing the entrance valve. Fresh reformate is diverted to a second, unsaturated adsorbent bed, where the same adsorption process occurs. At the same time, non-H_2 species are removed from the saturated bed via three regeneration steps: (1) depressurization, (2) purging, and (3) repressurization. The first step (depressurization) releases the non-H_2 species, because the adsorbent bed holds less material at lower pressures. The second step (purging) removes the non-H_2 species from the adsorbent vessel. The third step (repressurization) ensures the bed will be ready for the next batch of reformate. The two beds oscillate between adsorption and desorption such that reformate can be continuously purified [44]. The process of reducing the pressure of the bed to reduce its adsorptive ability and then repressurizing it is called the pressure swing mechanism [45]. The parasitic power required to operate a PSA is negligible; the PSA's control system consumes only a small fraction of the fuel processor subsystem's electronics load.

10.2.10 Palladium Membrane Separation

Palladium–silver alloy membranes filter out pure H_2. Different species in a gas can permeate a membrane at different rates. The H_2 molecules can diffuse through a palladium membrane at a faster rate than other species, such as CO, N_2, and CH_4, due to the lattice structure of palladium metal [46].

The H_2 yield from a palladium membrane depends on its (1) pressure differential, (2) operating temperature, and (3) thickness:

1. The hydrogen flux through the membrane can be increased by increasing the pressure drop across the membrane such that a higher density of hydrogen molecules permeates the membrane. A high pressure drop drives H_2 molecules through the membrane and produces low-pressure H_2.

2. Hydrogen flux can also be increased by increasing the operating temperature. Higher temperatures increase the permeation kinetics, because the rates of processes governed by activation energies change exponentially with temperature. The kinetics of permeation is controlled by bulk diffusion at low temperatures and surface chemisorption at high temperatures [47]. At higher temperatures, the palladium material changes to the α-phase, which has a substantially higher hydrogen solubility and therefore permits a higher percentage of hydrogen molecules to permeate.

3. In addition to the pressure differential and the operating temperature, the thickness of the membrane affects its performance. Hydrogen molecules need to do less work to diffuse through a thin membrane, although thinner membranes may be more delicate and susceptible to leaks. According to Sievert's law, which describes the bulk diffusion of species across a pressure differential through a thickness, the normalized flux (the product of the flux and the thickness) should be independent of the thickness if processes are controlled by bulk diffusion. In practice, Sievert's law does not usually hold.

High H_2 yield is limited by (1) purging and (2) leaks. Hydrogen yield is limited by the need to purge the gas stream, which releases some H_2. As the palladium membrane allows H_2 gas to filter through it, non-H_2 species that have not passed through the membrane build up at its surface. As a result, the concentration of H_2 at the surface declines. In most designs, to increase the concentration of H_2 at the surface, this gas stream is periodically purged just for a moment; both H_2 and non-H_2 species are intentionally released from the system. Periodic purging of the gas stream increases the concentration of H_2 at the palladium membrane's surface and therefore the partial pressure of H_2 and the hydrogen flux through it. Hydrogen yield is also limited by pinhole leaks in the membrane that reduce gas purity.

10.3   THERMAL MANAGEMENT SUBSYSTEM

Having learned about important components of the fuel processing subsystem, we now look in detail at a second primary subsystem, the thermal management subsystem. This subsystem is used to manage heat among the fuel cell, the chemical reactors in the fuel processing subsystem, and any source of thermal demand outside of the system. The thermal management subsystem incorporates a system of heat exchangers to heat or cool system components, channeling waste heat from exothermic reactors (such as the fuel cell and afterburner) to endothermic ones (such as a steam generator) and to external sinks (such as a CHP fuel cell system providing heat to a building). A CHP fuel cell system with optimized heat recovery can achieve an overall efficiency ε_O of 80% of the fuel energy. In this section, we will learn about a methodology for managing heat within a fuel cell system so as to maximize heat recovery.

We will learn about managing heat in a fuel cell system using the technique of pinch point analysis [48, 49]. The primary goal of pinch point analysis is to optimize the overall heat recovery within a process plant by minimizing the need to supply additional heating and/or cooling [50, 51]. In an ideal pinch point analysis solution, hot streams are used to heat cold ones with a minimum amount of additional heat transfer from an external source. Unnecessary external heat transfer increases fuel consumption and thereby decreases overall energy efficiency (ε_O) and profitability. The goals of maximum heat recovery and minimum supplemental energy can be met by designing a network of heat exchangers. Various permutations of these heat exchanger networks can be tested using scenario analysis and chemical engineering models of the fuel cell system.

10.3.1   Overview of Pinch Point Analysis Steps

Pinch point analysis is a heat transfer analysis methodology that follows several steps:

1. Identify hot and cold streams in the system.
2. Determine thermal data for these streams.
3. Select a minimum acceptable temperature difference ($dT_{\min,\text{set}}$) between hot and cold streams.
4. Construct temperature–enthalpy diagrams and check $dT_{\min} > dT_{\min,\text{set}}$.
5. If $dT_{\min} < dT_{\min,\text{set}}$, change heat exchanger orientation.
6. Conduct scenario analysis of heat exchanger orientation until $dT_{\min} > dT_{\min,\text{set}}$.

These steps are illustrated below using the fuel cell system design shown in Figure 10.1 as an example.

1. *Identify Hot and Cold Streams.* A hot stream is a flowing fluid that needs to be cooled (or can be cooled). A cold stream is one that needs to be heated. In reference to the system design of Figure 10.1, we will investigate three important hot streams that require cooling:

(a) The hot reformate stream exiting the WGS reactor and eventually entering the fuel cell's anode (labeled 4 through 2)

HOW HEAT EXCHANGERS WORK

A heat exchanger is a mechanical device that conveys thermal energy or heat (Q) from a hot-fluid stream on one side of a barrier to a cold-fluid stream on the other side without allowing the fluids to directly mix. An example of a heat exchanger is a car's radiator, which conveys heat from fluids inside the engine to the surrounding air by convection. Figure 10.7 illustrates one type of heat exchanger, a counterflow heat exchanger, in which the hot fluid flows in one horizontal direction and the cold fluid flows in the reverse horizontal direction. As the hot fluid flows across the top of the plate, heat (Q) is transferred through the conductive plate to the cold fluid below. As a result, the hot-stream temperature declines along the length of the plate, from its inlet ($T_{H,IN}$) to its outlet ($T_{H,OUT}$). This decline in temperature over the length of the heat exchanger is shown by a nonlinear temperature profile. Over the same length of heat exchanger, the cold-stream temperature increases from its inlet ($T_{C,IN}$) to its outlet ($T_{C,OUT}$), also shown by a temperature profile. The temperature difference between the hot and cold streams is dT. Heat only flows from hot to cold.

WHY IS HEAT RECOVERY IMPORTANT FOR FUEL CELLS?

We will now touch on two heat transfer design problems for fuel cells:

- *External Heat Transfer.* Say you have a 70°C PEMFC stack producing 6 kW of electricity and 9 kW of heat. Given the large percentage of heat released, you want to use this heat to heat water for a building up to 90°C. Because heat only flows from hot streams to cold streams, you might initially assume that the heat from the fuel cell stack is NOT transferable to the building. However, it is. You will see this in our example problems.
  Effective heat recovery becomes more challenging the smaller the difference between the hot- (T_H) and cold-temperature (T_C) streams and the lower the temperature of the hot stream (T_H). Because low-temperature fuel cell systems (such as PEMFCs and PAFCs) produce heat at low hot-stream temperatures (T_H), it is even more important to design their heat exchanger network carefully to capture this heat [52].
- *Internal Heat Transfer.* Say you are operating the fuel cell system shown in Figure 10.1. You want to design a heat exchanger system to extract heat dissipated by the fuel cell at 150°C (shown by 1 in Figure 10.1) and from the afterburner at 600°C (shown by 6 in Figure 10.2). You would like to use this heat for an upstream endothermic steam reformer which operates at 800°C (shown by 3 in Figure 10.2) and for the inlet gas preheater which operates at 500°C (shown by 2 in Figure 10.2). What is the optimal design?
  Pinch point analysis will help. Designing a heat exchanger network is especially important for fuel cell systems with complex fuel processors, where different parts may produce or consume heat.

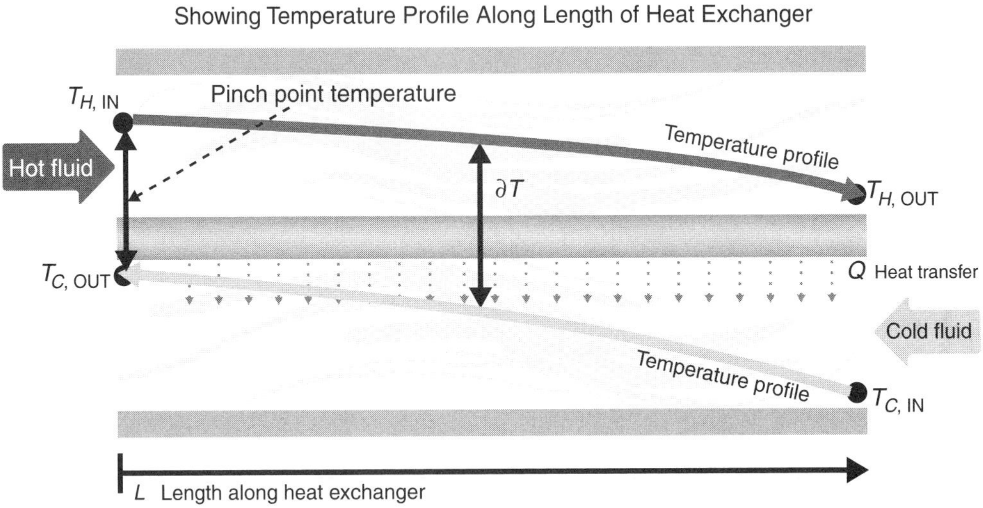

Figure 10.7. Temperature profiles of hot and cold streams in counterflow heat exchanger. The pinch point is the minimum temperature difference between hot and cold streams across the length (L) of the heat exchanger, located at $L = 0$ here.

(b) The cooling loop for the fuel cell stack (labeled 1)

(c) The hot anode and cathode exhaust stream exiting the afterburner and entering the condenser (labeled 5)

Heat management along each of these streams is important. (a) The hot reformate stream must remain within certain temperature ranges to avoid sintering catalysts in the CO clean-up reactor and at the fuel cell's anode. (b) The fuel cell stack also operates most effectively within a certain temperature range. Also, quite importantly, the stack produces a large portion of the recoverable heat from the system. (c) The condenser also releases a large portion of the recoverable heat from the system over a wide temperature. We also will investigate the coldest stream that requires heating: the building's heating loop (labeled 6). This loop provides heating for the air space in the building and for hot water.

2. *Determine Thermal Data for These Streams.* For each hot and cold stream identified, thermal data on the streams must be compiled. These data include the following:

(a) The supply temperature T_{in}, the initial temperature at which the stream is available before entering a heat exchanger

(b) The target temperature T_{out}, the desired outlet temperature for the stream upon exiting a heat exchanger

(c) The heat capacity flow rate $\dot{m}c_p$, the product of the stream's mass flow rate $\dot{m}$ (in kg/s) and the specific heat of the fluid in the stream, c_p (in kJ/kg · °C), whereby the specific heat of the stream is assumed constant over the temperature range

(d) The change in enthalpy $d\dot{H}$ in the stream passing through the heat exchanger

As discussed in Chapter 2, according to the first law of thermodynamics, at constant pressure,

$$d\dot{H} = \dot{Q} + \dot{W} \tag{10.65}$$

Since a heat exchanger performs no mechanical work ($\dot{W} = 0$), $d\dot{H} = \dot{Q} = \dot{m}c_p(T_{\text{in}} - T_{\text{out}})$, where $\dot{Q}$ represents the flow of heat into or out of a stream and $d\dot{H}$ represents the change in enthalpy flow across the stream. The supply temperature data may be measured from an operating system or may be calculated by chemical engineering modeling of reactors. Target temperatures (the desired outlet temperatures) may be determined this way or can be based on other system constraints. For the hot and cold streams identified in step 1, data are tabulated in Table 10.9.

> **Example 10.8.** The fuel cell system in Figure 10.1 produces 6 kWe of electricity with an electrical efficiency of 34% based on the HHV of the original natural gas fuel consumed by the system. (1) Estimate the maximum quantity of heat available from the system for heating the building. (2) Based on the first law of thermodynamics only, if it were possible to transfer all of the energy available in the hot streams of this fuel cell system to a cold stream heating the building via radiators, estimate the maximum flow rate of water for this stream. Assume this cold-water stream circulating through the building has a supply temperature of 25°C and a target temperature of 80°C. (3) If this heat was used for a building's hot water, how many hot showers could it provide?
>
> *Solution*
>
> 1. As we learned in Chapter 2, the real electrical efficiency of the fuel cell stack is described by
>
> $$\varepsilon_R = \frac{P_e}{\Delta \dot{H}_{(\text{HHV}),\text{fuel}}} \tag{10.66}$$
>
> where P_e is the electrical power output of the fuel cell stack. Assuming the parasitic power draw from pumps and compressors is negligible, the maximum heat recovery efficiency ε_H is given as
>
> $$\varepsilon_H = 1 - \varepsilon_R \tag{10.67}$$
>
> The maximum quantity of heat recoverable ($d\dot{H}_{\text{MAX}}$) from the system is then
>
> $$d\dot{H}_{\text{MAX}} = \frac{(1 - \varepsilon_R)P_e}{\varepsilon_R} \tag{10.68}$$
>
> $$= \frac{(1 - 0.34)\,6\ \text{kW}}{0.34} = 11.6\ \text{kW} \tag{10.69}$$

TABLE 10.9. Thermodynamic Data for Hot and Cold Streams in Fuel Cell System Design Shown in Figure 10.1

Stream Number	Source of Heat or Cooling	Stream Description	Hot or Cold?	Supply Temperature, T_{in} (°C)	Target Temperature, T_{out} (°C)	Heat Flow Capacity, $\dot{m}c_p$ (W/K)	Heat Flow $\dot{Q}$ (W)
1	Fuel cell stack	Heat extracted from fuel cell stack	Hot	70	60	276	2760
2	Aftercooler	Heat extracted from reformate stream after selective oxidation reactor	Hot	110	70	276/6	860
3	Selective oxidation	Heat extracted from reformate stream at exothermic selective oxidation reactor	Hot	120	110	6	60
4	Post-WGS reactor	Heat extracted from reformate stream after shift reactor	Hot	260	120	6	840
5	Condenser	Heat extracted from condensing water from anode and cathode exhaust	Hot	219	65	200/9.5	3370
6	Building heat loop	Domestic water-cooling loop exchanging heat between fuel cell system and building	Cold	25	80	143	7890

Note: Stream numbers refer to labeled streams in Figure 10.1. Data were used to construct $T-H$ diagrams. Streams 1–5 refer to hot streams within the fuel cell system. Stream 1 is the cooling stream for the fuel cell stack. Stream 2 is the reformate stream before it enters the fuel cell. Stream 3 is the reformate passing through a selective oxidation chemical reactor. Stream 4 is the reformate stream passing through the WGS reactor. Streams 2–4 are essentially the same contiguous streams passing through different stages. Stream 5 is the anode and cathode exhaust stream passing through a condenser. Stream 6 refers to a building's cold stream. This stream requires heating to provide hot water and space heating for the building. For each stream, thermodynamic data are listed, including (1) inlet and (2) outlet temperatures, (3) the heat flow capacity, and (4) the change in enthalpy or heat flow within the stream. Heat flow capacity is the product of the stream's mass flow rate $\dot{m}$ and its heat capacity c_p.

2. Assuming perfect heat exchange, the mass flow rate of water is

$$\dot{m} = \frac{\dot{Q}}{c_p(T_{in} - T_{out})} \tag{10.70}$$

The heat capacity of water is 4.19 kJ/kg · °C over this ΔT, such that

$$\dot{m} = \frac{11.6 \text{ kW}}{4.19 \text{ kJ/kg} \cdot °\text{C} (80°\text{C} - 25°\text{C})} = 0.05 \text{ kg/s} \tag{10.71}$$

3. The flow rate of hot water from a shower is estimated to be 0.20 kg/s for maximum flow. With a 100 l (100 kg) hot-water storage tank on site, this flow rate would be enough for a single 8-min shower every 30 min.

3. *Select a Minimum Temperature Difference* ($dT_{\min,\text{set}}$) *between Hot and Cold Streams.* While the first law of thermodynamics describes the conservation-of-energy equation for calculating changes in enthalpy, the second law of thermodynamics describes the direction of heat flow. Heat may only flow from hot streams to cold streams. As a result, within a heat exchanger, the temperature of the hot stream cannot dip below the temperature of a cold stream, and a cold stream cannot be heated to a temperature higher than the supply temperature of the hot stream. A minimum temperature difference $dT_{\min}$ must exist between the streams to drive heat transfer such that, for the hot-stream temperature T_H and the cold-stream temperature T_C,

$$T_H - T_C \geq dT_{\min} \tag{10.72}$$

at any length along the heat exchanger. For a set of streams, the minimum temperature difference observed between the streams at any length along the heat exchanger is referred to as the pinch point temperature. In the heat exchanger shown in Figure 10.7, the temperature difference between the hot and cold streams (dT) changes along the length of the heat exchanger, as shown by the difference in the hot- and cold-temperature profiles over its length L. In this heat exchanger, the minimum temperature difference $dT_{\min}$ is at $L = 0$, at the inlet of the hot-fluid stream and the outlet of the cold-fluid stream. For the purposes of pinch point analysis, $dT_{\min}$ is often set at a desired value, between 3 and 40°C, depending on the type of heat exchanger and the application. For example, while shell-and-tube heat exchangers require $dT_{\min,\text{set}}$ of 5°C or more, compact heat exchangers can achieve higher heat transfer rates due to their larger effective surface area and may require $dT_{\min,\text{set}}$ of only 3°C.

For our analysis of the heat streams of Figure 10.1, we select $dT_{\min,\text{set}} = 20°\text{C}$.

4. *Construct Temperature-Enthalpy Diagrams and Check* $dT_{\min} > dT_{\min,\text{set}}$. Temperature–enthalpy diagrams ($T–H$) show the change in temperature versus the change in enthalpy for hot and cold streams. On a $T–H$ diagram, any stream with a constant c_p should be represented by a straight line from T_{in} to T_{out}.

We use the thermal data we gathered in Table 10.9 to make the $T–H$ plots. Given the large quantity of energy available from the condenser (stream 5 in Table 10.9), we plot its data on a T–H diagram. The data plotted in Figure 10.8 are based on the condenser's $T_{in} = 219°C$, $T_{out} = 65°C$, and $\dot{Q} = 3370$ W. We also plot data for the building's cold-stream loop (stream 6 in Table 10.9), based on its $T_{in} = 25°C$ and $T_{out} = 80°C$. We assume this loop could absorb from the condenser up to 3370 W, a portion of the 7890 W of heat it could absorb from all five hot streams. We plot these data on a T–H diagram.

Figure 10.8 shows T–H diagrams that illustrate our best understanding of the hot stream of the condenser and the cold stream of the building's heating loop assuming these loops are separated, that is, not connected by a heat exchanger. Note the schematic at the bottom of the diagram which indicates how the hot and cold streams are separated in different pipes that do not intersect. If the two streams are separated, each must rely on an external source of heat transfer from outside the fuel cell system to cool or heat itself. For example,

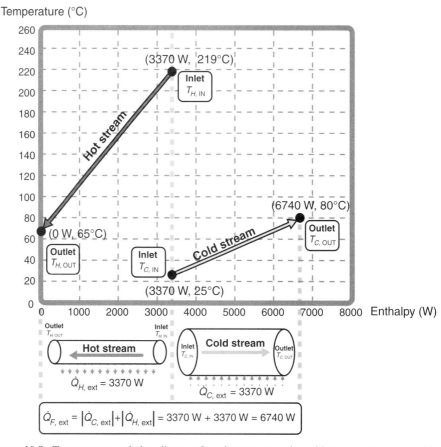

Figure 10.8. Temperature–enthalpy diagram for a hot stream and a cold stream not connected by a heat exchanger. External heat transfer is maximum. The hot stream rejects 3370 W to the environment. The cold stream absorbs 3370 W from an external heat source. Arrow heads on the T–H plots indicate the direction of stream flow. The schematic at the bottom illustrates the processes occurring by showing pipes carrying fluid and the heat transfer through these pipes. The change in enthalpy is roughly analogous to the change in the length along a pipe.

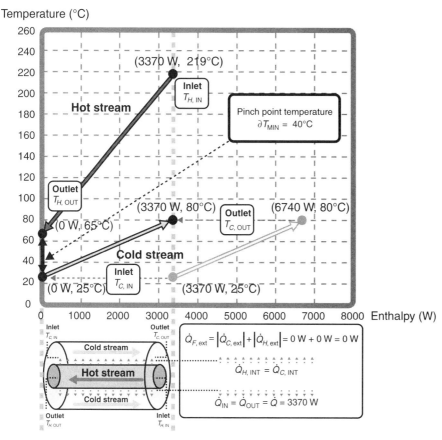

Figure 10.9. Temperature–enthalpy diagram for hot and cold streams in Figure 10.8 but connected by a heat exchanger (shown at bottom). External heat transfer $\dot{Q}_{ext}$ is zero. The hot stream rejects 3370 W to the cold stream. The pinch point, the minimum temperature difference between hot and cold streams, appears to be at the entrance to the cold stream and has a value of 40°C, based on our available data. The figure at bottom depicts the combined streams in an exchanger.

instead of relying on heat from the condenser, the building's heating loop would have to rely on burning natural gas as a source of heat.

Figure 10.9 shows the effect of incorporating a heat exchanger between the two streams in order to thermally connect them. For heat exchange to take place between the hot stream and the cold stream, the hot-stream $T–H$ curve must lie above the cold-stream $T–H$ curve. The $T–H$ diagram for the cold stream has been shifted to the left such that the cold stream now cools the hot stream and the hot stream heats the cold one. Less external heat transfer is necessary. The heat recovery efficiency ε_H of the system increases and therefore, according to

$$\varepsilon_O = \varepsilon_R + \varepsilon_H \tag{10.73}$$

the overall efficiency ε_O of the system increases.

When a hot $T-H$ diagram and a cold $T-H$ diagram are horizontally shifted on top of each other in these diagrams, the change in enthalpy along the x axis can be thought of in terms of the change in length along the heat exchanger. At a given length along a heat exchanger, the quantity of heat that leaves the hot stream to enter the cold stream (the cumulative change in enthalpy of the hot stream) must be equal to the quantity of heat absorbed by the cold stream from the hot stream (the cumulative change in enthalpy of the cold stream). The bottom of Figure 10.9 shows that the hot and cold streams of the two separate pipes have been merged together as two concentric pipes (a heat exchanger). In this way, the length along the heat exchanger is analogous to the cumulative change in enthalpy of the streams.

> **Example 10.9.** You would like to use the heat of the condenser (stream 6 in Table 10.9) to warm a cold stream of utility water from 25 to 80°C for a building's heat. (1) Report the quantity of heat available from this component as a percentage of the HHV fuel energy input. (2) Construct the appropriate $T-H$ diagrams and check the pinch point temperature. Ensure that $T_{min} > T_{min,set} = 20$°C.
>
> *Solution*
>
> 1. Based on data in Example 10.8,
>
> $$\Delta \dot{H}_{(HHV),fuel} = \frac{P_e}{\varepsilon_R} = \frac{6 \text{ kWe}}{0.34} = 17.6 \text{ kW} \tag{10.74}$$
>
> The maximum quantity of heat available from the hot stream is
>
> $$d\dot{H}_{MAX} = 3370 \text{ W} \tag{10.75}$$
>
> $$\frac{d\dot{H}_{MAX}}{\Delta \dot{H}_{(HHV),fuel}} = 19\% \tag{10.76}$$
>
> Almost 20% of the energy in the fuel is available as heat for recovery from this single component.
>
> 2. The $T-H$ diagrams are shown in Figures 10.8 and 10.9 for these assumptions:
>
> $$dT_{min} = 40°C > dT_{min,set} = 20°C \tag{10.77}$$

> **Example 10.10.** Given the large quantity of heat available from the condenser, you improve your understanding of it. You realize that the condenser's stream changes phase during the heat exchange process, as the water vapor condenses to liquid water. Because the heat capacity c_p of the stream changes between gas and liquid phases, $\dot{m}c_p$ is not constant across the heat exchanger. You more carefully measure the thermodynamic properties of the stream. You measure the mass flow rate and estimate the heat capacity for the vapor and gas phases based on the stream's constituent species. For the liquid phase $\dot{m}c_{p,1} = 200$ W/°C, and for the vapor phase $\dot{m}c_{p,2} = 9.5$ W/°C. (1) Calculate the temperature at which the stream changes

phase. (2) Reconstruct the appropriate T–H diagram and check the pinch point temperature.

Solution

1. Using

$$\dot{Q} = \dot{m}c_{p,1}(T_{cond} - T_{out}) + \dot{m}c_{p,2}(T_{in} - T_{cond}) \qquad (10.78)$$

we have

$$3370 \text{ W} = (200 \text{ W}/^\circ\text{C})(T_{cond} - 65^\circ\text{C}) + (9.5 \text{ W}/^\circ\text{C})(219^\circ\text{C} - T_{cond}) \qquad (10.79)$$

or

$$T_{cond} = 75^\circ\text{C} \qquad (10.80)$$

2. Figure 10.10 shows the appropriate T–H diagram. In this condenser example, the pinch point does not occur at either the entrance or the exit of the heat exchanger but rather occurs within the heat exchanger. The pinch point occurs during the phase change from gas to liquid and is 17°C. Because $dT_{min} = 17°\text{C}$ is not greater than $dT_{min,set} = 20°\text{C}$, we need to reconfigure the heat exchangers to meet the set pinch point temperature. To do this, for example, after partially heating the utility water-cooling loop, we may have the condenser heat a colder stream in the system and have a hotter stream heat the utility water loop the rest of the way. This example is extremely important because all fuel cell systems produce water vapor in the product stream and most will use cold streams from other components to condense it for heat recovery and water balance.

5. *If $dT_{min} < dT_{min,set}$, Change Heat Exchanger Orientation.* If the actual pinch point temperature is less than the set minimum pinch point temperature, the hot and cold streams must be reoriented. For the new orientation, a new T–H diagram is developed and the pinch point temperature and location within the heating network are recalculated. Additional streams may be included in the analysis to increase the number of options available.

6. *Conduct Scenario Analysis of Heat Exchanger Orientation Until $dT_{min} > dT_{min,set}$.* Different orientations of streams and heat exchangers can be evaluated using scenario analysis. Scenario analysis is greatly aided by computer software that incorporates the chemical engineering process descriptions and pinch point temperature analysis capability. Although beyond the scope of this brief introduction to pinch point analysis, these programs can be used to investigate better heat exchanger network designs. From these analyses, one can determine the number of heat exchangers required and conduct a cost–benefit analysis to compare the cost of different heat exchanger network scenarios with the financial benefits of higher fuel efficiency and heat recovery.

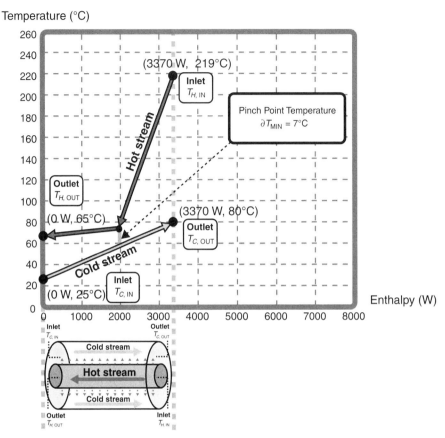

Figure 10.10. Temperature–enthalpy diagram for a hot and a cold stream connected by a heat exchanger, with hot stream changing phase from gas to liquid in the middle. The change in phase is marked by the hot stream's abrupt change in slope, where slope is the inverse of the heat flow capacity $\dot{m}c_p$. The change in phase causes a pinch point. Aggregate conservation of energy calculations would not have detected the pinch.

Example 10.11. You are designing the thermal management subsystem for the fuel cell system shown in Figure 10.1. You plan to capture heat from the fuel cell system to heat a building. Table 10.9 provides the thermal characteristics of some of the most important hot streams within the fuel cell system (streams 1–5). Figure 10.1 shows the arrangement of these five hot streams within the system (numbered 1–5). You plan to use heat from these five streams (a total of 7890 W) to heat the building. Table 10.9 also shows the thermal characteristic of the single stream you want to heat, the building's cold stream (steam 6). You would like to heat this cold stream from 25 to 80°C, as shown in Table 10.9. You would like to capture every single watt of heat from the five hot streams to warm the building. Capturing this heat will give the fuel cell system a very high heat recovery efficiency and therefore a high overall efficiency.

Conduct a pinch point analysis on one possible heating loop design. Assume the building's cold stream exchanges heat with the hot streams placed in series in this order: (1) the fuel cell stack, (2) the aftercooler, (3) the selective oxidation reactor, (4) the post-WGS reactor, and (5) the condenser.

1. Plot these hot and cold streams on a $T-H$ diagram and identify the location of the pinch point.
2. Calculate the pinch point temperature dT_{min}.
3. If $dT_{min} > dT_{min,set} = 10°C$, suggest another heating loop design to increase the pinch.

For the aftercooler, the heat flow capacity $\dot{m}c_{p,1,aft}$ for the liquid portion of the stream is 276 W/°C. The heat flow capacity $\dot{m}c_{p,2,aft}$ for the vapor portion of the stream is 6 W/°C.

Solution

1. To do this analysis, we realize that the heat capacity c_p of the condenser's stream does not remain constant. The same is true for the aftercooler's stream. In both of these streams, water condenses from a vapor to a liquid midstream.
   Using

$$\dot{Q} = \dot{m}c_{p,1,aft}(T_{cond,aft} - T_{out}) + \dot{m}c_{p,2,aft}(T_{in} - T_{cond,aft}) \tag{10.81}$$

we have

$$860 \text{ W} = (276 \text{ W/°C})(T_{cond,aft} - 70°C) + (6 \text{ W/°C})(110°C - T_{cond,aft}) \tag{10.82}$$

Thus, the aftercooler stream condenses at

$$T_{cond,aft} = 72.3°C \tag{10.83}$$

From Example 10.10, we know the fluid in the condenser will condense at $T_{cond} = 75°C$. The thermodynamic characteristics table 10.9 gives us the change in enthalpy $(d\dot{H} = \dot{Q})$ and the change in temperature (dT) for each of the five streams. We plot a curve of $d\dot{H}$ versus dT curve for each of the five stages consecutively from coldest to hottest, resulting in the $T-H$ curve in Figure 10.11, which shows us that the pinch point occurs in the condenser.

2. To find the value of the pinch point temperature dT_{min} at the condenser, we observe that the pinch point occurs just as the vapor condenses, at $T_{cond} = 75°C$. Also, $dT_{min} = T_{cond} - T_b$, where T_b is the building loop temperature. At $T_{cond} = 75°C$, we want to know the cumulative enthalpy transfer $d\dot{H}_{cum}$, the value on the x axis:

$$d\dot{H}_{cum} = \dot{Q}_{FC} + \dot{Q}_{AC} + \dot{Q}_{SO} + \dot{Q}_{PS} + \dot{Q}_{cond1,A} \tag{10.84}$$

where $\dot{Q}_{FC}$ is the heat flow at the fuel cell, $\dot{Q}_{AC}$ is the heat flow at the aftercooler, $\dot{Q}_{SO}$ is the heat flow at the selective oxidation reactor, $\dot{Q}_{PS}$ is the heat flow at the

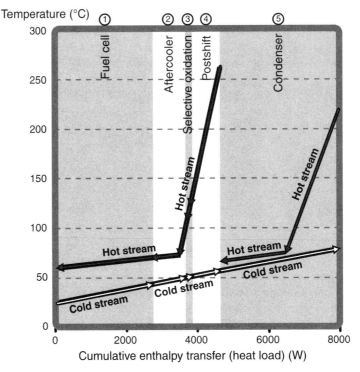

Figure 10.11. Temperature–enthalpy diagram for hot and cold streams from fuel cell system of Figure 10.1. The two separate hot streams are from two different parts of the system. They heat the cold stream in series. First, the cold stream absorbs heat from (1) the fuel cell stack, (2) the aftercooler, (3) a selective oxidation reactor, and (4) the reformate leaving the water gas shift reactor. Second, the cold stream absorbs heat from a condenser. The dT–dH data were plotted from Table 10.9.

postshift reactor, and $\dot{Q}_{\mathrm{cond1,A}}$ is the heat flow in the cold stage (liquid) of the condenser. From Example 10.11,

$$\dot{Q}_{\mathrm{cond1,A}} = \dot{m}c_{p,1}(T_{\mathrm{cond}} - T_{\mathrm{out}}) = (200 \text{ W}/^{\circ}\text{C})(75^{\circ}\text{C} - 65^{\circ}\text{C}) = 2000 \text{ W} \quad (10.85)$$

$$d\dot{H}_{\mathrm{cum}} = 2760 \text{ W} + 860 \text{ W} + 60 \text{ W} + 840 \text{ W} + 2000 \text{ W} = 6520 \text{ W} \quad (10.86)$$

where $d\dot{H}_{\mathrm{cum}} = 6520$ W is the value on the x axis where the pinch occurs. For the building's heating loop, the relationship between T_b and $d\dot{H}$ can be described by

$$T_b = \left(\frac{80^{\circ}\text{C} - 25^{\circ}\text{C}}{7890 \text{ W}}\right) d\dot{H} + 25^{\circ}\text{C} \quad (10.87)$$

$$= \left(\frac{80^{\circ}\text{C} - 25^{\circ}\text{C}}{7890 \text{ W}}\right) 6520 \text{ W} + 25^{\circ}\text{C} \quad (10.88)$$

$$= 70.5^{\circ}\text{C} \quad (10.89)$$

$$dT_{\mathrm{min}} = T_{\mathrm{cond}} - T_b = 75^{\circ}\text{C} - 70.5^{\circ}\text{C} = 4.5^{\circ}\text{C} < dT_{\mathrm{min,set}} = 10^{\circ}\text{C} \quad (10.90)$$

This pinch point temperature is extremely low. We need to propose another heat exchanger network to increase the pinch.

3. One option is to split the building's cooling stream into two separate but parallel streams. One stream extracts heat from the first four heat sources in series: (1) the fuel cell stack, (2) the aftercooler, (3) the selective oxidation reactor, and (4) the post-WGS reactor. The second stream extracts heat from the fifth heat source, the condenser. The ratio of flow rates between the building loop's two parallel streams could be optimized to maximize the pinch. Such a detailed analysis, performed by computer simulations, leads to a pinch greater than $dT_{\min,\text{set}}$ (10°C) over a range of molar flow ratios.

CHAPTER SUMMARY

In this chapter, we gained a solid understanding of the four primary fuel cell subsystems. We learned in detail about two of these main subsystems: (1) the fuel processor subsystem and (2) the thermal management subsystem.

- The overall efficiency ε_O of a fuel cell system is the sum of its net system electrical efficiency ε_R and its heat recovery efficiency ε_H.
- A fuel processor efficiency ε_{FP} is the ratio of the HHV of H_2 in the output gas ($\Delta H_{(\text{HHV}),H_2}$) compared with the HHV of fuel ($\Delta H_{(\text{HHV}),\text{fuel}}$) in the input, including fuel consumed to provide energy for the fuel processor itself, or

$$\varepsilon_{\text{FP}} = \frac{\Delta H_{(\text{HHV}),H_2}}{\Delta H_{(\text{HHV}),\text{fuel}}} \tag{10.91}$$

- Exothermic reactions release energy; endothermic ones consume it.
- Hydrogen yield y_{H_2} is the molar percentage of H_2 in a chemical stream:

$$y_{H_2} = \frac{n_{H_2}}{n} \tag{10.92}$$

where n_{H_2} is the number of moles of H_2 and n is the total number of moles of all gases in the stream.
- Hydrogen can be produced from a hydrocarbon (HC) fuel via three main processes: (1) steam reforming, (2) partial oxidation, and (3) autothermal reforming.
- Steam reforming is an endothermic reaction that combines a HC fuel with steam:

$$C_xH_y + xH_2O_{(g)} \leftrightarrow xCO + \left(\tfrac{1}{2}y + x\right)H_2 \tag{10.93}$$

- Partial oxidation is an exothermic reaction that combines a HC fuel with deficient O_2:

$$C_xH_y + \tfrac{1}{2}xO_2 \leftrightarrow xCO + \tfrac{1}{2}yH_2 \tag{10.94}$$

- Autothermal reforming is energy neutral and combines a HC fuel with H_2O and O_2:

$$C_xH_y + zH_2O_{(l)} + \left(x - \tfrac{1}{2}z\right)O_2 \leftrightarrow xCO + \left(z + \tfrac{1}{2}y\right)H_2 \qquad (10.95)$$

- The water gas shift reaction (1) increases H_2 yield and (2) decreases CO yield:

$$CO + H_2O_{(g)} \leftrightarrow CO_2 + H_2 \qquad \Delta\hat{h}_r(25°C) = -42.1 \text{ kJ/mol} \qquad (10.96)$$

- Pinch point analysis optimizes heat recovery by minimizing the need to supply additional heating and/or cooling. Temperature–enthalpy diagrams are constructed to locate the pinch point temperature dT_{min}, the minimum temperature difference between hot and cold streams. We rearrange heat exchangers to maximize (1) internal use of heating and (2) dT_{min}.

CHAPTER EXERCISES

Review Questions

10.1 What are the four primary subsystems of a fuel cell system? Give examples of subsystem components that depend on the operation of other subsystem components. How might these subsystem components be integrated?

10.2 Explain the purpose and operation of the pressure swing absorption unit, including the reason for its name.

10.3 Label the following processes as endothermic, exothermic, or neither: oxidation of hydrogen fuel in a fuel cell, steam reforming, partial oxidation, autothermal reforming, the water gas shift reaction, selective methanation, selective oxidation, hydrogen separation via palladium membranes, pressure swing adsorption, combustion of fuel cell exhaust gases, condensing water vapor to liquid, compression of natural gas, and expansion of hydrogen gas.

10.4 Sketch out a process diagram for a fuel cell system for a scooter. Some primary components of the system include a PEMFC stack, a hydrogen tank, an electrical storage device such as a battery or capacitor for buffering load, and an electric motor that fits into the hub of the scooter's wheel. Draw the primary system components, stream flows, and heat flows. Label the four subsystems. (One way to approach this problem is to begin with the process diagram shown in Figure 10.1 and decide which components are not needed.)

10.5 Resketch the process diagram in problem 10.4 assuming hydrogen is stored on the bike in a metal hydride (discussed in Chapter 9) that requires heating and cooling for hydrogen storage and release. Sketch $T–H$ diagrams for managing heat. Discuss important thermodynamic characteristics of metal hydrides.

Calculations

10.6   An idealized partial oxidation fuel reformer consumes isooctane fuel ($C_8H_{18(l)}$) and air, which is similar to gasoline. What is the maximum H_2 yield?

10.7   **(a)**   Based on Examples 10.3 and 10.6, what is the minimum quantity of methane fuel you must burn to provide enough heat for the steam reformer assuming that the efficiency of heat exchange is only 72%.

   (b)   Calculate the fuel reformer efficiency ε_{FR} in terms of HHV. Assume the reactants and products enter and leave the reformer at 1000 K.

10.8   You would like a hydrogen generator similar to the one discussed in Example 10.6 to operate on emergency back-up fuels such as propane ($C_3H_{8(g)}$) and to use an autothermal reformer (not a steam reformer). For a 100% efficient reformer, specify a reasonable steam-to-carbon ratio (S/C) and the quantity of hydrogen the reformer would produce per unit of fuel consumed. Assume the reactants and products enter and leave the reformer at 1000 K.

10.9   Assuming that the endothermic steam reformer attains its heat from the combustion of methane, compare the ratio of hydrogen produced per unit of methane consumed for (1) a steam reformer, (2) a partial oxidation reformer, and (3) an autothermal reformer. Assume, in all three cases, that the reactants enter the reactor at 1000 K, having been preheated, and the products leave at 1000 K.

10.10   If the fuel cell system described in Example 10.8 was used for space heating, estimate the air space it could heat during winter with an outside temperature of 0°C and a desired indoor temperature of 23°C. Assume a radiative heating system that is closed loop. How many rooms in a building can be heated? Assume a log-cabin structure made of 5-cm-thick wood with a thermal conductivity of 0.17 W/m°C, no windows, and no free convection of air along the outside.

10.11   You are designing a PEMFC scooter for use in a developing country where water resources are scarce. You design your fuel cell system to condense the product water in the outlet stream for reuse, taking advantage of the PEMFC stack's relatively low operating temperature. Sketch $T-H$ diagrams for capturing heat from such a condenser. Determine how the forced convection of air against the bike could provide enough cooling for the condenser, such that no additional air pumps or blowers would be needed. The stack's maximum electrical power output is 1 kW. Estimate the volume and mass of the onboard water tank. Assume half of the waste heat from the fuel cell system exits via the cathode exhaust gas, a 40% efficient fuel cell system, and the scooter stores enough hydrogen at minimum for a two-hour ride.

10.12   Continue the analysis of Example 10.11. Develop alternative heat exchanger network designs that increase the pinch. If the parallel stream network is implemented, calculate a range of mass flow rate ratios over which $dT_{min} \geq dT_{min,set} = 10°C$.

10.13   In Example 10.11, locate and determine the value of the pinch point considering all hot streams except the condenser.

CHAPTER 11

ENVIRONMENTAL IMPACT OF FUEL CELLS

In this chapter, we will learn how to quantify the potential environmental impact of fuel cells. We will calculate potential changes in emissions from their use. We will calculate how these changes in emissions affect global warming, air pollution, and human health. We will learn how to evaluate these changes not just at the vehicle or power plant level but also across the entire supply chain, from raw material extraction to end use.

First, we will learn a tool called life cycle assessment (LCA), which we can use to evaluate how a new energy technology (such as fuel cells) affects energy use, energy efficiency, and emissions. Second, to conduct an LCA thoroughly, we will need to quantify the most important global warming and air pollution emissions. Therefore, we will briefly discuss the theory behind global warming and detail the primary global warming emissions from conventional vehicles, power plants, and fuel cell systems. Third, we will review the primary air pollutants from fossil fuel combustion devices and fuel cell systems and their effects on human health. Finally, using LCA and our knowledge of emissions impacts, we will develop a complete "what if" scenario to look at how fuel cell implementation can change the global environmental context. After learning these tools and following these examples, you will be well equipped to quantify the impact of your own scenarios of possible fuel cell futures.

11.1 LIFE CYCLE ASSESSMENT

Life cycle assessment is a methodology for systematically analyzing the effect of changes in the implementation and use of energy-related technologies.[1] With a change in energy

---

[1]Life cycle assessment also may be referred to as wheel-to-wheel analysis, process chain analysis, or supply chain analysis, depending on the emphasis of the analysis, be it environmental or economic.

technology, LCA helps us evaluate changes in efficiency, emissions, and other environmental consequences [53, 54]. These environmental consequences include the economic costs of global warming and the human health impacts of air pollution.

11.1.1 Life Cycle Assessment as a Tool

Life cycle assessment consists of three primary stages:

1. Analyze the relevant energy and material inputs and outputs associated with the change in energy technology along the entire supply chain. The supply chain begins with raw material extraction, continues to processing, then to production and end use, and finally to waste management. Within this chain, it is important to focus on the most energy and emission intensive processes, the "process bottlenecks" [55].

2. Quantify the environmental impacts associated with these energy and material changes.

3. Rate the proposed change in energy technology against other scenarios.

Figure 11.1 shows an example of a supply chain for today's conventional gasoline internal combustion engine (ICE) vehicles. The figure shows primary energy and pollutant flows during petroleum fuel extraction, production, transport, storage, delivery, and use on a vehicle. Processes are depicted via boxes, emissions via wavy arrows at the top of the boxes, fuel flow via small arrows between boxes, and energy flows via thick arrows at the bottoms of the boxes. This supply chain could serve as a base case for comparing alternative vehicle supply chains.

Now that we understand the concepts of the supply chain and process bottlenecks, we will dig deeper into a detailed methodology for LCA. A useful methodology for LCA follows these steps:

1. *Research and develop an understanding of the supply chain from raw material production to end use.*

2. *Sketch a supply chain showing important processes and primary mass and energy flows.* Examples of processes include chemical and energy conversion, production and transport of fuels, and fuel storage. Mass flows include the flow of raw materials, fuels, waste products, and emissions. Energy flows include the use of electric power, additional chemical energy consumed in a process, and work done on a process.

3. *Identify the "bottleneck" processes*, which consume the largest amounts of energy or produce the largest quantities of harmful emissions (or both).

4. *Analyze the energy and mass flows in the supply chain using a control volume analysis and the principles of conservation of mass and energy.* A control volume is a volume of space into which (and from which) mass flows. The boundaries of the control volume are shown by a control surface. Draw a control surface around individual processes in the supply chain, with particular focus on bottleneck processes. Analyze the mass and energy flows entering and exiting these processes. Employ the

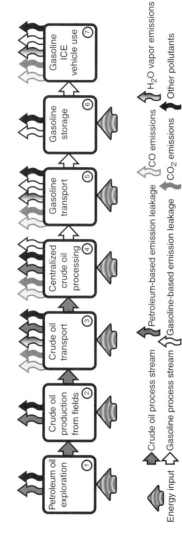

Figure 11.1. Supply chain for today's conventional gasoline ICE vehicles. Energy is consumed (bottom arrows) and emissions are produced (top arrows) during the primary processes (boxes 1–7) from petroleum fuel extraction to its use on a vehicle.

327

conservation-of-mass equation

$$m_1 - m_2 = \Delta m \tag{11.1}$$

where m_1 is the mass entering the control volume, m_2 is the mass leaving the control volume, and Δm is the mass accumulating within the control volume. Employ the conservation of energy equation for steady flow,

$$\dot{Q} - \dot{W} = \dot{m}\left[h_2 - h_1 + g(z_2 - z_1) + \tfrac{1}{2}(V_2^2 - V_1^2)\right] \tag{11.2}$$

where $\dot{Q}$ is the heat flow into the process, $\dot{W}$ is rate of work done by the process, $\dot{m}$ is the mass flow rate, $h_2 - h_1$ is the change in enthalpy between outgoing and incoming streams, g is the acceleration of gravity, $z_2 - z_1$ is the change in height, and $V_2^2 - V_1^2$ is the change in the square of the velocity. The last three terms refer to the change in the internal energy, potential energy, and kinetic energy of a flowing stream, respectively.

5. Having analyzed the individual processes within the supply chain, *evaluate the entire supply chain as a single control volume.* Aggregate net energy and emission flows for the chain.

6. *Quantify the environmental impacts of these net flows,* for example, in terms of human health impacts, external costs, and potential for global warming. We will discuss definitions of these terms and methods for conducting this analysis in subsequent sections.

7. *Compare the net change in energy flows, emissions, and environmental impacts of one supply chain with another.*

8. *Rate the environmental performance of each supply chain against the others.*

9. *Repeat the analysis* for an expanded, more detailed number of processes in the supply chain.

Each of these steps will be expanded upon throughout the rest of the chapter through detailed examples and explanations with a particular focus on fuel cell technologies. A significant amount of attention will be given to methods for quantifying environmental impacts.

11.1.2   Life Cycle Assessment Applied to Fuel Cells

Using the first three steps in this methodology for LCA, we will build and analyze a potential supply chain for fuel cell vehicles:

1. *Research and develop an understanding of the supply chain from raw material production to end use.* Using our knowledge from Chapter 10, we know that we can chemically process natural gas into a H_2-rich gas. Assume we will fuel our fuel cell vehicles with H_2 derived from steam reforming of natural gas. These steam reformers could be placed at similar locations as conventional gasoline refueling stations and could consume natural gas fuel piped in through the existing natural gas pipeline network. During these processes

some methane (CH_4) in the natural gas could leak into the surrounding environment. Hydrogen produced at the fuel processor could then be compressed into high-pressure tanks, stored at the station to buffer supply, and finally used to refuel high-pressure tanks onboard the vehicle. During these processes, some H_2 could leak into the environment.

2. *Sketch a supply chain showing important processes and primary mass and energy flows*. Figure 11.2 shows a sketch of this potential fuel cell vehicle supply chain. Processes include natural gas exploration (box 1); production from gas fields (box 2); storage in underground tanks and reservoirs (box 3); chemical processing into a refined gas, including the addition of sulfur as an odorant (box 4); and transmission through pipelines (box 5). Up to this point, this part of the chain is identical to the supply chain already in existence for natural gas used to supply homes and buildings with fuel for heating and gas turbine power plants with fuel for generating electric power. Remaining processes include the conversion of natural gas to H_2 at the fuel processor (box 6), H_2 compression (box 7), storage (box 8), and use onboard the vehicle (box 9). As shown in Figure 11.2, most of these processes require at least some additional energy or work input. The dark arrows show natural gas fuel flow and the light arrows show H_2 fuel flow. Emissions include leaked CH_4 in the natural gas stream; leaked H_2 in the H_2 stream; carbon dioxide (CO_2), carbon monoxide (CO), and other emissions produced during fuel processing and electricity production for powering the compression of hydrogen; and water vapor emissions (H_2O) at the vehicle.

3. *Identify the most energy intensive and most polluting portions of the chain, that is, "bottleneck" processes*. Think about the energy input arrows at the bottom of the process boxes. Approximately 0.7% of the HHV of natural gas is required for its exploration (box 1), about 5.6% for production (box 2), 1.0% for storage and processing (boxes 3 and 4), and 2.7% for transmission (box 5). Thus, about 10% of the HHV of natural gas is required to provide energy for the first five boxes in Figure 11.2. As shown in Chapter 10, approximately 30% of the HHV of natural gas is required for the operation of the fuel processor. As we learned in Chapter 9, the energy required to compress H_2 is approximately 10% of the HHV of H_2. Storage energy is a fraction of this. Therefore, the two single most energy intensive processes in the chain are (1) fuel processing of natural gas and (2) compression of H_2.

The most energy intensive processes are likely to produce the largest quantities of harmful emissions. Therefore, the most energy intensive processes should be examined first. However, this relationship may not always hold. Different types of emissions are more harmful than others. Therefore, the most energy intensive processes are an excellent starting point for determining the highest emitting processes, but other processes must also be investigated.

Think about the emission arrows at the top of the process boxes, beginning with the most energy intensive processes: (1) fuel processing of natural gas and (2) compression of H_2. Consider the first process bottleneck: fuel processing. Based on research of steam reformers used in conjunction with fuel cell systems, emission factors for a commercial natural gas steam reformer are shown in Table 11.1. For reference, Table 11.1 also benchmarks the steam reformer's emissions against emissions from another type of hydrogen generator, a coal gasification plant, and against emissions from electric power plants fueled by natural gas and coal. The steam reformer's emissions are quite low. For example, the steam

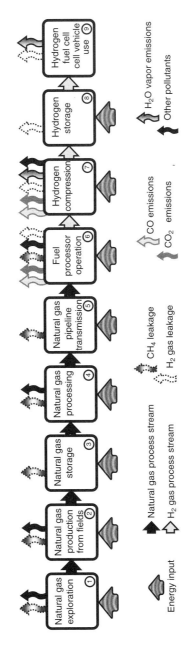

Figure 11.2. Supply chain for hydrogen fuel cell vehicle fleet that obtains its hydrogen fuel from steam reforming of natural gas. Approximately 30% of the HHV of natural gas is needed for the operation of the steam reformer (box 6). Approximately 10% of the HHV of H_2 is required for H_2 compression (box 7). These are the most energy intensive links in the supply chain.

330

TABLE 11.1. Emission Factors for Two Types of Hydrogen Generators and Two Types of Electricity Generators

	Hydrogen Generator Emission Factors		Electricity Plant Emission Factors			
	Natural Gas Steam Reformer	Coal Gasification	Natural Gas Combustion (Combined-Cycle Gas Turbine, low NO_x)		Coal Combustion (Coal Boiler, Steam Turbine, low NO_x)	
Emission	(kg emission /kg natural gas fuel)	(kg emission /kg coal fuel)	g emission /kWh electricity	kg emission /kg natural gas fuel	g emission /kWh electricity	kg emission /kg coal fuel
CO_2	2.6	2.37	390	2.5	850	2.4
CH_4	0.000048	Unknown	1.5	0.010	3.0	0.0084
Particulate matter	Negligible	0	0.074	0.00047	0.20	0.00056
SO_2	Negligible	0.000762	0.27	0.0017	1.0	0.0028
NO_x as NO_2	0.00046	0.000108	0.70	0.0045	2.0	0.0056
CO	0.0000033	0.00734	0.33	0.0021	0.12	0.00035
VOC	0.00000066	0	0.016	0.00010	0.013	0.000038

reformer produces negligible SO_x and particulate matter. Now consider the second process bottleneck: H_2 compression. Hydrogen compressors run on electric power from the surrounding electric grid. Although the energy required to compress H_2 is 10% of its HHV, this energy refers to the electric power drawn by the compressor. An additional energy penalty is paid due to the efficiency of the electric power plant. The average efficiency of all power plants connected to the grid is approximately 32% and their distribution by fuel type is shown in Figure 11.3. Over half of U.S. electric power plants are coal plants, which produce the most harmful emissions of any power plant per unit of electricity produced. Considering the relatively low emissions from the natural gas steam reformer and the efficiency penalty of the power plants, emissions from the use of electric power for H_2 compression may be the most significant contributor to air pollution.

Example 11.1. (1) Identify the bottleneck processes in the gasoline vehicle supply chain. (2) Estimate the energy required to complete some of the important processes in the chain from petroleum production from oil fields (box 2 in Figure 11.1) to the delivery of gasoline at the vehicle (box 6).

Solution

1. Bottleneck processes are those that consume the largest quantities of energy or that produce the largest quantities of harmful emissions in the supply chain. Based on background research on the petroleum industry and the supply chain shown in Figure 11.1, some of the bottleneck processes are (1) production of crude oil from fields (box 2), (2) centralized chemical processing of crude oil into gasoline (box 4), and (3) combustion of gasoline in the engine onboard the vehicle (box 7). Additional energy-intensive processes may include the transport of crude oil and gasoline (boxes 3 and 5), depending on the location of the vehicles relative to the oil fields. These

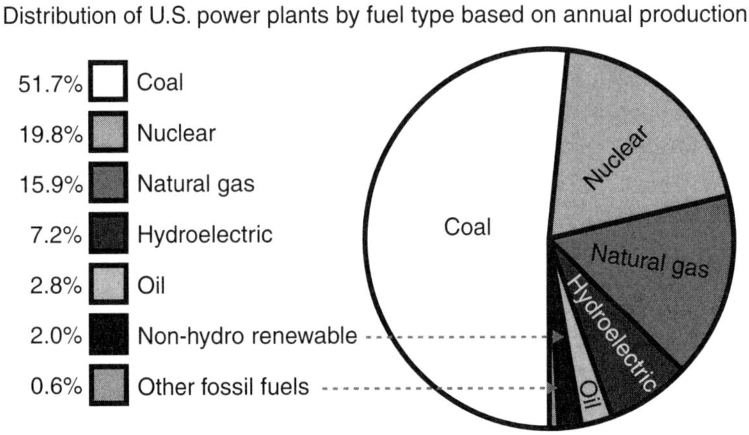

Distribution of U.S. power plants by fuel type based on annual production

51.7%	Coal
19.8%	Nuclear
15.9%	Natural gas
7.2%	Hydroelectric
2.8%	Oil
2.0%	Non-hydro renewable
0.6%	Other fossil fuels

Figure 11.3. Most U.S. electric power derives from conventional coal-fired power plants, which burn coal in a boiler to generate steam that runs through a steam turbine. The second largest portion of electric power comes from nuclear power plants, which extract heat from nuclear fission reactions to generate steam in a boiler that is then run through a steam turbine. The third most prevalent form of electric power is natural gas plants, which burn gas in a turbine.

bottleneck processes should be the focus of a further study of this supply chain via LCA.

2. Although estimates vary, approximately 12% of the HHV of gasoline fuel is required for its production, transport, and processing (boxes 2–5) [56]. The storage of gasoline (box 6) does not require a large quantity of energy because it remains a liquid at room temperature, with some evaporation. (Consider conducting additional research on the petroleum industry to quantify these estimates, which may vary by region due to differences in the distance to oil fields and in environmental legislation.)

Example 11.2. Having completed steps 1–3 of the LCA, we will now explore step 4 of LCA, analyzing the energy and mass flows in the supply chain using a control volume analysis and the principles of conservation of mass and energy. Imagine that the fuel cell vehicle fleet described in Figure 11.2 replaces the current on-road vehicle fleet shown in Figure 11.1. Emissions from this fuel cell fleet ultimately depend on the quantity of H_2 fuel it consumes. Assume this fuel cell fleet requires the same propulsive power as the current fleet—the total mass of the vehicles, their aerodynamic drag, rolling resistance, frontal area, and inertia are the same [57]. Based on fuel tax revenue records, the U.S. Environmental Protection Agency (EPA) estimates that on-road vehicles traveled 2.68 trillion miles (2.68×10^{12} miles) in 1999, and the average mileage of this fleet was 17.11 miles per gallon. The HHV of gasoline fuel is 47.3 MJ/kg and the HHV for H_2 fuel is 142.0 MJ/kg [58]. The density of gasoline is 750 kg/m^3. Having reviewed the relevant literature, you estimate that for the current vehicle fleet the average gasoline vehicle's efficiency (its motive energy to propel the vehicle/HHV of fuel) is 16%. Considering the performance of precommercial fuel cell vehicle prototypes, you estimate that the fuel cell vehicle's efficiency is 41.5% [59, 60]. Based on the conservation of energy, estimate the mass of H_2 needed to fuel this fleet.

Solution: We draw a control surface around box 7 in Figure 11.1 and box 9 in Figure 11.2 to compare mass and energy flows into and out of these processes. Based on the conservation of energy, we assume the work done by the current fleet ($\dot{W}_c$) equals the work done by the fuel cell fleet ($\dot{W}_f$), $\dot{W}_c = \dot{W}_f$. The required propulsive work of the average car in each fleet is the same. The propulsive work of the current fleet is

$$\dot{W}_c = \dot{m}_g \, \Delta H_{(\text{HHV}),g} \varepsilon_g \tag{11.3}$$

where $\dot{m}_g$ is the mass of gasoline fuel consumed by vehicles per year (kg/year), $\Delta H_{(\text{HHV}),g}$ is the HHV of gasoline fuel (MJ/kg), and ε_g is the gasoline vehicle's efficiency.

The mass of gasoline consumed per year is also

$$\dot{m}_g = \frac{\rho_g V_{MT}}{\bar{M}_{gvf} V_c} \tag{11.4}$$

where ρ_g is the density of gasoline (kg/m^3), V_{MT} the vehicle miles travelled per year (10^6 miles), $\bar{M}_{gvf}$ the average mileage of the conventional fleet (miles/gal), and V_c

the volumetric conversion (264.17 gallons/m^3). The propulsive work of the fuel cell fleet is

$$\dot{W}_f = \dot{m}_h \, \Delta H_{(HHV),h} \varepsilon_h \qquad (11.5)$$

where $\dot{m}_h$ is the mass of H$_2$ consumed by vehicles per year (kg/year), $\Delta H_{(HHV),h}$ is the HHV of H$_2$ fuel (MJ/kg), and ε_h is the fuel cell vehicle's efficiency. Setting $\dot{W}_c = \dot{W}_f$ and combining the last three equations, the mass of hydrogen consumed by the fleet is

$$\dot{m}_{H_2,C} = \frac{V_{MT}}{F_h} \qquad (11.6)$$

where

$$F_h = \frac{\bar{M}_{gvf} V_c \, \Delta H_{(HHV),h} \varepsilon_h}{\rho_g \, \Delta H_{(HHV),g} \varepsilon_g} \qquad (11.7)$$

is the mileage of hydrogen fuel cell vehicles (miles/kg H$_2$). Based on the information in our example,

$$F_h = \frac{(17.11 \text{ miles/gal})(264 \text{ gal/m}^3)(142 \text{ MJ/kg})(0.415)}{(750 \text{ kg/m}^3)(47.3 \text{ MJ/kg})(0.16)}$$

$$= 46.9 \text{ miles/kg H}_2 \qquad (11.8)$$

$$\dot{m}_{H_2,C} = \frac{V_{MT}}{F_h} = \frac{2.68 \times 10^{13} \text{ miles/yr}}{46.9 \text{ miles/kgH}_2} = 5.71 \times 10^{10} \text{ kg H}_2/\text{year}. \quad (11.9)$$

Based on this derivation, a fuel cell vehicle fleet would consume 57 megatonnes (MT) of H$_2$/year. Figure 11.4 shows a spatial distribution of hydrogen consumption by such a fuel cell fleet by county, based on gasoline consumption data by county recorded by the EPA [61].

11.2 IMPORTANT EMISSIONS FOR LCA

To conduct the next steps in LCA (especially steps 5 and 6), we first have to determine which types of emissions are important to evaluate in the supply chain. Important emissions fall into two categories: (1) those that influence global warming and (2) those that influence air pollution. In the two subsequent sections, we will discuss both of these. Emissions that influence global warming include CO$_2$ and CH$_4$. Important emissions that lead to air pollution include ozone (O$_3$),[2] CO, nitrogen oxides (NO$_x$), particulate matter (PM), sulfur oxides (SO$_x$), and volatile organic compounds (VOCs). In the sections that follow, we will

---

[2]In the upper atmosphere, ozone creates a protective layer around Earth by absorbing ultraviolet radiation that would otherwise harm life. However, ozone emitted at sea level causes smog and air pollution and damages human health.

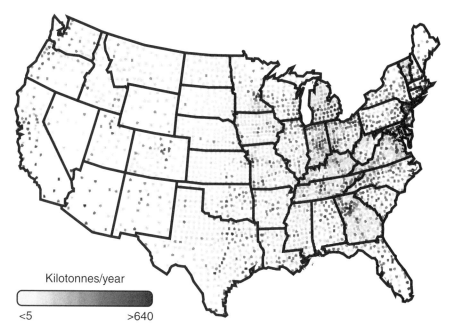

Figure 11.4. Annual hydrogen consumption by fuel cell vehicles by county, plotted at the center of each U.S. county, assuming a complete switch of fleet from conventional vehicles to fuel cell vehicles.

(a) discuss the importance of these emissions and (b) describe methods for quantifying their environmental impact.

11.3   EMISSIONS RELATED TO GLOBAL WARMING

11.3.1   Climate Change

Earth's climate has changed over time. Earth's average near-surface temperature is currently close to 15°C, but geological evidence suggests that in the past one million years it may have fluctuated to as high as 17°C and as low as 8°C. Climate scientists are now concerned that these natural fluctuations are being overtaken by warm-side temperature changes induced by human activity, specifically the combustion of fossil fuels that release gases and particles that have a warming effect on the atmosphere [62].

11.3.2   Natural Greenhouse Effect

The natural greenhouse effect is the process by which gases normally contained in the atmosphere, such as CO_2 and water vapor (H_2O), trap a portion of the Sun's energy in the form of infrared (IR) radiation. As a result, Earth's temperature is high enough to support life as we know it. When the Sun's light hits Earth's surface, some of this energy

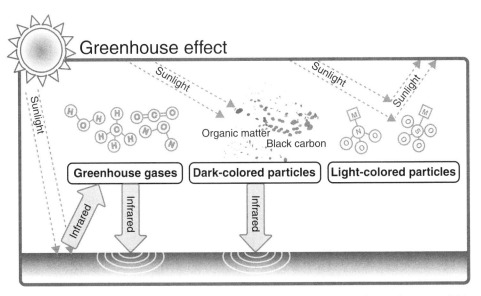

Figure 11.5. Left: Sunlight hits Earth's surface and is partly absorbed. Earth reemits some of this energy as IR radiation (thermal energy). Greenhouse gases, including H_2O, CH_4, CO_2, and N_2O, selectively absorb this IR radiation and reemit it out to space and back toward Earth's surface and thereby warm Earth's surface. Center: Sunlight hits dark-colored particles, such as black carbon, suspended in the Earth's atmosphere. These dark particles absorb the light and reemit this energy as IR radiation, some of which may reach Earth's surface and may warm it. Organic matter focuses light onto black carbon, thereby enhancing black carbon's warming effect. Right: Light-colored particles, including sulfates and nitrates, reflect sunlight and have a cooling effect.

is absorbed and warms Earth. Earth's surface then reemits some of this energy to the atmosphere as IR radiation or thermal energy. Greenhouse gases are special in that, unlike other molecules, they selectively absorb 80% of IR radiation, and then reemit this radiation back up to space and back towards Earth's surface. The left portion of Figure 11.5 shows the warming mechanism of greenhouse gases. In a process somewhat similar to heat trapping in a glass greenhouse, greenhouse gases absorb and reemit some IR radiation while remaining transparent to 50% of visible sunlight and other wavelengths. As a result, the more greenhouse gases present in the atmosphere, the more heat is trapped near Earth's surface. The natural greenhouse effect contributes 33 K of Earth's average near-surface air temperature of 288 K. Without this effect, Earth would be too cold to support life as we know it.

11.3.3 Global Warming

Most climate scientists concur that an increase in anthropogenic (i.e., man-made) emissions of greenhouse gases is contributing to an intensification of the greenhouse effect. Global warming refers to the increase in Earth's temperature above that caused by the natural greenhouse effect as a result of the addition of anthropogenic greenhouse gases and certain particles. Anthropogenic greenhouse gases include CO_2, CH_4, H_2O, and nitrous oxide (N_2O). In addition to these gases, certain particles also have a warming effect on Earth but through a different mechanism. Dark-colored particles, such as soot, absorb sunlight, reemit this energy as IR radiation, and therefore also may warm Earth's surface. Black carbon (BC) is a predominant global warming particle [63, 64]. The warming effect of black carbon is enhanced by organic matter (OM), which focuses additional light onto black carbon. The center portion of Figure 11.5 shows the warming mechanism of dark-colored particles. Figure 11.5 shows that these gases and particles reemit IR radiation toward Earth's surface to cause warming; they also reemit IR radiation away from Earth. In contrast, light-colored particles reflect sunlight and have a cooling effect. Light-colored particles that cool Earth include sulfates (SULF) and nitrates (NIT). SULF also attract water, which reflects light as well. Emitted gases that have a cooling effect include SO_x, NO_x, and nonmethane organic compounds or VOCs. These gases react in the atmosphere and convert to particles which are mostly light in color. Sulfur oxide converts to SULF, NO_x converts to NIT, and VOCs convert to light-colored organics. The right portion of Figure 11.5 shows the cooling mechanism of light-colored particles.

11.3.4 Evidence of Global Warming

Since the 1860s, the concentration of primary greenhouse gases—CO_2, CH_4, and N_2O—in the lower atmosphere has increased by 30%, 143%, and 14%, respectively. Figure 11.6

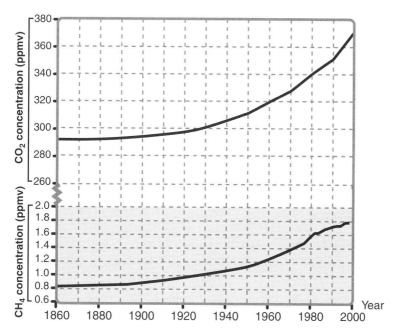

Figure 11.6. Since the 1860s, concentrations of the primary greenhouse gases CO_2 and CH_4 in the lower atmosphere have increased by 30% and 143%, respectively.

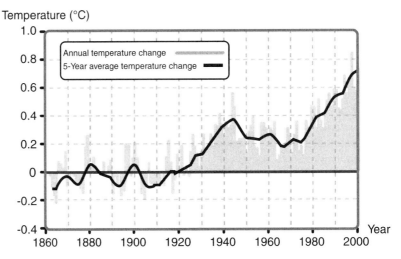

Figure 11.7. Since the 1860s, Earth's near-surface temperature has increased by 0.6°C on average.

shows the increase in CO_2 and CH_4 over the past century. With the start of the Industrial Revolution 200 years ago, people began to combust fossil fuels to provide energy for industrial processes and began releasing much larger quantities of CO_2 into the atmosphere than in previous times. At the start of the Industrial Revolution, CO_2 concentrations were close to 280 parts per million by volume (ppmv). Currently, they are close to 380 ppmv and are increasing at a rate of 2 ppmv/year. Over the same period, Figure 11.7 shows the change in Earth's near-surface temperature, which has increased by $0.6°C \pm 0.2°C$ over the past 100 years. Compared with historical records, this rate of temperature increase is unusually high. Further evidence of global warming includes the following:

1. An increase in temperature in the past four decades in the lowest 8 km of the atmosphere
2. A decrease in snow cover, ice extent, and glacier extent
3. A 40% reduction in the thickness of Arctic sea ice in summer and autumn in recent decades,
4. An increase in global average sea level by 10–20 cm due to warmer oceans expanding
5. An increase in the heat content of the ocean

Other pieces of evidence indicating anthropogenic climate change include flowers blooming earlier, birds hatching earlier, and a cooling of the middle portion of the atmosphere.

11.3.5 Hydrogen as a Potential Contributor to Global Warming

Since industrialization, the concentration of H_2 in the atmosphere is estimated to have increased via 200 parts per billion by volume (ppbv) [65] to 530 ppbv [66]. The majority of H_2 emissions originate from the oxidation of HCs, especially the incomplete combustion of gasoline and diesel fuels in automobiles, and the burning of biomass. When released,

H_2 most commonly does not combust with oxygen in the air because its concentration and its temperature are usually too low to facilitate the reaction. The self-ignition temperature of H_2 is 858 K and its ignition limits in air are between 4 and 75%. Once released into the atmosphere, H_2 is estimated to have a lifetime of between 2 and 10 years.

If fuel cells become widespread, H_2 release will likely accelerate. As we saw in Figure 11.2, H_2 may leak into the environment during its production, compression, storage, and use onboard vehicles (boxes 6–9). In addition, H_2 may leak during transport, especially if transmitted over long distances through pipelines, in much the same way natural gas leaks today (box 5). Because H_2 is one of the smallest molecules, it may be more likely than other fuels to escape from small openings. For example, the mass diffusion coefficient of H_2 is four times higher than that of natural gas. In addition to leakage, H_2 may also be intentionally released into the environment. For example, some fuel cell systems are designed to purge anode exhaust gas (containing H_2) from the stack periodically so as to prevent blockage of reaction sites at the anode by other species (such as water.) Also, liquid H_2 tanks require a periodic release of H_2 to avoid pressure buildup.

As a result, climate researchers are now trying to determine the potential effect of released H_2 on global warming. One mechanism through which released H_2 might increase global warming is by indirectly increasing the concentration of the greenhouse gas CH_4. In the troposphere (lower atmosphere), H_2 reacts with the hydroxyl radical (OH) according to the reaction

$$H_2 + OH \rightarrow H + H_2O \qquad (11.10)$$

If H_2 did not consume OH in this reaction, OH might otherwise reduce the presence of CH_4 via the reaction

$$CH_4 + OH \rightarrow CH_3 + H_2O \qquad (11.11)$$

However, numerous other chemical reactions must also be considered. The net effect of H_2 on global warming is still the subject of research.

> **Example 11.3.** You read an article that claims that fuel cell vehicles might increase global warming as a result of the additional water vapor they will produce. You decide to invoke LCA to make your own determination. You decide to compare two different scenarios, one being the current vehicle fleet (shown in Figure 11.1) and the other being a fuel cell vehicle fleet (shown in Figure 11.2). You decide to calculate the water vapor emitted in each of these scenarios to compare them to see if there would be a genuine increase in water vapor emissions between a current fleet and a fuel cell fleet. The 1999 vehicle fleet consumed approximately 450 MT/yr of combined gasoline ($C_nH_{1.87n}$) and light diesel fuel ($C_nH_{1.8n}$) [67]. Gasoline and light diesel fuels represented 78 and 22% of fuel consumption in vehicles [68], respectively.
>
> 1. Locate the sources of H_2O emission in each supply chain.
> 2. Identify the bottleneck processes for H_2O emission.
> 3. Based on the conservation-of-mass equation, calculate the quantity of water vapor emitted in the bottleneck processes.
> 4. Is the article's assertion valid?

Solution

1. Locate the source of H_2O vapor emission in the supply chain. In the current fleet, water vapor is emitted as a product of combustion. As shown in Figure 11.1, water vapor is emitted during the transport of petroleum fuel by truck, railroad, or ship (boxes 3 and 5) and during ICE vehicle use (box 7). As shown in Figure 11.2, in the fuel cell fleet scenario, water vapor is emitted as a product of the electrochemical oxidation of hydrogen at the exhaust of the fuel cell vehicle (box 9). Water vapor is also emitted indirectly because hydrogen compressors consume electric power from power plants (box 7), and some of these power plants (coal and natural gas) produce water as a product of combustion.

2. Identify the bottleneck processes for H_2O emission. As a first approximation, we assume that the majority of H_2O emissions occur in the last step of each process chain (box 7 in Figure 11.1 and box 9 in Figure 11.2) during vehicle use.

3. Calculate the quantity of water vapor emitted in the bottleneck processes. Within an internal combustion engine, combustion can be described by

$$CH_{1.85} + 1.4625O_2 \rightarrow CO_2 + 0.925H_2O + \text{work} + \text{heat} \qquad (11.12)$$

where $CH_{1.85}$ is a chemical formula representing gasoline ($C_nH_{1.87n}$) and light diesel ($C_nH_{1.8n}$) fuels weighted by their consumption in the vehicle fleet (78 and 22%, respectively). The molecular weight of $CH_{1.85}$ is 13.85 g/mol. The molecular weight of H_2O is 18 g/mol. Every kilogram of $CH_{1.85}$ consumed produces 1.2 kg H_2O (18 kg/mol $H_2O \times 0.925$ mol H_2O/13.85 g/mol $CH_{1.85}$). For every 450 MT/yr fuel consumed, approximately 540 MT H_2O/yr is produced. Within a fuel cell, every mole of hydrogen consumed produces 1 mole H_2O, according to

$$H_2 + 0.5O_2 \rightarrow H_2O + \text{electricity} + \text{heat} \qquad (11.13)$$

The molecular weight of H_2 is 2 g/mole. Thus, every kilogram of H_2 consumed produces 9 kg H_2O. In Example 11.2, we calculated that a fuel cell fleet would consume 57 MT H_2/year. The fleet would then produce about 510 MT/yr of H_2O. Based on these estimates, a fuel cell vehicle fleet would produce approximately the same quantity of water vapor as the current fleet [69]. (This calculation may over-estimate the amount of water vapor produced by the fuel cell vehicles because it assumes all water is emitted in vapor form, when it could actually condense as a liquid, especially given the low operating temperature of hydrogen fuel cells.)

4. Is the article's assertion valid? The quantity of water vapor produced by either the current fleet or a fuel cell fleet is one million times smaller than the emission rate of water vapor from natural sources—5×10^8 MT/yr. Based on these considerations, the water vapor emitted by either fleet will have a negligible effect on the atmosphere. Thus, the article's assertion does not appear to be valid.

11.3.6   Quantifying Environmental Impact—Carbon Dioxide Equivalent

One important method for quantifying the environmental impact of emissions related to global warming is the calculation of the carbon dioxide equivalent ($CO_{2\,equivalent}$) of a mixture of emitted gases and particles. To estimate the potential for a mixture of gases and particles to contribute to global warming, one can calculate the $CO_{2\,equivalent}$ of these gases. The $CO_{2\,equivalent}$ is the mass of CO_2 gas that would have an equivalent warming effect on Earth as the mixture of different gases. The $CO_{2\,equivalent}$ helps us quantify and compare the warming effect of different types and quantities of emissions. One equation for measuring the $CO_{2\,equivalent}$ of gases over a 100-year period is [69, 70]

$$CO_{2\,equivalent} = m_{CO_2} + 23m_{CH_4} + 296m_{N_2O} + \alpha(m_{OM,2.5} + m_{BC,2.5}) \qquad (11.14)$$

$$- \beta(m_{SULF,2.5} + m_{NIT,2.5} + 0.40m_{SO_x} + 0.10m_{NO_x} + 0.05m_{VOC})$$

where m is the mass of each species emitted, with, for example, $m_{OM,2.5}$ indicating the mass of organic matter 2.5 μm in diameter and less. The coefficient α can range between 95 and 191. The coefficient β can range between 19 and 39. The logic of this formula follows from our description of the various gases and particles that contribute to global warming or cooling, as shown in Figure 11.5. In the formula, gases or particles with a warming effect are preceded by a plus sign and those with a cooling effect are preceded by a minus sign. The coefficients in front of the masses (23, 296, α, and β, respectively) represent the global warming potential (GWP) of each of the species over a 100-year period. The GWP is an index for estimating the relative global warming contribution of a unit mass of a particular greenhouse gas or particle emitted compared to the emission of a unit mass of CO_2. For example, a GWP of 23 for CH_4 indicates that it is 23 times more efficient at absorbing radiation than CO_2. The GWP for H_2 is not included in the equation above because its value is still being determined by climate researchers. Values of the GWP are calculated for different time horizons due to the different lifetimes of gases in the atmosphere. Anthropogenic H_2O emission is not usually considered in $CO_{2\,equivalent}$ calculations because, as we learned in Example 11.3, natural sources of H_2O are five orders of magnitude higher than anthropogenic sources. The $CO_{2\,equivalent}$ equation above is only an estimate of the potential for global warming of some of the important gases and particles and must be periodically updated with further climate research findings. More accurate results than ones derived using the above equation can be obtained through the use of global-scale computer models of the atmosphere.

11.3.7   Quantifying Environmental Impact—External Costs
of Global Warming

A second important method for quantifying the environmental impact of emissions related to global warming is the calculation of external costs of global warming. The potential effects of global warming include the following:

1. An increase in sea level, resulting in flooding of some low-lying areas
2. An intensification of the hydrological cycle, resulting in both more drying and more flooding due to an increase in extreme precipitation events

3. Shifts in regions with arable land and changes in agricultural regions
4. Damage to ecosystems

Researchers estimate the external cost of global warming at between \$0.026 and \$0.067 per kilogram of CO_2 equivalent emission in 2004 dollars [71, 72]. This external cost is the damage cost of an additional unit of mass of CO_2 (or equivalent gas) into the atmosphere.[3] An external cost arises when all of the costs of a good are not included into its free-market price [73]. An example of an externality is the cost of damage to a piece of real-estate due to flooding resulting from a sea-level rise related to global warming. By definition, the external costs of global warming related to land use are not incorporated into the free-market prices for property. Researchers' estimates of the economic value of externalities vary over a large range because these costs are difficult to quantify precisely. However, to ignore external costs is to incorrectly assume that their value is zero.

> ***Example 11.4.*** (1) The EPA tabulates emissions from vehicles, power plants, and all other sources in a National Emission Inventory (NEI). You check the NEI for emissions from on-road fossil fuel vehicles in 1999 and create Table 11.2. In Table 11.2, PM_{10} refers to particulate matter that is 10 μm in diameter and less; $PM_{2.5}$ refers to matter 2.5 μm in diameter and less. Calculate the CO_2 equivalent of this fleet. Compare this with only the CO_2 released by the fleet. (2) Now imagine instantaneously replacing this fossil fuel vehicle fleet with a hydrogen fuel cell vehicle fleet. Calculate the CO_2 equivalent of this fleet, considering only the change in vehicles. What is this percentage reduction in terms of total anthropogenic CO_2 equivalent in the United States? (3) To make this comparison more even-handed, what might you also consider? (4) What is the reduction in external costs (costs of the damage to society from global warming that is not incorporated into free-market prices)?

TABLE 11.2. U.S. Emissions from All Man-Made Sources, 1999 (metric tonnes/year)

Species	On-Road Vehicles[a]	Total All Sources[b]
Gases		
Carbon monoxide (CO)	6.18×10^7	1.12×10^8
Nitrogen oxides (NO_x) as NO_2	7.57×10^6	2.19×10^7
Sulfur oxides (SO_x) as SO_2	2.72×10^5	1.81×10^7
Ammonia (NH_3)	2.39×10^5	4.53×10^6
Hydrogen (H_2)	1.55×10^5	2.79×10^5
Carbon dioxide (CO_2)	1.37×10^9	5.30×10^9
Water (H_2O)	5.19×10^8	1.99×10^9

(continued)

---

[3]External costs are referred to also as damage, societal, and/or environmental costs, depending on the source of the costs.

TABLE 11.2. (*continued*)

Species	On-Road Vehicles[a]	Total All Sources[b]
Organics		
Paraffins (PAR)	3.53×10^6	1.40×10^7
Olefins (OLE)	1.61×10^5	5.21×10^5
Ethylene (C_2H_4)	2.27×10^5	9.12×10^5
Formaldehyde (HCHO)	4.43×10^4	2.23×10^5
Higher aldehydes (ALD2)	1.72×10^5	3.39×10^5
Toluene (TOL)	3.29×10^5	2.60×10^6
Xylene (XYL)	4.66×10^5	2.25×10^6
Isoprene (ISOP)	4.86×10^3	9.92×10^3
Total nonmethane organics	*4.93×10^6*	*2.09×10^7*
Methane (CH_4)	7.91×10^5	6.31×10^6
Particulate matter		
Organic matter ($OM_{2.5}$)	5.04×10^4	2.64×10^6
Black carbon ($BC_{2.5}$)	9.07×10^4	5.92×10^5
Sulfate ($SULF_{2.5}$)	1.88×10^3	3.10×10^5
Nitrate ($NIT_{2.5}$)	2.47×10^2	2.67×10^4
Other ($OTH_{2.5}$)	2.40×10^4	8.26×10^6
Total $PM_{2.5}$	*1.67×10^5*	*1.18×10^7*
Organic matter (OM_{10})	7.19×10^4	5.77×10^6
Black carbon (BC_{10})	1.07×10^5	9.62×10^5
Sulfate ($SULF_{10}$)	2.99×10^3	4.91×10^5
Nitrate (NIT_{10})	3.15×10^2	7.10×10^4
Other (OTH_{10})	3.66×10^4	3.75×10^7
Total PM_{10}	*2.19×10^5*	*4.48×10^7*

[a] Conventional on-road fossil fuel vehicles.
[b] All man-made sources including industrial facilities and power plants.

Solution

1. Based on the $CO_{2\,equivalent}$ formula and the data in Table 11.2, we can calculate high and low values for the range of $CO_{2\,equivalent}$ gases and particles emitted by on-road vehicles.

$$CO_{2\,equivalent,\,LOW} \qquad\qquad\qquad (11.15)$$
$$= m_{CO_2} + 23m_{CH_4} + 296m_{N_2O} + 95(m_{OM_{2.5}} + m_{BC_{2.5}})$$
$$- 39(m_{SULF_{2.5}} + m_{NIT_{2.5}} + 0.40m_{SO_X} + 0.10m_{NO_X} + 0.05m_{VOC})$$

$$CO_{2\,equivalent,\,HIGH} \qquad\qquad\qquad (11.16)$$
$$= m_{CO_2} + 23m_{CH_4} + 296m_{N_2O} + 191(m_{OM_{2.5}} + m_{BC_{2.5}})$$
$$- 19(m_{SULF_{2.5}} + m_{NIT_{2.5}} + 0.40m_{SO_x} + 0.10m_{NO_x} + 0.05m_{VOC})$$

Because the NEI does not tabulate N_2O, as an estimate, consider only the other terms. This range is between 1.36×10^9 and 1.39×10^9 tonnes/year. These values differ from the total CO_2 fleet emissions by -0.87% and 1.75%. Thus, in this example, the primary contributor to $CO_{2\,equivalent}$ is CO_2 itself.

2. Considering only the change in the fleet and no upstream fuel production sources, the hydrogen fuel cell vehicle fleet would produce no CO_2. Its $CO_{2\,equivalent}$ would also be zero.

Based on the $CO_{2\,equivalent}$ formula and the data in Table 11.2, we can calculate high and low values for the range of $CO_{2\,equivalent}$ gases and particles emitted by all sources in the United States, 5.33×10^9 and 5.86×10^9 tonnes/year. This change represents an approximate reduction in $CO_{2\,equivalent}$ in a range of 23.21–26.17%.

3. To make this analysis more even handed, one might also consider the change in $CO_{2\,equivalent}$ gases and particles from upstream sources, including fuel production in both the fossil fuel and hydrogen supply chain.

4. Based on a range of external costs of global warming of between \$0.026 and \$0.067 per kilogram of $CO_{2\,equivalent}$, a reduction in $CO_{2\,equivalent}$ in a range of 1.36×10^9–1.39×10^9 tonnes/year translates to a reduction in external costs of between \$35.3 and \$93.5 billion/year due to global warming.

11.4 EMISSIONS RELATED TO AIR POLLUTION

To conduct the later steps in LCA related to emissions, in addition to emissions that influence global warming, we have to determine which emissions in the supply chain influence air pollution. The primary source of air pollution is combustion in power plants, furnaces, and vehicles. This air pollution can harm the health of humans, animals, and vegetation and can damage materials. Six primary emissions that create air pollution are O_3, CO, NO_x, PM, SO_x, and VOCs. Volatile organic compounds are nonmethane organic compounds, such as the higher HCs (C_xH_y). Some of these compounds are air pollutants themselves. Others react with chemicals to produce air pollution. Effects of air pollution on human health can include respiratory illness, pulmonary illness, damage to the central nervous system, cancer, and increased mortality.

11.4.1 Hydrogen as a Potential Contributor to Air Pollution

Because an increase in the use of fuel cells might increase the quantity of H_2 released into the atmosphere, climate researchers are now trying to determine the potential effect of released H_2 on air pollution. One mechanism through which released H_2 might increase one type of air pollutant is through a series of chemical reactions that enhance the concentration of O_3. In the troposphere, H_2 might increase O_3 by increasing the concentration of atomic hydrogen (H). After several years in the atmosphere, molecular hydrogen decays to atomic hydrogen in the presence of the hydroxyl radical (OH), via the reaction

$$H_2 + OH \rightarrow H_2O + H \tag{11.17}$$

Atomic hydrogen (H) could then react with oxygen (O_2) in air in the presence of photon energy ($h\nu$) from light to increase O_3 through the following set of reactions:

$$H + O_2 + M \rightarrow HO_2 + M \tag{11.18}$$

$$NO + HO_2 \rightarrow NO_2 + OH \tag{11.19}$$

$$NO_2 + h\nu \rightarrow NO + O \tag{11.20}$$

$$O + O_2 + M \rightarrow O_3 + M \tag{11.21}$$

where M represents any molecule in the air that is neither created nor destroyed during the reaction but that absorbs energy from the reaction. However, other sets of reactions must also be considered, with a focus on their net effect on air pollution. The net effect of these reactions might be determined with computer simulations of chemical reactions in the atmosphere (atmospheric models). As we learned in LCA, to be accurate, these simulations should model, not the mere addition or subtraction of an individual chemical component, but rather the net change in emissions among different scenarios.

11.4.2 Quantifying Environmental Impact—Health Effects of Air Pollution

Table 11.3 summarizes some of the most important emissions and the ambient air pollutants that evolve from them via chemical reactions with other compounds [74]. The table also lists some important health effects from these pollutants. For example, emissions of both CO and PM increase the human death rate (mortality). Finally, the table shows estimates of the number of cases of each health effect per unit mass of ambient pollutant.[4] The estimates in Table 11.3 primarily apply to vehicles rather than power plants; vehicles tend to be used in population centers where they are close to people. Therefore, their emissions have a stronger impact on human health per unit mass of emission than power plants, which

---

[4]Estimates were derived from the number of U.S. cases of each health effect stemming from automotive pollution and total U.S. emissions of each type from automobiles (NEI).

TABLE 11.3. Health Effects of Air Pollution

Emission	Ambient Pollutant	Health Effect	Health Effect Factor (thousands of cases/tonne ambient pollutant)		Change in Health Effects (thousands of cases) with a Fleet Change from Conventional to Fuel Cell	
			Low	High	Low	High
CO	CO	Headache	1.22	1.45	-7.53×10^4	-8.95×10^4
		Hospitalization	0.000572	0.000164	-3.54	-10.2
		Mortality	0.00000357	0.0000107	-0.221	-0.663
NO$_x$	NO$_2$	Sore throat	11.5	11.6	-8.68×10^4	-8.81×10^4
		Excess phlegm	5.26	5.34	-3.98×10^4	-4.04×10^4
		Eye irritation	4.73	4.81	-3.58×10^4	-3.64×10^4
VOC + NO$_x$	O$_3$	Asthma attacks	0.0811	0.255	-1.01×10^3	-3.19×10^3
		Eye irritation	0.752	0.830	-9.40×10^3	-1.04×10^4
		Low respiratory illness	1.08	1.80	-1.35×10^4	-2.25×10^4
		Upper respiratory illness	0.328	0.548	-4.10×10^3	-6.85×10^3
		Any symptom or condition (ARD2)	0	6.13	0	-7.67×10^4
PM$_{10}$, SO$_2$,	PM$_{10}$	Asthma attacks	0.147	0.155	-188	-199
NO$_x$, VOC		Respiratory restricted activity days (RRAD)	4.33	5.87	-5566	-7540
		Chronic illness	0.00190	0.00454	-2	-6
		Mortality	0.00391	0.00669	-5	-9

Note: Emissions from vehicles (column 1) evolve by chemical reaction to ambient pollutants (column 2). These ambient pollutants lead to various health effects in people (column 3). The health effects are estimated primarily for automotive pollutants in terms of the number of cases of each health effect per unit mass of ambient pollutant, with low values (column 4) and high values (column 5). An example is shown for the change in health effects cases with a switch in vehicle fleet from the conventional one to a hydrogen fuel cell fleet (columns 6 and 7).

tend to be located further from population centers. Table 11.3 lists incidents of health effects as a function of ambient pollutant levels. To calculate the health effects per tonne of emission, one can estimate that every tonne of VOC or NO_x emitted yields, via chemical reaction, 1 tonne of O_3 as an ambient pollutant,

$$m_{O_3, \text{AMB}} = m_{\text{VOC}} + m_{\text{NO}_x} \qquad (11.22)$$

where m is the mass of each type of emission, and that ambient pollution of PM_{10} can be calculated as

$$m_{\text{PM10,AMB}} = m_{\text{PM2.5}} + 0.1(m_{\text{PM10}} - m_{\text{PM2.5}}) + 0.4 m_{\text{SO}_2} + 0.1 m_{\text{NO}_2} + 0.05 m_{\text{VOC}}$$

$$(11.23)$$

where the coefficients in front of the m refer to the percentage of emitted mass that converts to ambient PM_{10} pollution via reaction with other species [69].

11.4.3 Quantifying Environmental Impact—External Costs of Air Pollution

If people are less healthy, they require more medical services and miss more productive working days. Additional medical services and a decrease in labor productivity incur a financial cost on society. Therefore, the health effects of air pollution can be quantified in financial terms. Based on the health effects data shown in Table 11.3, Table 11.4 estimates the financial costs of these and other emissions on human health [74]. Interestingly, the majority of the health costs in Table 11.4 are due to automotive emissions. The health costs per unit mass of emissions are estimated to be about an order of magnitude lower for power plants due to their greater distance from people. The financial costs related to human health are the dominant source of external costs of air pollution. As with the external costs of global warming, the external costs of air pollution are not incorporated into free-market prices. Although these costs are difficult to quantify, ignoring them incorrectly assumes their value is zero.

> ***Example 11.5.*** (1) Based on the scenario of fuel cell vehicle adoption outlined in Example 11.4, calculate the change in health effects for a replacement of conventional vehicles with fuel cell vehicles. For simplicity, in this LCA comparison, focus on the change in emissions at the vehicle, ignoring upstream changes in emissions. (2) Calculate the change in external costs (the financial costs of health damage born by society). (3) Compare the change in external costs due to air pollution with the change due to global warming.
>
> *Solution*
>
> 1. The change in health effects is shown in the last column of Table 11.3. Volatile organic compounds include all of the organics listed in Table 11.2 except methane. One can calculate the quantity of ambient ozone pollution from the emitted VOCs and NO_x based on $m_{O_3, \text{AMB}} = m_{\text{VOC}} + m_{\text{NO}_x}$ and the quantity of ambient pollution of PM_{10} from several emissions based on

TABLE 11.4. Financial Costs of Air Pollution

Emission	Ambient Pollutant	Health Cost of Air Pollution ($2004/tonne of emission)		Change in Health Costs Due to Air Pollution ($2004) with a Fleet Change from Conventional to Fuel Cell	
		Low	High	Low	High
CO	CO	12.7	114	-7.87×10^8	-7.08×10^9
NO$_x$	Nitrate–PM$_{10}$	1.30×10^3	2.11×10^4	-9.83×10^9	-1.60×10^{11}
	NO$_2$	191	929	-1.45×10^9	-7.03×10^9
PM$_{2.5}$	PM$_{2.5}$	1.33×10^4	2.03×10^5	-2.22×10^9	-3.39×10^{10}
PM$_{2.5}$–PM$_{10}$	PM$_{2.5}$–PM$_{10}$	8.52×10^3	2.25×10^4	-4.38×10^8	-1.16×10^9
SO$_x$	Sulfate–PM$_{10}$	8.78×10^3	8.33×10^4	-2.39×10^9	-2.27×10^{10}
VOC	Organic–PM$_{10}$	127	1.46×10^3	-6.27×10^8	-7.21×10^9
VOC + NO$_x$	O$_3$	12.7	140	-1.59×10^8	-1.75×10^9
Total				-1.79×10^{10}	-2.40×10^{11}

Note: Emissions from vehicles (column 1) evolve by chemical reaction to ambient pollutants (column 2). These ambient pollutants lead to health effects in people and therefore a human health cost to society (columns 3 and 4). An example is shown for the change in health costs with a switch in vehicle fleet from the conventional one to a hydrogen fuel cell fleet (columns 5 and 6).

$$m_{\text{PM10, AMB}} = m_{\text{PM2.5}} + 0.1(m_{\text{PM10}} - m_{\text{PM2.5}})$$
$$+ 0.4m_{\text{SO}_2} + 0.1m_{\text{NO}_2} + 0.05m_{\text{VOC}} \quad (11.24)$$

The reduction in health effects shown in Table 11.3 is an upper bound estimate. A more developed analysis takes into account the net change in emissions all along each supply chain.

2. The change in health costs is shown in the last column of Table 11.4. The external costs shown in Table 11.4 are per-unit mass of emission (not per unit mass of ambient pollutant as in Table 11.3.) With a switch in the vehicle fleet, health costs decrease by between $18 and $240 billion per year.

3. With a switch in fleet, we have seen that global warming costs decrease by between $35.3 and $93.5 billion per year. The reduction in health costs is in a similar range.

11.5 ANALYZING ENTIRE SCENARIOS WITH LCA

We have now seen several examples of different segments of LCA. We have also learned important tools for quantifying the environmental impact of different supply chains. We will now combine these tools to analyze an additional scenario on electric power production through the lens of energy efficiency.

11.5.1 Electric Power Scenario

Having read so much about fuel cells, you are interested in exploring the possibility of installing a fuel cell system on your local university's campus. You would like this system to provide electricity to nearby buildings. Because you live in an area of the country rich in coal reserves, you would like to explore the possibility of using coal as the original fuel. Your local university currently gets most of its electricity from a nearby coal power plant. You decide to compare (1) the current scenario with electricity derived from a coal power plant against (2) a possible process chain of a fuel cell system fueled by hydrogen derived from coal. You would like to determine whether it would be more efficient to use a fuel system. You decide to compare the overall electrical efficiency across the process chain to see which scenario might be more efficient.

1. *Research and develop an understanding of the supply chain.* First, think about the current supply chain for electricity. Coal is extracted from coal mines, processed from chunks into smaller pieces, and transported via railroad or barge to power plants that are usually within close proximity to the mine. The coal power plant produces electricity that is transmitted across high-voltage transmission lines long distances and later at reduced voltages over low-voltage distribution lines to the university's buildings.

Second, think about the potential H_2 supply chain. Based on our knowledge of fuel processing from Chapter 10 and some additional reading, we learn how we can chemically process coal into a H_2-rich gas, a process called coal gasification. Coal gasification is a chemical conversion process that transforms solid coal and steam into a gaseous mixture of H_2 and CO at elevated pressures and temperatures. Because coal contains little H_2, much of the H_2 originates from the added steam. Emissions for a coal gasification plant optimized for H_2 production are shown in Table 11.1. This plant has a HHV efficiency of 60%.

Assume our coal gasification plants are placed at similar locations as conventional coal power plants. They rely on the same upstream processes as traditional coal plants, including coal mining, processing, and transport. After H_2 production, H_2 is transmitted through large hydrogen transmission pipelines over long distances and then through smaller distribution lines to local areas. Then H_2 is stored and consumed in fuel cell systems located throughout your university campus. Each fuel cell system provides electricity to one or more buildings.

2. *Sketch a supply chain.* Figures 11.8 and 11.9 describe these two separate supply chains. The first three boxes of Figure 11.8 are the same as for Figure 11.9.

3. *Identify the "bottleneck" processes.* In Figure 11.8, think about the energy input arrows at the bottom of the process boxes in terms of efficiency. The HHV efficiency of the first three combined processes—extraction, processing, and transport—is approximately 90%; about 10% of the original energy in the coal fuel is required for its combined mining (box 1), processing (box 2), and transport (box 3). The HHV efficiency of a typical coal plant (box 4) is approximately 32%; for every 100 units of coal energy entering the plant, 32 units leave as electricity and 68 leave as heat dissipated to the environment. The efficiency of electricity transmission (box 5) is 97%; about 3% of the electricity transmitted over the high-voltage wires from the coal plant to urban areas is dissipated as heat. The

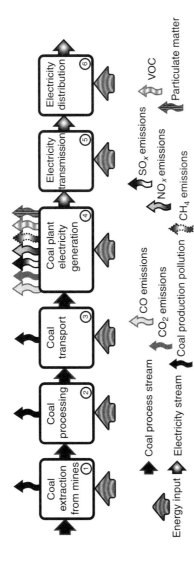

Figure 11.8. Supply chain for conventional electricity generation from coal. The most energy and emission intensive process in the chain is electricity generation (box 4).

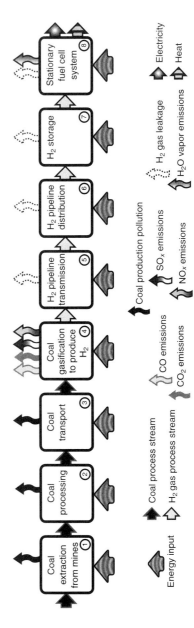

Figure 11.9. Supply chain for coal gasification plant. The most energy intensive processes in the chain are coal gasification (box 4) and electricity generation at the stationary fuel cell system (box 8).

efficiency of electricity distribution is about 93%; about 7% of electricity conveyed over the low-voltage wires around local areas is lost to the environment as heat. Therefore, for the scenario in Figure 11.8, the most energy intensive process is by far electricity generation at the coal plant.

In Figure 11.9, think about the energy input arrows at the bottom of the process boxes in terms of efficiency. The HHV efficiency of the first three combined processes (boxes 1, 2, and 3) is the same as in the supply chain of Figure 11.8, approximately 90%. The HHV efficiency of the coal gasification plant (box 4) is approximately 60%; that is, for every 100 units of coal energy entering the plant, 60 units leave as hydrogen energy. The efficiencies of hydrogen transmission (box 5) and distribution (box 5) are both 97%, similar to natural gas. The HHV efficiency of hydrogen storage not at pressure is about 100%. The HHV electrical efficiency of the fuel cell system is 50%. Therefore, for the scenario in Figure 11.9, the most energy intensive processes are by far coal gasification and electricity generation at the fuel cell system.

4. *Analyze the energy and mass flows in the supply chain.* Focusing on the bottleneck processes, emissions for the coal plant and the coal gasification plant are shown in Table 11.1 Emissions at the fuel cell system are only water vapor.

5. *Aggregate net energy and emission flows for the chain.* The supply chain in Figure 11.8 has an overall efficiency across the entire chain of 26%. The supply chain in Figure 11.9 has an overall efficiency across the entire chain of 25%. Therefore, there might be no gain in overall efficiency from switching to fuel cell power in this scenario.

However, a comparison of the emissions per unit mass of fuel in Table 11.1 shows a potential reduction in emissions with a switch to the supply chain of Figure 11.9. Therefore, you continue to think about how a fuel cell scenario might work for your campus. You realize that the fuel cell system you were interested in installing can also recover heat. The HHV heat recovery efficiency of the fuel cell system is 20%. Across the entire supply chain, the heat recovery efficiency ($\varepsilon_{H,SC}$) is then 10%; that is, 10% of the original energy in the coal mined can be used as heat on your university campus. Therefore, the overall (electrical and heat recovery) efficiency across the entire supply chain is

$$\varepsilon_{O,SC} = \varepsilon_{R,SC} + \varepsilon_{H,SC} = 25\% + 10\% = 35\% \tag{11.25}$$

in this scenario.

To make a fair comparison, you also investigate heat recovery for the supply chain of Figure 11.8. You discover that coal plants are almost always located close to coal mines because of the high cost of transporting a solid fuel. As a result, coal plants are not often located near large population centers where there is a source of demand for electricity or heat. The coal plant that serves your university is no different; it is located 20 miles away from your university and 50 miles away from the nearest city. As a result, it would not be practical to try to recover heat from it. The practical heat recovery efficiency of this supply chain is zero. The overall electrical and thermal efficiency of the supply chain in Figure 11.8 is then

$$\varepsilon_{O,SC} = \varepsilon_{R,SC} + \varepsilon_{H,SC} = 26\% + 0\% = 26\% \tag{11.26}$$

You thus decide to investigate more seriously the prospect of installing a fuel cell system with heat recovery on your university campus.

CHAPTER SUMMARY

The purpose of this chapter was to understand the potential environmental impact of fuel cells by applying quantitative tools to help us calculate changes in emissions, energy use, and efficiency with their adoption. We learned a tool called life cycle assessment (LCA).

- To compare a change in energy technology from one to another, the entire supply chain associated with each technology is considered.
- The supply chain begins with the extraction of raw materials, continues on to the processing of materials, then on to energy production and end use, and finally to waste management.
- Within a chain, attention focuses on the most energy and emission intensive processes, the "process bottlenecks."
- Scenarios are compared by analyzing the relevant energy and material inputs and outputs along the entire supply chain based on the conservation-of-mass equation $m_1 - m_2 = \Delta m$ and the conservation-of-energy equation

$$\dot{Q} - \dot{W} = \dot{m}\left[h_2 - h_1 + g(z_2 - z_1) + \tfrac{1}{2}(V_2^2 - V_1^2)\right] \qquad (11.27)$$

- Aggregate emissions and energy use for one supply chain are compared with aggregate emissions and energy for another.
- The environmental impact of emissions related to global warming is quantified by (1) calculating the $CO_{2\,\text{equivalent}}$ of emitted gases and (2) the external costs of these emissions.
- $CO_{2\,\text{equivalent}}$ is the mass of CO_2 gas that would have an equivalent warming effect on Earth as a mixture of different types of gases and particles. One equation for measuring the $CO_{2\,\text{equivalent}}$ of gases and particles over a 100-year period is

$$CO_{2\,\text{equivalent}} = m_{CO_2} + 23m_{CH_4} + 296m_{N_2O} + \alpha(m_{OM_{2.5}} + m_{BC_{2.5}})$$
$$- \beta[m_{SULF_{2.5}} + m_{NIT_{2.5}} + 0.40m_{SO_x} + 0.10m_{NO_x} + 0.05m_{VOC}]$$
$$(11.28)$$

- An external cost is the cost of a good that is not included in its free-market price.
- The environmental impact of emissions related to air pollution can be quantified by (1) calculating the impacts on human health and (2) the external costs of these emissions.
- By comparing these quantities, the environmental performance of various supply chains can be rated against one another.
- The analysis can be repeated to incorporate greater detail along the various segments of the chain.

CHAPTER EXERCISES

Review Questions

11.1   What are the primary steps of life cycle assessment (LCA)?

11.2   What are some of the gases and particles that have a warming effect on Earth? How? What are some of the gases and particles that have a cooling effect on Earth? How?

11.3   What are some of the most important air pollutants that affect human health?

11.4   What is a national emissions inventory? Describe the type of information it contains.

11.5   When might leaked hydrogen combust with oxygen in air?

11.6   How might hydrogen contribute to global warming and air pollution?

11.7   Develop an abstract for a research proposal to answer a question you feel is important that relates to the environmental impact of fuel cells. You plan to use data from a national emissions inventory and to conduct an LCA.

Calculations

11.8   Estimate the $CO_{2\,equivalent}$ of the following mixture of gases and particles: all organic gases and particulate matter from all sources listed in the 1999 NEI.

11.9   Based on Example 11.2, estimate the mass flow rate of natural gas that must be produced at the gas field to supply enough fuel to the downstream steam reformers. Assume the ratio of fuel cell vehicle efficiency to gasoline vehicle efficiency is 2, 2% of total hydrogen production is leaked in the supply chain, and 1% of methane in natural gas is leaked. How does this quantity of natural gas compare with current annual natural gas production, as a percentage? Calculate the $CO_{2\,equivalent}$ and the external cost of the leaked methane.

11.10 Based on U.S. emissions listed in Table 11.2 and Example 11.4, compare the $CO_{2\,equivalent}$ emissions from the fossil fuel vehicle fleet with a hydrogen vehicle fleet taking into account changes in upstream emissions during the production of hydrogen and fossil fuels. Assume all hydrogen is produced via a high-efficiency steam reformer. Assume that half of the U.S. total VOC emissions are related to the transportation sector and are emitted during gasoline and diesel production. Rely on the 1999 U.S. NEI, available at the EPA's website, for additional data on emissions.

11.11 Imagine replacing current U.S. electrical power with stationary hydrogen fuel cell power plants. Conduct an LCA to evaluate the change in efficiency and emissions across the supply chain.

11.12 Imagine the same scenario as in problem 11.11 except that heat is also recovered from the fuel cell systems. Heat recovered from the fuel cell systems replaces heat that would otherwise be produced by combusting natural gas and oil in furnaces.

Assume that, on average through the seasons, 30% of the HHV of natural gas fuel is recovered by the fuel cell systems as useful heat and consumed in surrounding buildings for space heating or industrial applications. Assume the same emissions profile as shown in Table 11.1 for a steam reformer matches that of a fuel cell system. The original emissions data shown in Table 11.1 are from a United Technologies Corporation PAFC 200-kWe system. Conduct an LCA to evaluate the change in efficiency and emissions across both the electricity supply chain and the heating supply chain.

11.13 Building on Examples 11.4–11.5, for the same LCA comparison recalculate the change in health effects and in external costs due to air pollution taking into account changes in upstream emissions. Also, for the entire supply chain, calculate the change in $CO_{2\,equivalent}$ and in external costs due to global warming. Rely on the 1999 U.S. NEI, available at the EPA's website, for additional data on emissions.

11.14 Conduct an LCA for a scenario in which hydrogen is derived from coal gasification. Assume that the coal gasification plant has the emissions profile shown in Table 11.1.

11.15 Building on Example 11.2, estimate the quantity of hydrogen leaked into the environment by a fuel cell vehicle fleet. Assume the hydrogen leakage rate is similar to that for natural gas (approximately 1% of production). How does this quantity of released hydrogen compare with the amount released by conventional on-road vehicles, shown in Table 11.2?

APPENDIXES

APPENDIX A

CONSTANTS AND CONVERSIONS

Physical Constants

Avogadro's number	N_A	6.02×10^{23} atoms/mol
Universal gas constant	R	0.08205 L $\cdot$ atm/mol $\cdot$ K
		8.314 J/mol $\cdot$ K
		0.08314 bar $\cdot$ m^3/mol $\cdot$ K
		8.314 kPa $\cdot$ m^3/mol $\cdot$ K
Planck's constant	h	6.626×10^{-34} J $\cdot$ s
		4.136×10^{-15} eV $\cdot$ s
Boltzmann's constant	k	1.38×10^{-23} J/K
		8.61×10^{-5} eV/K
Electron mass	m_e	9.11×10^{-31} kg
Electron charge	q	1.60×10^{-19} C
Faraday's constant	F	$96{,}485.34$ C/mol

Conversions

Weight	$2.20 \text{ lb} = 1 \text{ kg}$
Distance	$0.622 \text{ miles} = 1 \text{ km}$
	$3.28 \times 10^{-2} \text{ ft} = 1 \text{ cm}$
Volume	$1000 \text{ L} = 1 \text{ m}^3$
	$0.264 \text{ gal} = 1 \text{ L}$
	$3.53 \times 10^{-2} \text{ ft}^3 = 1 \text{ L}$
Pressure	$1.013250 \times 10^5 \text{ Pa} = 1 \text{ atm}$
	$1.013250 \text{ bars} = 1 \text{ atm}$
	$10^5 \text{ Pa} = 1 \text{ bar}$
	$14.7 \text{ psi} = 1 \text{ atm}$
Energy	$6.241506 \times 10^{18} \text{ eV} = 1 \text{ J}$
	$4.186800 \text{ calorie} = 1 \text{ J}$
	$9.478134 \times 10^{-4} \text{ btu} = 1 \text{ J}$
	$2.777778 \times 10^{-7} \text{ kWh} = 1 \text{ J}$
Power	$1 \text{ J/s} = 1 \text{ W}$
	$1.34 \cdot 10^{-3} \text{ horsepower} = 1 \text{ W}$
	$3.415 \text{ BTU/hour} = 1 \text{ W}$

THERMODYNAMIC DATA

This appendix lists thermodynamic data for H_2, O_2, $H_2O_{(g)}$, $H_2O_{(l)}$, CO, CO_2, CH_4, and N_2 as a function of temperature at $P = 1$ bar.

TABLE B.1. H₂ Thermodynamic Data

T (K)	$\hat{g}(T)$ (kJ/mol)	$\hat{h}(T)$ (kJ/mol)	$\hat{s}(T)$ (J/mol · K)	$C_p(T)$ (J/mol · K)
200	−26.66	−2.77	119.42	27.26
220	−29.07	−2.22	122.05	27.81
240	−31.54	−1.66	124.48	28.21
·260	−34.05	−1.09	126.75	28.49
280	−36.61	−0.52	128.87	28.70
298.15	−38.96	0.00	130.68	28.84
300	−39.20	0.05	130.86	28.85
320	−41.84	0.63	132.72	28.96
340	−44.51	1.21	134.48	29.04
360	−47.22	1.79	136.14	29.10
380	−49.96	2.38	137.72	29.15
400	−52.73	2.96	139.22	29.18
420	−55.53	3.54	140.64	29.21
440	−58.35	4.13	142.00	29.22
460	−61.21	4.71	143.30	29.24
480	−64.08	5.30	144.54	29.25
500	−66.99	5.88	145.74	29.26
520	−69.91	6.47	146.89	29.27
540	−72.86	7.05	147.99	29.28
560	−75.83	7.64	149.06	29.30
580	−78.82	8.22	150.08	29.31
600	−81.84	8.81	151.08	29.32
620	−84.87	9.40	152.04	29.34
640	−87.92	9.98	152.97	29.36
660	−90.99	10.57	153.87	29.39
680	−94.07	11.16	154.75	29.41
700	−97.18	11.75	155.61	29.44
720	−100.30	12.34	156.44	29.47
740	−103.43	12.93	157.24	29.50
760	−106.59	13.52	158.03	29.54
780	−109.75	14.11	158.80	29.58
800	−112.94	14.70	159.55	29.62
820	−116.14	15.29	160.28	29.67
840	−119.35	15.89	161.00	29.72
860	−122.58	16.48	161.70	29.77
880	−125.82	17.08	162.38	29.83
900	−129.07	17.68	163.05	29.88
920	−132.34	18.27	163.71	29.94
940	−135.62	18.87	164.35	30.00
960	−138.91	19.47	164.99	30.07
980	−142.22	20.08	165.61	30.14
1000	−145.54	20.68	166.22	30.20

TABLE B.2. O$_2$ Thermodynamic Data

T (K)	$\hat{g}(T)$ (kJ/mol)	$\hat{h}(T)$ (kJ/mol)	$\hat{s}(T)$ (J/mol · K)	$C_p(T)$ (J/mol · K)
200	−41.54	−2.71	194.16	25.35
220	−45.45	−2.19	196.63	26.41
240	−49.41	−1.66	198.97	27.25
260	−53.41	−1.10	201.18	27.93
280	−57.45	−0.54	203.27	28.48
298.15	−61.12	0.00	205.00	28.91
300	−61.54	0.03	205.25	28.96
320	−65.66	0.62	207.13	29.36
340	−69.82	1.21	208.92	29.71
360	−74.02	1.81	210.63	30.02
380	−78.25	2.41	212.26	30.30
400	−82.51	3.02	213.82	30.56
420	−86.80	3.63	215.32	30.79
440	−91.12	4.25	216.75	31.00
460	−95.47	4.87	218.14	31.20
480	−99.85	5.50	219.47	31.39
500	−104.25	6.13	220.75	31.56
520	−108.68	6.76	221.99	31.73
540	−113.13	7.40	223.20	31.89
560	−117.61	8.04	224.36	32.04
580	−122.10	8.68	225.48	32.19
600	−126.62	9.32	226.58	32.32
620	−131.17	9.97	227.64	32.46
640	−135.73	10.62	228.67	32.59
660	−140.31	11.27	229.68	32.72
680	−144.92	11.93	230.66	32.84
700	−149.54	12.59	231.61	32.96
720	−154.18	13.25	232.54	33.07
740	−158.84	13.91	233.45	33.19
760	−163.52	14.58	234.33	33.30
780	−168.21	15.24	235.20	33.41
800	−172.93	15.91	236.05	33.52
820	−177.66	16.58	236.88	33.62
840	−182.40	17.26	237.69	33.72
860	−187.16	17.93	238.48	33.82
880	−191.94	18.61	239.26	33.92
900	−196.73	19.29	240.02	34.02
920	−201.54	19.97	240.77	34.12
940	−206.36	20.65	241.51	34.21
960	−211.20	21.34	242.23	34.30
980	−216.05	22.03	242.94	34.40
1000	−220.92	22.71	243.63	34.49

TABLE B.3. $H_2O_{(l)}$ Thermodynamic Data

T (K)	$\hat{g}(T)$ (kJ/mol)	$\hat{h}(T)$ (kJ/mol)	$\hat{s}(T)$ (J/mol · K)	$C_p(T)$ (J/mol · K)
273	−305.01	−287.73	63.28	76.10
280	−305.46	−287.20	65.21	75.81
298.15	−306.69	−285.83	69.95	75.37
300	−306.82	−285.69	70.42	75.35
320	−308.27	−284.18	75.28	75.27
340	−309.82	−282.68	79.85	75.41
360	−311.46	−281.17	84.16	75.72
373	−312.58	−280.18	86.85	75.99

TABLE B.4. H$_2$O$_{(g)}$ Thermodynamic Data

T (K)	$\hat{g}(T)$ (kJ/mol)	$\hat{h}(T)$ (kJ/mol)	$\hat{s}(T)$ (J/mol · K)	$C_p(T)$ (J/mol · K)
280	−294.72	−242.44	186.73	33.53
298.15	−298.13	−241.83	188.84	33.59
300	−298.48	−241.77	189.04	33.60
320	−302.28	−241.09	191.21	33.69
340	−306.13	−240.42	193.26	33.81
360	−310.01	−239.74	195.20	33.95
380	−313.94	−239.06	197.04	34.10
400	−317.89	−238.38	198.79	34.26
420	−321.89	−237.69	200.47	34.44
440	−325.91	−237.00	202.07	34.62
460	−329.97	−236.31	203.61	34.81
480	−334.06	−235.61	205.10	35.01
500	−338.17	−234.91	206.53	35.22
520	−342.32	−234.20	207.92	35.43
540	−346.49	−233.49	209.26	35.65
560	−350.69	−232.77	210.56	35.87
580	−354.91	−232.05	211.82	36.09
600	−359.16	−231.33	213.05	36.32
620	−363.43	−230.60	214.25	36.55
640	−367.73	−229.87	215.41	36.78
660	−372.05	−229.13	216.54	37.02
680	−376.39	−228.39	217.65	37.26
700	−380.76	−227.64	218.74	37.50
720	−385.14	−226.89	219.80	37.75
740	−389.55	−226.13	220.83	37.99
760	−393.97	−225.37	221.85	38.24
780	−398.42	−224.60	222.85	38.49
800	−402.89	−223.83	223.83	38.74
820	−407.37	−223.05	224.78	38.99
840	−411.88	−222.27	225.73	39.24
860	−416.40	−221.48	226.65	39.49
880	−420.94	−220.69	227.56	39.74
900	−425.51	−219.89	228.46	40.00
920	−430.08	−219.09	229.34	40.25
940	−434.68	−218.28	230.21	40.51
960	−439.29	−217.47	231.07	40.76
980	−443.92	−216.65	231.91	41.01
1000	−448.57	−215.83	232.74	41.27

TABLE B.5. CO Thermodynamic Data

T (K)	$\hat{g}(T)$ (kJ/mol)	$\hat{h}(T)$ (kJ/mol)	$\hat{s}(T)$ (J/mol · K)	$C_p(T)$ (J/mol · K)
200	−150.60	−113.42	185.87	30.20
220	−154.34	−112.82	188.73	29.78
240	−158.14	−112.23	191.31	29.50
260	−161.99	−111.64	193.66	29.32
280	−165.89	−111.06	195.83	29.20
298.15	−169.46	−110.53	197.66	29.15
300	−169.83	−110.47	197.84	29.15
320	−173.80	−109.89	199.72	29.13
340	−177.81	−109.31	201.49	29.14
360	−181.86	−108.72	203.16	29.17
380	−185.94	−108.14	204.73	29.23
400	−190.05	−107.56	206.24	29.30
420	−194.19	−106.97	207.67	29.39
440	−198.36	−106.38	209.04	29.48
460	−202.55	−105.79	210.35	29.59
480	−206.77	−105.20	211.61	29.70
500	−211.01	−104.60	212.83	29.82
520	−215.28	−104.00	214.00	29.94
540	−219.57	−103.40	215.13	30.07
560	−223.89	−102.80	216.23	30.20
580	−228.22	−102.19	217.29	30.34
600	−232.58	−101.59	218.32	30.47
620	−236.95	−100.98	219.32	30.61
640	−241.35	−100.36	220.29	30.75
660	−245.77	−99.75	221.24	30.89
680	−250.20	−99.13	222.17	31.03
700	−254.65	−98.50	223.07	31.17
720	−259.12	−97.88	223.95	31.31
740	−263.61	−97.25	224.81	31.46
760	−268.12	−96.62	225.65	31.60
780	−272.64	−95.99	226.47	31.74
800	−277.17	−95.35	227.28	31.88
820	−281.73	−94.71	228.07	32.01
840	−286.30	−94.07	228.84	32.15
860	−290.88	−93.43	229.60	32.29
880	−295.48	−92.78	230.34	32.42
900	−300.09	−92.13	231.07	32.55
920	−304.72	−91.48	231.79	32.68
940	−309.37	−90.82	232.49	32.81
960	−314.02	−90.17	233.18	32.94
980	−318.69	−89.51	233.86	33.06
1000	−323.38	−88.84	234.53	33.18

TABLE B.6. CO_2 Thermodynamic Data

T (K)	$\hat{g}(T)$ (kJ/mol)	$\hat{h}(T)$ (kJ/mol)	$\hat{s}(T)$ (J/mol · K)	$C_p(T)$ (J/mol · K)
200	−436.93	−396.90	200.10	31.33
220	−440.95	−396.25	203.16	32.77
240	−445.04	−395.59	206.07	34.04
260	−449.19	−394.89	208.84	35.19
280	−453.39	−394.18	211.48	36.24
300	−457.65	−393.44	214.02	37.22
320	−461.95	−392.69	216.45	38.13
340	−466.31	−391.92	218.79	39.00
360	−470.71	−391.13	221.04	39.81
380	−475.15	−390.33	223.21	40.59
400	−479.63	−389.51	225.31	41.34
420	−484.16	−388.67	227.35	42.05
440	−488.73	−387.83	229.32	42.73
460	−493.33	−386.96	231.23	43.38
480	−497.98	−386.09	233.09	44.01
500	−502.66	−385.20	234.90	44.61
520	−507.37	−384.31	236.66	45.20
540	−512.12	−383.40	238.38	45.76
560	−516.91	−382.48	240.05	46.30
580	−521.72	−381.54	241.69	46.82
600	−526.59	−380.60	243.28	47.32
620	−531.46	−379.65	244.84	47.80
640	−536.37	−378.69	246.37	48.27
660	−541.31	−377.72	247.86	48.72
680	−546.28	−376.74	249.32	49.15
700	−551.29	−375.76	250.75	49.57
720	−556.31	−374.76	252.15	49.97
740	−561.37	−373.76	253.53	50.36
760	−566.45	−372.75	254.88	50.73
780	−571.56	−371.73	256.20	51.09
800	−576.71	−370.70	257.50	51.44
820	−581.86	−369.67	258.77	51.78
840	−587.05	−368.63	260.02	52.10
860	−592.26	−367.59	261.25	52.41
880	−597.50	−366.54	262.46	52.71
900	−602.76	−365.48	263.65	53.00
920	−608.05	−364.42	264.82	53.28
940	−613.35	−363.35	265.97	53.55
960	−618.68	−362.27	267.10	53.81
980	−624.04	−361.19	268.21	54.06
1000	−629.41	−360.11	269.30	54.30

TABLE B.7. CH₄ Thermodynamic Data

T (K)	$\hat{g}(T)$ (kJ/mol)	$\hat{h}(T)$ (kJ/mol)	$\hat{s}(T)$ (J/mol · K)	$C_p(T)$ (J/mol · K)
200	−112.69	−78.25	172.23	36.30
220	−116.17	−77.53	175.63	35.19
240	−119.71	−76.83	178.67	34.74
260	−123.32	−76.14	181.45	34.77
280	−126.97	−75.44	184.03	35.12
298.15	−130.33	−74.80	186.25	35.65
300	−130.68	−74.73	186.48	35.71
320	−134.43	−74.01	188.80	36.47
340	−138.23	−73.27	191.04	37.36
360	−142.07	−72.52	193.20	38.35
380	−145.95	−71.74	195.31	39.40
400	−149.88	−70.94	197.35	40.50
420	−153.85	−70.12	199.36	41.64
440	−157.86	−69.27	201.32	42.80
460	−161.90	−68.41	203.25	43.98
480	−165.99	−67.51	205.15	45.16
500	−170.11	−66.60	207.01	46.35
520	−174.27	−65.66	208.86	47.54
540	−178.46	−64.70	210.67	48.73
560	−182.69	−63.71	212.47	49.90
580	−186.96	−62.70	214.24	51.07
600	−191.26	−61.67	215.99	52.23
620	−195.60	−60.61	217.72	53.37
640	−199.97	−59.53	219.43	54.50
660	−204.38	−58.43	221.13	55.61
680	−208.82	−57.31	222.80	56.71
700	−213.29	−56.16	224.46	57.79
720	−217.79	−55.00	226.10	58.85
740	−222.33	−53.81	227.73	59.90
760	−226.90	−52.60	229.34	60.93
780	−231.51	−51.37	230.94	61.94
800	−236.14	−50.13	232.52	62.93
820	−240.81	−48.86	234.08	63.90
840	−245.50	−47.57	235.64	64.85
860	−250.23	−46.26	237.17	65.79
880	−254.99	−44.94	238.70	66.70
900	−259.78	−43.60	240.20	67.60
920	−264.60	−42.23	241.70	68.47
940	−269.45	−40.86	243.18	69.33
960	−274.33	−39.46	244.65	70.17
980	−279.23	−38.05	246.11	70.99
1000	−284.17	−36.62	247.55	71.79

TABLE B.8. N$_2$ Thermodynamic Data

T (K)	$\hat{g}(T)$ (kJ/mol)	$\hat{h}(T)$ (kJ/mol)	$\hat{s}(T)$ (J/mol · K)	$C_p(T)$ (J/mol · K)
200	−38.85	−2.83	180.08	28.77
220	−42.48	−2.26	182.82	28.72
240	−46.16	−1.68	185.31	28.72
260	−49.89	−1.11	187.61	28.76
280	−53.66	−0.53	189.75	28.81
298.15	−57.11	0.00	191.56	28.87
300	−57.48	0.04	191.74	28.88
320	−61.33	0.62	193.60	28.96
340	−65.22	1.20	195.36	29.05
360	−69.15	1.78	197.02	29.14
380	−73.10	2.37	198.60	29.25
400	−77.09	2.95	200.11	29.35
420	−81.11	3.54	201.54	29.46
440	−85.15	4.13	202.91	29.57
460	−89.22	4.72	204.23	29.68
480	−93.32	5.32	205.50	29.79
500	−97.44	5.92	206.71	29.91
520	−101.59	6.51	207.89	30.02
540	−105.76	7.12	209.02	30.13
560	−109.95	7.72	210.12	30.24
580	−114.16	8.33	211.19	30.36
600	−118.40	8.93	212.22	30.47
620	−122.65	9.54	213.22	30.58
640	−126.92	10.16	214.19	30.69
660	−131.22	10.77	215.14	30.80
680	−135.53	11.39	216.06	30.91
700	−139.86	12.01	216.96	31.02
720	−144.21	12.63	217.83	31.13
740	−148.57	13.25	218.69	31.24
760	−152.96	13.88	219.52	31.34
780	−157.35	14.51	220.34	31.45
800	−161.77	15.14	221.13	31.55
820	−166.20	15.77	221.91	31.66
840	−170.64	16.40	222.68	31.76
860	−175.11	17.04	223.43	31.86
880	−179.58	17.68	224.16	31.96
900	−184.07	18.32	224.88	32.06
920	−188.58	18.96	225.58	32.16
940	−193.10	19.61	226.28	32.25
960	−197.63	20.25	226.96	32.35
980	−202.17	20.90	227.63	32.44
1000	−206.73	21.55	228.28	32.54

APPENDIX C

STANDARD ELECTRODE POTENTIALS AT 25°C

Electrochemical Half Reaction			E^0
$Li^+ e^-$	$\rightarrow$	Li	-3.04
$2H_2O + 2e^-$	$\rightarrow$	$H_2 + 2OH^-$	-0.83
$Fe^{2+} + 2e^-$	$\rightarrow$	Fe	-0.440
$CO_2 + 2H^+ + 2e^-$	$\rightarrow$	$CHOOH_{(aq)}$	-0.196
$2H^+ + 2e^-$	$\rightarrow$	H_2	$+0.00$
$CO_2 + 6H^+ + 6e^-$	$\rightarrow$	$CH_3OH + H_2O$	$+0.03$
$\frac{1}{2}O_2 + H_2O + 2e^-$	$\rightarrow$	$2OH^-$	$+0.40$
$O_2 + 4H^+ + 4e^-$	$\rightarrow$	$2H_2O$	$+1.23$
$H_2O_2 + 2H^+ + 2e^-$	$\rightarrow$	$2H_2O$	$+1.78$
$O_3 + 2H^+ + 2e^-$	$\rightarrow$	$O_2 + H_2O$	$+2.07$
$F_2 + 2e^-$	$\rightarrow$	$2F^-$	$+2.87$

APPENDIX D

QUANTUM MECHANICS

A number of key discoveries in the early part of the twentieth century led to the foundation of modern quantum mechanics. We will highlight some of these discoveries and describe a few of the underlying assumptions in mostly qualitative terms. Readers are encouraged to broaden their knowledge in this area by studying relevant quantum mechanics and chemistry texts [75–76].

Before the emergence of modern quantum mechanics, Bohr [77], an early pioneer in atom physics, proposed in 1913 a model for the hydrogen atom in which the electron encircles the nucleus in only one of a number of allowed orbits. He assumed that the energy of the electron is quantized and that the change in energy of the electron, associated with transitioning from one orbit to the other, is accompanied by the absorption or emission of discrete light quanta. Balancing the attracting forces between the nucleus and the electron with the centrifugal forces of the electron, the Bohr model was able to predict the radius of the hydrogen atom quite accurately as 0.529×10^{-10} m. Nevertheless, Bohr's model is fundamentally based on Newtonian mechanics for which the quantization of energy levels does not occur naturally.

About a decade later, de Broglie [78] was the first to propose that electrons have both a particle and a wave nature. The electron diffraction experiments in atomic crystal structures of Davisson and Germer [79] in 1928 confirmed de Broglie's view that electrons may be indeed assigned a wavelength.

Schrödinger was able create the formalism of modern quantum mechanics by combining the wave nature of electrons following de Broglie and their quantized energy states in hydrogen according to Bohr. In 1926 Schrödinger [80] wrote in the journal *Annalen der Physik*:[1]

[1] Translation from German appears in Ref. 81.

The usual rule of quantization can be replaced by another postulate, in which there occurs no mention of whole numbers. Instead, the introduction of integers arises in the same natural way as, for example, in a vibrating string, for which the number of nodes is integral. The new conception can be generalized, and I believe that it penetrates deeply into the true nature of quantum rules.

In vibrating strings with fixed ends, the location of nodes does not change over time. More importantly, the number of nodes in vibrating strings with stationary ends can only be changed in discrete steps, that is, integer numbers $(1, 2, 3, \ldots, n)$. In other words, one cannot add a portion of a wave to a vibrating string with given length and fixed ends; only whole waves can be added. In analogy to string waves, quantum mechanics assumes that matter can be described with wave functions of amplitude $\Psi(t, x, y, z)$. These are "material" waves rather than electromagnetic waves. For the present considerations, we are interested in stationary waves only. In stationary waves, the nodes do not change as a function time; stationary waves depend on spatial coordinates only. We define the amplitudes of stationary waves as $\psi(x, y, z)$. We are focusing on electrons with stationary boundaries such as an electron in a box, or an electron wrapped around a positively charged nucleus, or electrons in an array of positively charged atoms, as found in any crystal structure. The wave function ψ cannot be directly observed or measured. But one can measure $|\psi(x, y, z)|^2$, which corresponds to the probability of finding the particle in (x, y, z), that is, the density of the material in that location.

It is important to realize that quantum mechanics is based on a number of postulates, such as: There exists a wave function which contains all possible information about the system considered. (A more detailed description of the postulates is given later in this appendix.) Postulates or axioms are underlying assumptions that cannot be further explained and cannot be further questioned. Their justification stems from the practicality of their results. The wave function cannot be measured; however, the absolute square can be. If experimental results are consistent with the assumptions of the theory, the theory is considered valid, at least until proven wrong.

Let's ask how one can calculate $\Psi(t, x, y, z)$ for a given atomic structure. One of the postulates of quantum mechanics is that $\Psi(t, x, y, z)$ can be obtained by solving the Schrödinger equation. The Schrödinger equation describes the evolution of a particle (wave function) over time. In classical mechanics, the time evolution of any particle system is described by its kinetic and potential energy. Similarly, the Schrödinger equation involves the kinetic and potential energy of the particles involved. In fact, it is a further postulate in quantum mechanics that the kinetic and potential energy in the Schrödinger equation are similar to that of the particles in classical mechanics.

For our purposes, we are focusing only on stationary waves; hence we are interested in the solution of the so-called time-independent part of the Schrödinger equation. All terms dependent on time are constant, like the nodes in stationary waves. If we take the absolute square of the stationary, or time-independent, solutions of the Schrödinger equation, we obtain a picture of the location and shape of the particles (in our case the electrons) and how they rearrange during different stages of a chemical reaction.

Following decades of research in quantum mechanics and the availability of modern numerical methods, a broad community of scientists and engineers is now able to study and

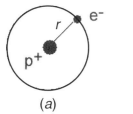

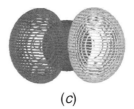

(a)   (b)   (c)

Figure D.1. (*a*) Electron circling proton according to Bohr. (*b*) Stationary electron density (1*s*) around proton, and (*c*) (2*p*) electrons in oxygen. Note that (*b*) and (*c*) are not drawn to the same scale

visualize the particle densities, quantify chemical bond formation, charge transfer reactions and diffusion phenomena. For example, the quantum simulation figures in Chapter 3 used a commercially available tool called Gaussian,[2] which is capable of determining the electron density and the minimum energy of the quantum system considered. Gaussian is based on density functional theory (DFT). Kohn [82], a pioneer of the DFT method, helped initiate a revolution that made quantum mechanical tools available for routine research in chemistry, electrochemistry, and physics.

D.1 ATOMIC ORBITALS

Using Gaussian we can illustrate the shape of an electron by considering the simplest atom there is: the hydrogen atom. Figure D.1*a* shows the hydrogen atom from Bohr's perspective, a proton being encircled by an electron; Figure D.1*b* describes the same atom by plotting $|\Psi^2|$, the proton surrounded by a stationary electron cloud, spherically symmetric but with varying electron density along the radius r. It just so happens that the radius of the electron orbit in the Bohr model turns out to be the same as the location of maximum electron density calculated by the time-independent Schrödinger equation. The space in which the electron may reside is called the *orbital*. The more electrons there are in an atom, the more orbitals exist. Orbital geometry is not easy to visualize. We can comfortably imagine stationary waves of a string since deflections occur in one dimension. We can also imagine that in a string with fixed ends the number of waves can be increased in incremental steps of whole numbers only (compare above remarks by Schrödinger). Yet, we have a hard time imagining 3D waves, especially 3D waves of higher order, interacting with electrically charged nuclei.

Computer tools such as Gaussian help in visualizing the complexity of 3D orbitals. Analogies to mechanical scenarios such as the buckling of a column also help our intuition. In fact, the 1D Schrödinger equation of a 1D particle in a box is identical to the differential equation leading to the calculation of the Euler buckling load. Engineers know there is a first-, second-, and higher order buckling load. Due to the 3D nature of orbitals, not only one quantum number n (as in buckling) exists to describe the possible states of

---

[2]Gaussian is a computational tool predicting energies, molecular structures, and vibrational frequencies of molecular systems by Gaussian Inc.

an electron in an atom. Instead, there are several quantum numbers describing the possible solutions of the Schrödinger equation and their respective energy levels. The quantum numbers commonly used in the solution for the Schrödinger equation are called n (principal quantum number), l (angular momentum number), and m (quantum number of z component of angular momentum). The following relations hold between the integers l, m, and n: $0 \leq l \leq n-1$, $-l \leq m \leq l$, and for a given n there are $\sum (2l+1) = n^2$ different states which happen to have the same energy. Two electrons (one with spin up, the other with spin down) may occupy the same set of quantum numbers (n, m, l).

It is helpful to make the link to the popular notation for electrons in most periodic systems: s, p, d, f. Historically, this notation came from the optical spectroscopy literature and means s (sharp), p (principal), d (diffuse), and f (fundamental). The orbital s stands for $l = 0$, p corresponds to $l = 1$, d to $l = 2$, and f to $l = 3$. Optical spectroscopy led to the first observations that electrons reside in discrete orbital states around the nucleus and was crucial for the earlier mentioned atomic hydrogen model of Bohr. Figure D.1b shows the 1s electrons in hydrogen, and Figure D.1c illustrates 2p electrons in oxygen.

Overlapping of orbitals between different atoms causes the formation of chemical bonds. Examples are hydrogen (H–H) or oxygen (O–O) or the formation of bonds between catalyst and H_2 or O_2. Needless to say that "molecular orbitals" may be quite complex. Only numerical tools provide quantitative insight into the strength of chemical bonds.

D.2   POSTULATES OF QUANTUM MECHANICS

The postulates, also referred to as axioms, of quantum mechanics were articulated by generations of physicists after Schrödinger's initial paper. Postulates or axioms are assumptions that cannot be further explained. They should be accepted as stated since they were shown to be useful and practical, but they sure sound abstract and not necessarily intuitive. They allow deriving results, which can be experimentally verified. In that sense they can be indirectly checked for their truth and practicality.

1. The first axiom in quantum mechanics says that there exists a wave function Ψ depending on time and space which contains all possible information about the system considered. In this book we consider the wave functions for electrons only.

2. The wave function Ψ has certain mathematical properties: It is differentiable finite, unique, and continuous. It is also important to realize that Ψ is complex, and it can be separated into a product of functions depending on time and space:

$$\Psi(t, x, y, z) = f(t)\psi(x, y, z) \tag{D.1}$$

3. The wave function Ψ cannot be measured. Only the function $|\Psi|^2$ can be observed, and it represents the probability of the particle to be in the location (x, y, z) at time t. For electrons the expression $|\Psi|^2$ is a measure of the electron density that can be observed in a variety of ways. Given the fact that the electron exists somewhere, it is reasonable to assume that the probability to find it in space is equal to 1. In equation

form

$$\int |\Psi|^2 \, dV = 1 \tag{D.2}$$

This property of the wave function is being referred to as normalizable.

4. An operator exists, the so-called Hamiltonian H, which, when applied to the wave function, describes the change of the wave function over time:

$$H\Psi = -i\hbar \frac{\partial}{\partial t} \Psi \tag{D.3}$$

This equation is called the Schrödinger equation where $\hbar = h/2\pi$ (h = Planck's constant). For the steady-state case, or the time-independent case, the Schrödinger equation can be reduced to

$$H\psi_n = \varepsilon_n \psi_n \tag{D.4}$$

where ε_n represents the energy of the system in state n. The ψ_n are the eigenfunctions of the operator H and ε_n the corresponding eigenvalue.

5. The Hamiltonian H is equivalent to the energy of classical mechanics, that is, $H = T + V$, kinetic energy plus potential energy. More specifically, the kinetic energy is

$$T = \frac{1}{2}mv^2 = \frac{p^2}{2m} \tag{D.5}$$

and m is the mass of the electron. The linear momentum p is, in contrast to classical mechanics, now an operator. In one dimension,

$$p = -i\hbar \frac{\partial}{\partial x} \tag{D.6}$$

and for three dimensions,

$$p_x = -i\hbar \frac{\partial}{\partial x} \qquad p_y = -i\hbar \frac{\partial}{\partial y} \qquad p_z = -i\hbar \frac{\partial}{\partial z} \tag{D.7}$$

For convenience the gradient vector is frequently defined as

$$\nabla = \left(\frac{\partial}{\partial x}, \frac{\partial}{\partial y}, \frac{\partial}{\partial z} \right) \tag{D.8}$$

The potential energy is a function of the three dimensions $V = V(x, y, z)$.

One should not attempt to understand these axioms but rather should become familiar with them or, better yet, memorize them. We need to mention that the axiom list as stated above is not quite complete but captures the essence of what we need for the present section.

D.3   ONE-DIMENSIONAL ELECTRON GAS

We will illustrate the quantum mechanical axioms by describing the behavior of the simplest system: one "free" electron in a 1D box of length L. *Free* means that there is no potential acting on the electron.

Consequently, the Schrödinger equation for the free electron reads

$$H\Psi = -\frac{\hbar^2}{2m}\frac{d^2\Psi}{dx^2} = \varepsilon_n \psi \tag{D.9}$$

The "box" of length L means that the wave function of the electrons is constrained at either end of the box. In other words,

$$\psi_n(0) = 0 \qquad \psi_n(L) = 0 \tag{D.10}$$

A solution to this equation is obviously harmonic in nature. We guess the solution

$$\psi_n = A \sin\left(\frac{n\pi}{L}x\right) \tag{D.11}$$

To check the guess, we take derivatives with respect to x of Equation D.11, yielding

$$\frac{d\psi_n}{dx} = A\left(\frac{n\pi}{L}\right)\cos\left(\frac{n\pi}{L}x\right) \tag{D.12}$$

$$\frac{d^2\psi_n}{dx^2} = -A\left(\frac{n\pi}{L}\right)^2 \sin\left(\frac{n\pi}{L}x\right) \tag{D.13}$$

The resulting levels for the energy are

$$\varepsilon_n = \frac{\hbar^2}{2m}\left(\frac{n\pi}{L}\right)^2 \tag{D.14}$$

The wave functions ψ_n are referred to as orbitals. The electron can be in any of the n orbitals. The important insight we obtain from this solution is that there are discrete, time-independent "stationary" states in which the electron may reside. The energy levels change incrementally; they are proportional to n^2. Transitions of the electron from one orbital to the other are accompanied by the emission or absorption of light quanta. Clearly, multiple electrons may be in the same box and may reside in available orbitals. Following the Pauli principle, which we accept without further explanation, a maximum of only two electrons may have the same orbital number n. However, the two electrons with the same n will differ in their spin, one is to be spin "up," the other one "down." In addition, the presence of multiple electrons in the same system (box) will modify the Hamiltonian in the Schrödinger equation since the presence of one electron will influence the others in the form of a nonzero potential energy term. The details of this problem go well beyond the introductory nature of this appendix and we refer to other texts [83].

D.4   ANALOGY TO COLUMN BUCKLING

Since this book is largely targeted for the engineering audience, we would like to draw attention to an analogy between the Schrödinger equation of the 1D electron gas and the mechanics of a buckling column. Consider a simple column of length L with pinned ends (Figure D.2) subject to an applied force P. The differential equation describing the bending moment in a column is formally identical to that of the Schrödinger equation of the electron in the box:

$$EI\frac{\partial^2 y}{\partial x^2} = Py \tag{D.15}$$

where E stands for Young's modulus, I is the cross-sectional moment of inertia, and y is the lateral deflection of the beam from the neutral position. The boundary conditions for the column and the solution for y are the same as the ones for the wave function; so are the solutions $y_n(x)$:

$$y_n(0) = 0 \qquad y_n(L) = 0 \qquad y_n = A \sin\left(\frac{n\pi}{L}x\right) \tag{D.16}$$

Interestingly, the discrete levels of energy ε_n resulting from the Schrödinger equation can now be interpreted as the discrete loads for column buckling, also called Euler buckling load:

$$P_n = EI\frac{n\pi}{L}^2 \tag{D.17}$$

We know that Euler buckling only happens above a critical threshold load, in analogy to the discrete levels of energy required to move an electron from one shell to another. The mathematical expressions in both cases are the same.

Figure D.2. Pinned column of length L subjected to force P buckles according to discrete modes.

D.5 HYDROGEN ATOM

The hydrogen atom is the only physical quantum mechanical system for which an analytical solution can be found. It consists of the nucleus, that is, one proton, and one electron surrounding the nucleus. The earlier discussed 1D free-electron gas is hypothetical in nature, but it gives insight into the methodology used below for the hydrogen atom. The solution of the Schrödinger equation for hydrogen is of significant historical importance since it shaped the thinking of generations of physicists. It provides qualitative insights into the behavior of more complex, multielectron systems for which analytical solutions are not available.

The Schrödinger equation of hydrogen can be established as follows. We are interested in the position of the electron relative to the proton only. Hence, the motion of the entire atom is unimportant. Establishing the Hamiltonian is the crucial step. The rest is mathematics and algebra. The kinetic energy in quantum mechanics is the square of the momentum divided by the mass (axiom 5). From axiom 5 we also know that the momentum is a differential operator acting on the wave function.

For the electron in the box there was no potential energy. For the interaction between the two electrically charged particles, the proton and electron, we know from classical electrostatics that there exists an attractive force of interaction which is inversely proportional to the distance square. Accordingly, the potential energy is inversely proportional to distance between the particles:

$$V(r) = -\frac{e^2}{4\pi\varepsilon_0 r} \tag{D.18}$$

with $\varepsilon_0 = 8.854 \times 10^{-12}$ C/V $\cdot$ m. Since $e^2/4\pi\varepsilon_0$ has the dimension of action times velocity (the units of action are energy times time), we can rewrite this term involving the Planck constant, which has the dimension of action and the speed of light c. In other words $e^2/4\pi\varepsilon_0 = \alpha\hbar c$ with $\alpha \approx 1/137$. We can now write the Schrödinger equation for hydrogen:

$$\left(-\frac{\hbar}{2m}\nabla^2 - \hbar c\frac{\alpha}{r}\right)\psi = E\psi \tag{D.19}$$

A hydrogen atom is completely spherical; there is no preferred orientation. Therefore, it is convenient to express all functions in spherical coordinates:

$$-\frac{\hbar^2}{2m}\left[\frac{\partial}{r^2\partial r}\left(r^2\frac{\partial\psi}{\partial r}\right) + \frac{1}{r^2\sin^2\theta}\frac{\partial^2\psi}{\partial\theta^2} + \frac{1}{r^2\sin\theta}\frac{\partial}{\partial\theta}\left(\sin\theta\frac{\partial\psi}{\partial\theta}\right) - \hbar c\frac{\alpha}{r}\right]\psi = E\psi \tag{D.20}$$

Partial differential equations like this one are frequently solved by a separation "Ansatz":

$$\psi(r,\theta,\varphi) = R(r)\Theta(\theta)\Phi(\varphi) \tag{D.21}$$

This separation leads to three differential equations. The discrete energy levels E_n [eigenvalues of $R(r)$] can be found as

$$E_n = -\tfrac{1}{2}Mc^2 \frac{\alpha^2}{n^2} \tag{D.22}$$

Without proof we give the solutions of these three differential equations:

$$R_{nl}(r) = -\left[\left(\frac{2}{na}\right)^3 \frac{(n-l-1)!}{2n[(n+l)!]^3}\right]^{1/2} e^{-\rho/2}\rho^l L_{n+l}^{2l+1}(\rho) \tag{D.23}$$

$$\Theta_{lm}(\theta) = \left[\frac{(2l+1)(l-|m|)!}{2(l+|m|)!}\right]^{1/2} P_l^{|m|}(\cos\theta) \tag{D.24}$$

$$\Phi_m(\varphi) = \frac{1}{\sqrt{2\pi}}e^{im\varphi} \tag{D.25}$$

By solving the three differential equations, one finds that, similar to the case of column buckling, there are discrete solutions or modes which we assign the indices (l, m, n). Accordingly, the stationary solution of the Schrödinger equation is of the form

$$\phi_{nlm}(r, \theta, \varphi) = R_{nl}(r)\Theta_{lm}(\theta)\Phi_m(\varphi) \tag{D.26}$$

The following polynomial expressions were used: L and P. The so-called Laguerre polynomial used by Schrödinger can be expressed as

$$L_{n+l}^{2l+1}(\rho) = \sum_{k=0}^{n-l-1}(-1)^{k+1}\frac{[(n+l)!]^2}{(n-l-1-k)!(2l+1+k)!k!}\rho^k \tag{D.27}$$

and the associated Legendre function P is recursively defined as

$$P_l^{|m|}(\cos\theta) = (1-\cos^2\theta)^{|m|/2}\frac{d^{|m|}}{dz^{|m|}}P_l(\cos\theta) \tag{D.28}$$

The Legendre polynomial P_l is given by

$$P_l(x) = \frac{1}{2^l l!}\frac{d^l}{dx^l}(x^2-1)^l \tag{D.29}$$

Furthermore we used the notation $\rho = [2/(na)]r$ and, more importantly,

$$a = \frac{\hbar^2}{\alpha m e^2} = 0.5292 \times 10^{-10}\ \text{m} \tag{D.30}$$

which is the radius of the innermost orbital of the electron in hydrogen, which coincides with Bohr's calculation of the size of the hydrogen atom.

We provided the solutions of the Schrödinger equation for hydrogen to give the students a perspective of the mathematical complexity of a relatively simple quantum system. From that sheer complexity, it appears obvious that more comprehensive systems than hydrogen can be solved with numerical means only by using computer tools such as Gaussian. Unfortunately, today's available computational tools take a long time to solve quantum mechanical problems. Algorithms tend to possess complexity of the order of at least n^3, where n is the number of electrons in the system. Therefore, doubling the size of a system may require eight times the computational resources. The study of the path of catalytic reactions requires the involvement of tens, if not hundreds, of atoms. Computer clusters available at the beginning of the twenty-first century may take days or weeks to solve the wave equations of hundreds of atoms.

Further improvement in the algorithmic efficiency and significant advances in computational speed are likely to lead to the discovery of radically new material combinations. A range of the challenges posed by catalysis might become resolved, including the discovery of platinum alternatives.

APPENDIX E

GOVERNING EQUATIONS OF CFD FUEL CELL MODEL

Because of their extensive computational flexibility, CFD methods can use an extensive set of governing equations. This provides greater realism to a fuel cell model. The governing equations of the CFD model we presented in Section 6.3 are summarized in Table E.1. Note that these equations consider all three dimensions. In this model, we do not make any of the assumptions that we made for the 1D model except one: We retain the single-phase flow assumption (no liquid water). Also, note the model presented here is just one of several popular models that fuel cell researchers employ [84–89]. Complete CFD models including two-phase flow and nonisothermal behavior are still an area of furious development.

Like the 1D model presented in Section 6.2, a CFD fuel cell model is based on the *conservation laws*. In our 1D model, we used simplified mass conservation and charge conservation laws. A CFD model can include a more complete set of conservation laws.

TABLE E.1. Governing Equations of CFD Fuel Cell Models.[a]

Category	Equations
1. Mass conservation	$\dfrac{\partial}{\partial t}(\epsilon \rho) + \nabla \cdot (\epsilon \rho \mathbf{U}) = 0$
2. Momentum conservation	$\dfrac{\partial}{\partial t}(\epsilon \rho \mathbf{U}) + \nabla \cdot (\epsilon \rho \mathbf{U}\mathbf{U}) = -\epsilon \nabla p + \nabla \cdot (\epsilon \zeta) + \dfrac{\epsilon^2 \mu \mathbf{U}}{\kappa}$
3. Species conservation	$\dfrac{\partial}{\partial t}(\epsilon \rho X_i) + \nabla \cdot (\epsilon \rho \mathbf{U} X_i) = \nabla \cdot \mathbf{J}_i + S_i$
4. Charge conservation	$\nabla \cdot \mathbf{i}_{\text{elec}} = -\nabla \cdot \mathbf{i}_{\text{ion}}$

[a]Symbols in boldface represent vectors.

The governing equations are similar to the usual conservation equations (e.g., Navier–Stokes equation) and include the following:

1. *Mass Conservation.* Mass conservation equations (or continuity equations) simply require that the change of mass in a unit volume must be equal to the sum of all the species entering (exiting) the volume in a given time period. Equation E.1 formulates the concept mathematically:

$$
\underbrace{\frac{\partial}{\partial t}(\varepsilon\rho)}_{\substack{\text{rate of mass change}\\\text{per unit volume}}} + \underbrace{\nabla\cdot(\varepsilon\rho\mathbf{U})}_{\substack{\text{net rate of mass change}\\\text{per unit volume by convection}}} = 0
\tag{E.1}
$$

Here, ρ and $\mathbf{U}$ stand for density and the velocity vector of the fluid respectively. Note that the porosity ε is implemented in this equation to account for porous domains such as the electrode and catalyst. By setting the correct value for porosity in each domain, the equation is globally valid for the entire physical fuel cell structure. For example, we can choose $\varepsilon = 0.4$ for the electrode, $\varepsilon = 1$ for the flow channels, and $\varepsilon = 0$ for the electrolyte. Porosity is similarly incorporated in all the other governing equations as well.

2. *Momentum Conservation.* Similar to mass conservation, we can set up an equation for momentum conservation as

$$
\underbrace{\frac{\partial}{\partial t}(\varepsilon\rho\mathbf{U})}_{\substack{\text{rate of}\\\text{momentum}\\\text{change}\\\text{per unit}\\\text{volume}}} + \underbrace{\nabla\cdot(\varepsilon\rho\mathbf{U}\mathbf{U})}_{\text{convection}} = \underbrace{-\varepsilon\nabla p}_{\text{pressure}} + \underbrace{\nabla\cdot(\varepsilon\zeta)}_{\substack{\text{viscous}\\\text{friction}}} + \underbrace{\frac{\varepsilon^2\mu\mathbf{U}}{\kappa}}_{\substack{\text{pore}\\\text{structure}}}
\tag{E.2}
$$

with labels: net rate of momentum change per unit volume by.

Here ζ and μ stand for the shear stress tensor and the viscosity of the fluid, respectively. Note that the last term on the right-hand side (RHS) is known as *Darcy's law*, which quantifies the viscous drag of fluids in porous media, such as pore wall–fluid interactions. The second to the last term on the RHS accounts for only fluid–fluid interaction. Permeability $\kappa\,[m^{-2}]$ quantifies the strength of this interaction depending on the pore structure configuration. A low permeability indicates greater interaction. Obviously, we may use an extremely large value of κ in the flow channels to ignore it.

3. *Species Conservation.*

$$
\underbrace{\frac{\partial}{\partial t}(\varepsilon\rho X_i)}_{\substack{\text{rate of a species}\\\text{mass change per}\\\text{unit volume}}} + \underbrace{\nabla\cdot(\varepsilon\rho\mathbf{U}X_i)}_{\text{convection}} = \underbrace{\nabla\cdot\mathbf{J}_i}_{\text{diffusion}} + \underbrace{S_i}_{\substack{\text{electrochemical}\\\text{reaction}}}
\tag{E.3}
$$

with label: net rate of species mass change per unit volume.

Here, X_i stands for the species mass fraction and J_i stands for a species diffusion mass flux that can be represented by any diffusion equation, such as Fick's law, the Maxwell–Stefan equation, and so on. S_i stands for a species source or sink. In fuel cells, electrochemical reactions act as species sources and sinks (e.g., hydrogen and oxygen consumption or water generation):

$$S_i = M_i \frac{j}{n_i F} \tag{E.4}$$

where n_i is the number of valence electrons associated with species i and M_i is the molecular weight of species i. Molecular weight is used to convert molar flux rate to mass flux rate.

4. *Charge Conservation.* From the continuity of current in a conducting material,

$$\nabla \cdot \mathbf{i} = 0 \tag{E.5}$$

where $\mathbf{i}$ stands for the current flux vector. Two types of charges are present in fuel cell systems—electrons and ions. Since both types of charge are generated from originally neutral species (hydrogen and/or oxygen), overall charge neutrality must be conserved:

$$\nabla \cdot \mathbf{i}_{\text{elec}} + \nabla \cdot \mathbf{i}_{\text{ion}} = 0 \tag{E.6}$$

where $\mathbf{i}_{\text{ion}}$ stands for the ionic current through an ion-conducting phase such as a catalyst layer or membrane and $\mathbf{i}_{\text{elec}}$ stands for an electronic current in an electron-conducting phase such as a catalyst layer or electrode. We rearrange Equation E.6, and relate it to local current density as

$$-\nabla \cdot \mathbf{i}_{\text{ion}} = \nabla \cdot \mathbf{i}_{\text{elec}} = j \tag{E.7}$$

By incorporating Ohm's law into Equation E.7, we get

$$\nabla \cdot (\sigma_{\text{ion}} \nabla \Phi_{\text{ion}}) = -\nabla \cdot (\sigma_{\text{elec}} \nabla \Phi_{\text{elec}}) = j \tag{E.8}$$

where Φ_{ion} and Φ_{elec} are the electric potentials in the ion conductor and electronic conductor, respectively, and σ is the conductivity. Note that this equation can be made valid for all the domains in a fuel cell by simply setting a proper value for σ in each domain. For example, we may use $\sigma_{\text{elec}} = \sigma_{\text{ion}} = 0$ in the flow channel and $\sigma_{\text{elec}} = 0$ in the membrane (no electronic conduction). The catalyst layer has both ionic and electronic conduction and so both conductivities may be considered. At the anode or cathode, this equation is coupled with a simplified Butler–Volmer equation to obtain j as

$$j = j_0 \exp\left[\frac{n_i \alpha F}{RT}(\Phi_{\text{ion}} - \Phi_{\text{elec}})\right] \frac{c_i}{c_i^0} \tag{E.9}$$

In this equation, c_i and c_i^0 stand for the local concentration and reference concentration of species i. A slight difference exists between this equation and the original Butler–Volmer Equation 3.33. The overvoltage η is replaced by $\Phi_{\text{ion}} - \Phi_{\text{elec}}$ to include the effect of the electronic potential drops and the ionic potential drops. Using this approach, now we can calculate the electronic ohmic voltage drop in the electrode. This loss was previously ignored in our 1D model.

APPENDIX F

PERIODIC TABLE OF THE ELEMENTS

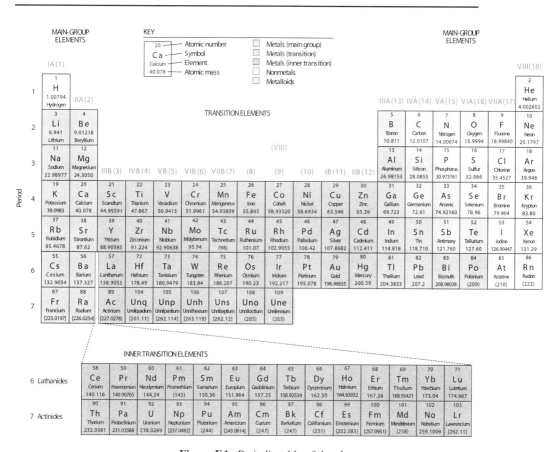

Figure F.1. Periodic table of the elements.

APPENDIX G

SUGGESTED FURTHER READING

The following references are suggested for further reading on the subject of fuel cells or electrochemistry (please see the bibliography for the detailed citations):

Fuel Cells:

- Fuel Cell Handbook [90]
- Fuel Cell Systems Explained [91]
- Handbook of Fuel Cell Technology [5]
- Springer Model of the PEMFC [8]

Electrochemistry:

- Electrochemical Methods [7]
- Electrochemistry [92]

Other:

- Basic Research Needs for the Hydrogen Economy [93]
- Transport Phenomena [11]
- Flow and Transport in Porous Formations [94]
- CFD Research Corporation User Manual [88]

BIBLIOGRAPHY

[1] R. P. Feynman, R. B. Leighton, and M. Sands, *The Feynman Lectures on Physics*, Section 4-1, Addison-Wesley, Reading, MA, 1963.

[2] D. V. Schroeder. *An Introduction to Thermal Physics*. Addison-Wesley, Reading, MA, 2000.

[3] G. H. J. Broers and J. A. A. Ketelaar. In *Fuel Cells*, G. J. Young (Ed.). Reinhold, New York, 1960.

[4] W. Vielstich, A. Lamm, and H. A. Gasteiger. *Handbook of Fuel Cells*, Vol. 2. Wiley, New York, 2003.

[5] C. Berger (Ed.). *Handbook of Fuel Cell Technology*. Prentice-Hall, Englewood Cliffs, NJ, 1968.

[6] A. Damjanovic, V. Brusic, and J. O'M Bockris. Electrode kinetics of oxygen reduction on oxide-free platinum electrodes. *Electrochimica Acta*, 12:615, 1967.

[7] A. J. Bard and L. R. Faulkner. *Electrochemical Methods*, 2nd ed. Wiley, New York, 2001.

[8] T. E. Springer, T. A. Zawodzinski, and S. Gottesfeld. Polymer electrolyte fuel cell model. *Journal of the Electrochemical Society*, 138(8):2334–2342, 1991.

[9] *Handbook of Chemistry and Physics*, 62nd ed. CRC Press, Boca Raton, FL, 1981.

[10] A. I. Ioffe, D. S. Rutman, and S. V. Karpachov. On the nature of the conductivity maximum in zirconia-based solid electrolyte. *Electrochimica Acta*, 23:141, 1978.

[11] R. Bird, W. Stewart, and E. Lightfoot. *Transport Phenomena*, 2nd ed. Wiley, New York, 2002.

[12] R. E. De La Rue and C. W. Tobias. *Journal of the Electrochemical Society*, 33(3):253–286, 1999.

[13] E. L. Cussler. *Diffusion: Mass Transfer in Fluid Systems*. Cambridge University Press, Cambridge, 1995.

[14] W. Sutherland. The viscosity of gases and molecular force. *Philosophical Magazine*, 5:507–531, 1893.

[15] J. Hilsenrath et al. Tables of thermodynamic and transport properties. National Bureau of Standards (U.S.) Circular 564.

[16] C. R. Wilke. *Journal of Chemical Physics*, 18:517–519, 1950.

[17] R. K. Shah and A. L. London. Laminar flow forced convection. In *Supplement 1 to Advance in Heat Transfer*, T. F. Irvine and J. P. Hartnett (Eds.), Academic, New York, 1978.

[18] W. M. Rohsenow, J. P Hartnett, and Y. I. Cho (Eds.). *Handbook of Heat Transfer*, 3rd ed. McGraw-Hill, New York, 1998.

[19] R. L. Borup and N. E. Vanderborgh. Design and testing criteria for bipolar plate materials for pem fuel cell applications. *Material Research Society Symposium Proceedings*, 393, p. 151–155, 1995.

[20] P. Adcock. *Development of Cost Effective, High Performance PEM Fuel Cells for Automotive Applications*. IQPC, London, 1998.

[21] D. P Davies, P. L. Adcock, M. Turpin, and S. J. Rowen. Stainless steel as a bipolar plate material for solid polymer fuel cells. *Journal of Power Source*, 86(1):237–242, 2000.

[22] R. C. Makkus, A. H. H. Janssen, F. A. de Bruijn, and R. K. A. M. Mallant. Use of stainless steel for cost competitive bipolar plates in the spfc. *Journal of Power Source*, 86(1):274–282, 2000.

[23] P. L Hentall, J. B. Lakeman, G. O. Mepsted, P. L. Adcock, and J. M. Moore. New materials for polymer electrolyte membrane fuel cell current collectors. *Journal of Power Source*, 802:235–241, 1999.

[24] D. R. Hodgson, B. May, P. L. Adcock, and D. P. Davies. New lightweight bipolar plate system for polymer electrolyte membrane fuel cells. *Journal of Power Source*, 96(1):233–235, 2001.

[25] H. Lee, C. Lee, T. Oh, S. Choi, I. Park, and K. Baek. Development of 1 kw class polymer electrolyte membrane fuel cell power generation system. *Journal of Power Sources*, 107(1):110–119, 2002.

[26] D. L Wood III, J. S. Yi, and T. V. Nguyen. Effect of direct liquid water injection and interdigitated flow field on the performance of proton exchange mebrane fuel cells. *Electrochimica Acta*, 43(24):3795–3809, 1998.

[27] J. S. Yi and T. V. Nguyen. Multicomponent transport in porous electrodes of proton exchange membrane fuel cells using the interdigitated gas distributors. *Journal of the Electrochemical Society*, 146(1):38–45, 1999.

[28] A. Kumar and R. G. Reddy. Effect of channel dimensions and shape in the flow-field distributor on the performance of polymer electrolyte membrane fuel cells. *Journal of Power Sources*, 113:11–18, 2003.

[29] E. Hontanon, M. J. Escudero, C. Bautista, P. L. Garcia-Ybarra, and L. Daza. Optimization of flow-field in polymer electrolyte membrane fuel cells using computational fluid dynamics techniques. *Journal of Power Sources*, 86:363–368, 2000.

[30] H. Naseri-Neshat, S. Shimpalee, S. Dutta, W. K. Lee, and J. W. Van Zee. Predicting the effect of gas-flow channels spacing on current density in pem fuel cells. *Proceedings of the ASME Advanced Energy Systems Division*, 39:337–350, 1999.

[31] D. M. Bernardi and M. W. Verbrugge. Mathematical model of the solid-polymer-electrolyte fuel cell. *Journal of the Electrochemical Society*, 139:2477, 1992.

[32] T. F. Fuller and J. Newman. Water and thermal management in solid-polymer-electrolyte fuel cells. *Journal of the Electrochemical Society*, 140:1218, 1993.

[33] V. Gurau, F. Barbir, and H. Liu. An analytical solution of a half-cell model for pem fuel cells. *Journal of the Electrochemical Society*, 147:2468–2477, 2000.

[34] T. V. Nguyen and R. E. White. Water and heat management model for proton-exchange-membrane fuel cells. *Journal of the Electrochemical Society*, 140:2178, 1993.

[35] J. W. Kim, A. V. Virkar, K. Z. Fung, K. Mehta, and S. C. Singhal. Polarization effects in intermediate temperature, anode-supported solid oxide fuel cells. *Journal of the Electrochemical Society*, 146(1):69–78, 1999.

[36] S. H. Chan, K. A. Khor, and Z. T. Xia. A complete polarization model of a solid oxide fuel cell and its sensitivity to the change of cell component thickness. *Journal of Power Sources*, 93:130–140, 2001.

[37] C. F. Curtiss and J. O. Hirschfelder. *Journal of Chemical Physics*, 17:550–555, 1949.

[38] S. Ahn and J. Tatarchuk. Air electrode: Identification of intraelectrode rate phenomena via ac impedance. *Journal of the Electrochemical Society*, 142(12):4169–4175, 1995.

[39] T. E. Springer, T. A. Zawodzinski, M. S. Wilson, and S. Gottesfeld. Characterization of polymer electrolyte fuel cell using impedance spectroscopy. *Journal of the Electrochemical Society*, 143(2):587–599, 1996.

[40] J. R. MacDonald. *Impedance Spectroscopy; Emphasizing Solid Materials and Systems*. Wiley-Interscience, New York, 1987.

[41] H. Ghezel-Ayagh, A. J. Leo, H. Maru, and M. Farooque. Overview of direct carbonate fuel cell technology and products development. Paper presented at the ASME First International Conference on Fuel Cell Science, Energy and Technology, Rochester, NY, Apr, 2003, p. 11.

[42] J. O. Bockris and A. K. N. Reddy. *Modern Electrochemistry*. Plenum, New York, 2000.

[43] M. V. Twigg. *Catalyst Handbook*. Manson, London, 1996.

[44] A. I. LaCava and S. V. Krishnan. Thermal effect of compression and expansion of gas in a pressure swing adsorption process. In *Fundamentals of Adsorption*, Vol. 6, F. Meunier (Ed.). Elsevier, New York, 1998.

[45] M. H. Chahbani and D. Tondeur. Compression, decompression and column equilibration in pressure swing-adsorption. In *Fundamentals of Adsorption*, Vol. 6, F. Meunier (Ed.). Elsevier, New York, 1998.

[46] A. G. Knapton. Palladium alloys for hydrogen diffusion membranes. *Platinum Metals Review*, 22(2):44–50, 1977.

[47] I. B. Elkina and J. H. Meldon. Hydrogen transport in palladium membranes. *Desalination*, 147:445–448, 2002.

[48] B. Linnhoff and P. Senior. Energy targets clarify scope for better heat integration. *Process Engineering*, 118:29–33, 1983.

[49] B. Linnhoff and J. Turner. Heat recovery networks: New insights yield big savings. *Chemical Engineering*, Nov. 1981, pp. 56–70.

[50] C. B. Snowdon. Pinch technology: Heat exchanger Networks. In *Process Design and Economics C5A*. Department of Engineering Sciences, Oxford University, 2002, p. 21.

[51] D. E. Winterbone. Pinch technology. In *Advanced Thermodynamics for Engineers*. Butterworth-Heinemann, New York, 1996, p. 47.

[52] W. Colella. Modelling results for the thermal management sub-system of a combined heat and power (CHP) fuel cell system (FCS). *Journal of Power Sources*, 118:129–149, 2003.

[53] *U.S. LCI Database Project—Phase 1 Final Report*, NREL/SR-550-33807. National Renewable Energy Laboratory, Golden, CO, Aug. 2003; http://www.nrel.gov/lci/pdfs/.

[54] G. Rebitzer, T. Ekvall, R. Frischnecht, D. Hunkeler, G. Norris, T. Rydberg, W.-P. Schmidt, S. Suh, B. P. Weidema, and D. W. Pennington. Life cycle assessment part 1: Framework, goal and scope definition, inventory analysis, and applications. *Environment International*, 30:701–720, 2004.

[55] E. M. Goldratt and J. Cox. *The Goal*. North River Press, New York, 1992.

[56] *Toyota FCHV—The First Step toward the Hydrogen Society of Tomorrow*, Toyota Special Report. http://www.toyota.co.jp/en/special/fchv/fchv_1.html.

[57] G. Sovran and D. Blaser. A contribution to understanding automotive fuel economy and its limits. SAE Technical Paper Series, 2003-01-2070:24, 2003.

[58] UTC Power, United Technology Corporation, 195 Governor's Highway, South Windsor, CT.

[59] *Ballard Transportation Products Xcellsis HY-80 Light Duty Fuel Cell Engine*. Ballard Power Corporation, Vancouver, BC, 2004; http://www.ballard.com/pdfs/XCS-HY-80_Trans.pdf.

[60] *Ballard Transportation Products A 600V300 MS High Power Electric Drive System*. Ballard Power Corporation, Vancouver, BC, 2004. http://www.ballard.com/pdfs/ballardedpc600v300ms.pdf.

[61] W. G. Colella, M. Z. Jacobson, and D. M. Golden. Switching to a hydrogen fuel cell vehicle fleet: The resultant change in emissions, energy use, and global warming gases. *Journal of Power Sources* (in review).

[62] Intergovernmental Panel on Climate Change 2001. *Climate Change 2001: The Scientific Basis*. Cambridge University Press, Cambridge, 2001.

[63] M. Z. Jacobson. A physically-based treatment of elemental carbon optic: Implications for global direct forcing of aerosols. *Geophysical Research Letters*, 27:217–220, 2000.

[64] M. Z. Jacobson. Strong radiative heating due to the mixing state of black carbon in atmospheric aerosols. *Nature*, 409:695–697, 2001.

[65] M. A. K. Khalil and R. A. Rasmussen. Global increase of atmospheric molecular hydrogen. *Nature*, 347:743–745, 1990.

[66] P. C. Novelli, P. M. Lang, K. A. Masarie, D. F. Hurst, R. Myers, and J. W. Elkins. Molecular hydrogen in the troposphere: Global distribution and budget. *Journal of Geophysical Research*, 104(30):427–430, 1999.

[67] J. B. Heywood. *Internal Combustion Engine Fundamentals*. McGraw-Hill, New York, 1988.

[68] http://www.eia.doe.gov/cneaf/alternate/page/datatables/table10.html.

[69] M. Z. Jacobson, W. G. Colella, and D. M. Golden. Air pollution and health effects of switching to hydrogen fuel cell and hybrid vehicles. *Science*, 308:1901, 2005.

[70] M. Z. Jacobson. The climate response of fossil-fuel and biofuel soot, accounting for soot's feedback to snow and sea ice albedo and emissivity. *Journal of Geophysical Research*, 109:D21201, 2004.

[71] ExternE, Externalities of energy. *Methodology*, 7, 1998; http://www.externe.info/reports.html.

[72] A. Rabl and J. Spadaro. Public health impact of air pollution and implications for the energy system. *Annual Review of Energy and the Environment*, 25:601–627, 2000.

[73] W. J. Baumol and A. S. Blinder. *Microeconomics: Principles and Policy*, 9th ed. South-Western College Publishing, Mason, OH, 2003.

[74] D. R. McCubbin and M. A. Delucchi. The health costs of motor-vehicle-related air pollution. *Journal of Transport Economics and Policy*, 33(3):253–286, 1999.

[75] L. Pauling and E. B. Wilson. *Introduction to Quantum Mechanics with Applications to Chemistry*. Dover, Mineola, NY, 1985.

[76] S. Brandt and H. D. Dahmen. *The Picture Book of Quantum Mechanics*. Springer, 2001.

[77] N. Bohr. On the constitution of atoms and molecules, *Philosophical Magazine*, 26:1, 1913.

[78] L. de Broglie. Researches on the Quantum Theorey, Thesis, Sorbonne University, Paris, 1924.

[79] C. J. Davisson. Are electrons waves? *Franklin Institute Journal*, 205:597, 1928.

[80] E. Schrödinger. Quantization as an eigenvalue problem, *Annalen der Physik*, 79:361, 1926.

[81] E. Schrödinger. *Collected Papers on Wave Mechanics*. Blackie and Son, London, 1928.

[82] W. Kohn and L. J. Sham. Self-consistent equations including exchange and correlation effects, *Physical Review*, 140:A1133, 1965.

[83] I. Levine. *Quantum Chemistry*. Allyn and Bacon, Boston, MA, 1983.

[84] S. Um, C. Y. Wang, and K. S. Chen. Computational fluid dynamics modeling of proton exchange membrane fuel cells. *Journal of the Electrochemical Society*, 147:4485, 2000.

[85] V. Gurau, S. Kakac, and H. Liu. Mathematical model for proton exchange membrane fuel cells. *American Society of Mechanical Engineers, Advanced Energy Systems Division (Publication) AES*, 38:205, 1998.

[86] W. He, J. S. Yi, and T. V. Nguyen. Two-phase flow model of the cathode of pem fuel cells using interdigitated flow fields. *AIChE Journal*, 46:2053, 2000.

[87] D. Natarajan and T. Nguyen. Three-dimensional effects of liquid water flooding in the cathode of a pem fuel cell. *Journal of the Electrochemical Society*, 148:A1324, 2001.

[88] CFD Research Corp. *CFD-ACE(U)TM User Manual version 2002*. CFD Research Corp., Huntsville, AL, 2002.

[89] Z. H. Wang, C. Y. Wang, and K. S Chen. Two-phase flow and transport in the air cathode of proton exchange membrane fuel cells. *Journal of Power Sources*, 94:40, 2001.

[90] J. H. Hirschenhofer, D. B. Stauffer, R. R. Engleman, and M. G. Klett. *Fuel Cell Handbook*, 6th ed. U.S. Department of Energy, Morgantown, WV, 2003.

[91] J. Larminie and A. Dicks. *Fuel Cell Systems Explained*. Wiley, New York, 2000.

[92] C. H. Hamann, A. Hamnet, and W. Vielstich. *Electrochemistry*. Wiley-VCH, Weinheim, 1998.

[93] M. Dresselhaus (Chair). Basic research needs for the hydrogen economy: Report of the basic energy sciences workshop on hydrogen production, storage, and use. Technical report, Workshop on Hydrogen Production, Storage, and Use, Rockville, MD, 2003.

[94] G. Dagan. *Flow and Transport in Porous Formations*. Springer-Verlag, Berlin, 1989.

IMPORTANT EQUATIONS

Thermodynamics

$$dU = dQ - dW = dQ - p\,dV$$

$$dS = k \ln \Omega = \frac{dQ}{T}$$

$$H = U + pV$$

$$G = H - TS$$

$$\Delta G = \Delta H - T\,\Delta S \text{ (isothermal process)}$$

$$\Delta G = -nFE$$

$$\mu = \mu^0 + RT \ln a$$

$$E = E^0 + \frac{\Delta S}{nF}(T - T_0) - \frac{RT}{nF} \ln \frac{\prod a_{\text{prod}}^{\nu_i}}{\prod a_{\text{react}}^{\nu_i}}$$

$$\varepsilon_{\text{real}} = \varepsilon_{\text{thermo}} \varepsilon_{\text{voltage}} \varepsilon_{\text{fuel}}$$

$$\varepsilon_{\text{thermo}, fc} = \frac{\Delta G}{\Delta H}$$

$$\varepsilon_{\text{voltage}} = \frac{V}{E}$$

Reaction Kinetics

$$j_0 = nFC^* f e^{-\Delta G^{\ddagger}/(RT)}$$

$$j = j_0^0 \left(\frac{C_R^*}{C_R^{0*}} e^{\alpha n F \eta/(RT)} - \frac{C_P^*}{C_P^{0*}} e^{-(1-\alpha)nF\eta/(RT)} \right)$$

$$j = j_0 \frac{nF\eta_{\text{act}}}{RT} \quad \text{(small overpotential/current)}$$

$$\eta_{\text{act}} = \frac{RT}{\alpha n F} \ln \frac{j}{j_0} \quad \text{(large overpotential/current)}$$

Charge Transport

$$\eta_{\text{ohmic}} = j(ASR_{\text{ohmic}}) = j\frac{L}{\sigma}$$

$$ASR_{\text{ohmic}} = A_{\text{fuel cell}} R_{\text{ohmic}} = \frac{L}{\sigma}$$

$$\sigma = zFcu$$

$$u = \frac{nFD}{RT}$$

$$D = D_0 e^{-\Delta G/(RT)}$$

Mass Transport

$$j_L = nFD^{\text{eff}} \frac{c_R^0}{\delta}$$

$$\eta_{\text{conc}} = \frac{RT}{\alpha n F} \ln \frac{j_L}{j_L - j} = c \ln \frac{j_L}{j_L - j}$$

Modeling

$$V = E_{\text{thermo}} - \eta_{\text{act}} - \eta_{\text{ohmic}} - \eta_{\text{conc}}$$

$$V = E_{\text{thermo}} - [a_A + b_A \ln(j + j_{\text{leak}})] - [a_C + b_C \ln(j + j_{\text{leak}})]$$

$$- (jASR_{\text{ohmic}}) - \left(c \ln \frac{j_L}{j_L - (j + j_{\text{leak}})} \right)$$

Characterization

$$Z_\Omega = R_\Omega$$

$$Z_C = \frac{1}{j\omega C}$$

$$Z_{\text{series}} = Z_1 + Z_2$$

$$Z_{\text{parallel}}^{-1} = Z_1^{-1} + Z_2^{-1}$$

$$Z_{\text{infinite Warburg}} = \frac{\sigma_1}{\sqrt{\omega}}(1 - j)$$

$$Z_{\text{finite Warburg}} = \frac{\sigma_1}{\sqrt{\omega}}(1 - j)\tanh\left(\delta\sqrt{\frac{j\omega}{D_i}}\right)$$

$$A_c = \frac{Q_h}{Q_m * A_{\text{geometric}}}$$

Systems

$$\text{Gravimetric energy storage density} = \frac{\text{stored enthalpy of fuel}}{\text{system mass}}$$

$$\text{Volumetric energy storage density} = \frac{\text{stored enthalpy of fuel}}{\text{system volume}}$$

$$\text{Carrier system effectiveness} = \frac{\%\text{ conversion of carrier to electricity}}{\%\text{ conversion of neat } H_2 \text{ to electricity}}$$

Fuel Cell Systems

$$\varepsilon_O = \varepsilon_R + \varepsilon_H$$

$$\varepsilon_R = \varepsilon_{FP} \times \varepsilon_{R,\text{SUB}} \times \varepsilon_{R,\text{PE}} = \frac{\Delta \dot{H}_{(\text{HHV}),H_2}}{\Delta \dot{H}_{(\text{HHV}),\text{fuel}}} \times \frac{P_{e,\text{SUB}}}{\Delta \dot{H}_{(\text{HHV}),H_2}} \times \frac{P_{e,\text{SYS}}}{P_{e,\text{SUB}}}$$

$$\frac{H}{P} = \frac{d\dot{H}}{P_{e,\text{SYS}}}$$

$$y_{H_2} = \frac{n_{H_2}}{n}$$

$$\frac{S}{C} = \frac{n_{H_2O}}{n_c}$$

Environmental Impact

$$\dot{Q} - \dot{W} = \dot{m}\left[h_2 - h_1 + g(z_2 - z_1) + \frac{1}{2}(V_2^2 - V_1^2)\right]$$

$$CO_{2\text{ equivalent}} = m_{CO_2} + 23m_{CH_4} + 296m_{N_2O} + \alpha(m_{OM2.5} + m_{BC2.5})$$
$$- \beta[m_{SULF2.5} + m_{NIT2.5} + 0.40m_{SO_X} + 0.10m_{NO_X} + 0.05m_{VOC}]$$

INDEX